Multibody Systems with Flexible Elements

Multibody Systems with Flexible Elements

Editors

Marin Marin
Dumitru Baleanu
Sorin Vlase

MDPI • Basel • Beijing • Wuhan • Barcelona • Belgrade • Manchester • Tokyo • Cluj • Tianjin

Editors

Marin Marin
Mathematics and Informatics
Transilvania University of
Brasov
Brasov
Romania

Dumitru Baleanu
Mathematics
Cankaya Universitesi
Ankara
Turkey

Sorin Vlase
Mechanical Engineering
Transilvania of Brasov
Brasov
Romania

Editorial Office
MDPI
St. Alban-Anlage 66
4052 Basel, Switzerland

This is a reprint of articles from the Special Issue published online in the open access journal *Symmetry* (ISSN 2073-8994) (available at: www.mdpi.com/journal/symmetry/special_issues/Muldtibody_Systems_Flexible_Elements).

For citation purposes, cite each article independently as indicated on the article page online and as indicated below:

LastName, A.A.; LastName, B.B.; LastName, C.C. Article Title. *Journal Name* **Year**, *Volume Number*, Page Range.

ISBN 978-3-0365-5258-3 (Hbk)
ISBN 978-3-0365-5257-6 (PDF)

Contents

About the Editors

Marin Marin

Marin Marin is a Professor Dr. hab. at Transilvania University of Brasov, Department of Mathematics and Informatics, Romania. He is a specialist in Applied Mechanics with more 200 papers published in international journals. He has a Hirsch factor of 47.

Dumitru Baleanu

Dumitru Baleanu is a Professor Dr. hab. in the Department of Mathematics at Cancaya University, Ankara, Turkey. He is a prominent specialist in Mathematics, Physics, and Applied Mechanics with more 1800 papers published in international journals. He has a Hirsch factor of 90.

Sorin Vlase

Sorin Vlase is a Professor at Transilvania University of Brasov, a member of the Romanian Academy of Technical Sciences, and Head of the Department of Mechanical Engineering. He published more than 200 scientific papers and 5 monographs in the field of Applied Mechanics.

Preface to "Multibody Systems with Flexible Elements"

Multibody systems with flexible elements are a field of great contemporary interest in multiple fields, with various industrial applications. Their study and the development of appropriate models represent a desire of current research in the field of applied mechanics. The high interest in the field is also indicated by the publications dealing with multibody systems and the large amount of scientific literature and numerous published monographs. This book aims to contribute to this large flow of scientific information in a representative field of current research. Mechanical systems with elastic elements are considered where the displacements of each or some of the elements of the system are generally large and cannot be neglected in mechanical modeling. The study of these multibody systems covers many industrial fields, but also has applications in medicine, sports, and art. The systematic treatment of the dynamic behavior of interconnected bodies has led to an important number of formalisms for multibody systems within mechanics. This formalism is currently used in many engineering fields, especially robotics and vehicle dynamics. The formalism of multibody systems offers a means of algorithmic analysis, assisted by computers, and a means of simulating and optimizing an arbitrary movement of a possibly high number of elastic bodies in the connection. The domain where researchers apply these methods are robotics, simulations of the dynamics of vehicles, biomechanics, aerospace engineering (helicopters and the behavior of cars in a gravitational field), internal combustion engines, gearboxes, transmissions, mechanisms, the cellulose industry, simulations of particle behavior (granulated particles and molecules), dynamic simulations, military applications, computer games, medicine, and rehabilitation.

We hope that the present volume will present a small but useful contribution in the field of multibody systems.

Marin Marin, Dumitru Baleanu, and Sorin Vlase
Editors

symmetry MDPI

Editorial

Multibody Systems with Flexible Elements

Marin Marin [1,*], Dumitru Băleanu [2,3,4,*] and Sorin Vlase [5,6,*]

1 Department of Mathematics and Computer Science Transilvania, University of Brașov, 500036 Brașov, Romania
2 Department of Mathematics, Faculty of Art and Sciences, Cankaya University, Ankara 0630, Turkey
3 Institute of Space Sciences, 077125 Bucharest-Magurele, Romania
4 Department of Medical Research, China Medical University Hospital, Taichung 40402, Taiwan
5 Department of Mechanical Engineering, Faculty of Mechanical Engineering, Transilvania University of Brașov, 500036 Brașov, Romania
6 Romanian Academy of Technical Science, Calea Victoriei, 700506 Bucharest, Romania
* Correspondence: m.marin@unitbv.ro (M.M.); dumitru.baleanu@gmail.com (D.B.); svlase@unitbv.ro (S.V.); Tel.: +40-722-643020 (M.M.)

1. Introduction

The formalism of multibody systems offers a means of computer-assisted algorithmic analysis and a means of simulating and optimizing an arbitrary movement of a possible high number of elastic bodies in the connection. The domains where researchers apply these methods are robotics, simulation of the dynamics of vehicles, biomechanics, aerospace engineering (helicopters, the behavior of cars in the gravitational field), internal combustion engines, gearboxes, transmissions, mechanisms, the cellulose industry, simulation of particle behavior (granulated particles and molecules), dynamic simulation, military applications, computer games, medicine, and rehabilitation. As a result, multibody systems have become widely used in all industries, such as in automotive engineering, airspace engineering, construction, and manufacturing [1–8]. It is for these reasons that there is continuous research into the development of the field. Some of this research is presented in this volume, in which a large group of researchers will present their latest findings. We hope that researchers will find an interesting and useful volume of information for their future work, but that the results will also be used by engineers for practical applications.

Citation: Marin, M.; Băleanu, D.; Vlase, S. Multibody Systems with Flexible Elements. *Symmetry* **2021**, *13*, 1359. https://doi.org/10.3390/sym13081359

Received: 5 July 2021
Accepted: 6 July 2021
Published: 27 July 2021

2. Statistics of the Special Issue

The statistics of papers called for this Special Issue, related to published or rejected items, are as follows [9–23]: 26 total submissions, of which 15 were published (57.6%) and 11 rejected (42.3%). The authors' geographical distribution is shown in Table 1, and it can be seen that the 38 authors are from 9 different countries. Note that it is usual for a paper to be written by more than one author, and for authors to collaborate with authors with different affiliations or multiple affiliations.

Table 1. Geographic distribution of authors by country.

Country	Number of Authors
Romania	22
Saudi Arabia	3
India	2
China	5
Egypt	3
Vietnam	2
Iran	2
UK	1
Poland	1

3. Authors of the Special Issue

The authors of this Special Issue and their main affiliations are summarized in Table 2, and it can be seen that there are three authors on average per manuscript.

Table 2. Affiliations and bibliometric indicators for authors.

Author	Affiliation	References
Marilena Ghitescu	Transilvania University of Brasov, Romania	[9,12,13]
Ioan-Marius Ghitescu	Transilvania University of Brasov, Romania	[9,12,13]
Sorin Vlase	Transilvania University of Brasov, Romania Technical Sciences Academy of Romania, B-dul Dacia 26, 030167 Bucharest, Romania	[9,13,18,19,22]
Paul Nicolae Borza	Transilvania University of Brasov, Romania	[9,12,13]
Octavian Postavaru	Center for Research and Training in Innovative Techniques of Applied Mathematics in Engineering, University Politehnica of Bucharest, 060042 Bucharest, Romania	[10]
Antonela Toma	Center for Research and Training in Innovative Techniques of Applied Mathematics in Engineering, University Politehnica of Bucharest, 060042 Bucharest, Romania	[10]
Joseph Dianavinnarasi	Department of Mathematics, Alagappa University, Karaikudi 630 004, India	[11]
Ramachandran Raja	Ramanujan Centre for Higher Mathematics, Alagappa University, Karaikudi 630 004, India	[11]
Jehad Alzabut	Department of Mathematics and General Sciences, Prince Sultan University, Riyadh 12435, Saudi Arabia	[11]
Michał Niezabitowski	Department of Automatic Control and Robotics, Faculty of Automatic Control, Electronics and Computer Science, Silesian University of Technology, Akademicka 16, 44-100 Gliwice, Poland	[11]
Ovidiu Bagdasar	Department of Electronics, Computing and Mathematics, University of Derby, Derby DE22 1GB, UK	[11]
Marin Marin	Transilvania University of Brasov, Romania	[12,20,22]
Maria Luminita Scutaru	Transilvania University of Brasov, Romania	[12,19,22]
Calin Itu	Transilvania University of Brasov, Romania	[14]
Polidor Bratu	ICECON SA, Bucharest, Romania	[14,15]
Dorin Lixandroiu	Transilvania University of Brasov, Romania	[14]
Chen Wang	Key Laboratory of Exploration Mechanism of the Deep Space Planet Surface, Ministry of Industry and Information Technology, Nanjing University of Aeronautics and Astronautics, Nanjing 211100, China	[16]
Jinbao Chen	Key Laboratory of Exploration Mechanism of the Deep Space Planet Surface, Ministry of Industry and Information Technology, Nanjing University of Aeronautics and Astronautics, Nanjing 211100, China	[16]
Shan Jia	Key Laboratory of Exploration Mechanism of the Deep Space Planet Surface, Ministry of Industry and Information Technology, Nanjing University of Aeronautics and Astronautics, Nanjing 211100, China	[16]
Heng Chen	Field Engineering College, Army Engineering University of PLA, Nanjing 210001, China	[16]
Attila Gerocs	Doctoral School of Mechanical Engineering, "Eftimie Murgu" University of Resita, 320085 Resita, Romania	[17]
Gilbert-Rainer Gillich	Doctoral School of Mechanical Engineering, "Eftimie Murgu" University of Resita, 320085 Resita, Romania	[17]
Dorian Nedelcu	Doctoral School of Mechanical Engineering, "Eftimie Murgu" University of Resita, 320085 Resita, Romania	[17]

Table 2. *Cont.*

Author	Affiliation	References
Zoltan-Iosif Korka	Doctoral School of Mechanical Engineering, "Eftimie Murgu" University of Resita, 320085 Resita, Romania	[17]
Gina Diana Musca (Anghelache)	Engineering and Agronomy Faculty in Braila, Research Center for Mechanics of Machines and Technological Equipments, "Dunarea de Jos" University of Galati, 810017 Braila, Romania	[18]
Carmen Debeleac	Engineering and Agronomy Faculty in Braila, Research Center for Mechanics of Machines and Technological Equipments, "Dunarea de Jos" University of Galati, 810017 Braila, Romania	[18]
Gabriel Leonard Mitu	COMAT, SA, str. Zizinului, nr.111, 500002 Brasov, Romania	[19]
Eliza Chircan	Department of Mechanical Engineering, Transilvania University of Brașov, B-dulEroilor 20, 500036 Brașov, Romania	
Aatef Hobiny	Nonlinear Analysis and Applied Mathematics Research Group (NAAM), Mathematics Department, King Abdulaziz University, Jeddah 21521, Saudi Arabia	[20]
Faris Alzahrani	Nonlinear Analysis and Applied Mathematics Research Group (NAAM), Mathematics Department, King Abdulaziz University, Jeddah 21521, Saudi Arabia	[20]
Ibrahim Abbas	Mathematics Department, Faculty of Science, Sohag University, Sohag 82524, Egypt	[20]
Ahmed A. El-Deeb	Department of Mathematics, Faculty of Science, Al-Azhar University, Nasr City, Cairo 11884, Egypt	[21]
Samer D. Makharesh	Department of Mathematics, Faculty of Science, Al-Azhar University, Nasr City, Cairo 11884, Egypt	[21]
Dumitru Baleanu	Cankaya University, Ankara, Turkey Institute of Space Sciences, Bucharest-Magurele, Romania Department of Medical Research, China Medical University Hospital, China Medical University, Taichung 40402, Taiwan, China	[10,12,21,23]
Iuliu Negrean	Technical Sciences Academy of Romania; B-dul Dacia 26, 030167 Bucharest, Romania Department of Mechanical Systems Engineering, Technical University of Cluj-Napoca, Str. Memorandumului 28, 400114 Cluj-Napoca, Romania	[22]
Mohammad Reza Mahmoudi	Institute of Research and Development, Duy Tan University, Da Nang 550000, Vietnam Department of Statistics, Faculty of Science, Fasa University, Fasa, Fars 7461686131, Iran	[23]
Roya Nasirzadeh	Department of Statistics, Faculty of Science, Fasa University, Fasa, Fars 7461686131, Iran	[23]
Kim-Hung Pho	Fractional Calculus, Optimization and Algebra Research Group, Faculty of Mathematics and Statistics, Ton Duc Thang University, Ho Chi Minh City 72915, Vietnam	[23]

4. Brief Overview of the Contributions to the Special Issue

The analysis of the topics identifies or summarizes the research undertaken. This section classifies the manuscripts according to the topics proposed in the Special Issue. There are three topics that are dominant, namely: modeling of the multibody system with symmetries, symmetry in applied mathematics and analytical methods in the symmetric multibody systems.

Author Contributions: Conceptualization, M.M., D.B. and S.V.; methodology, M.M., D.B. and S.V.; software, M.M., D.B. and S.V.; validation, M.M., D.B. and S.V.; formal analysis, M.M., D.B. and S.V.; investigation, M.M., D.B. and S.V.; resources, M.M., D.B. and S.V.; data curation, M.M., D.B. and S.V.; writing—original draft preparation, S.V.; writing—review and editing, M.M., D.B. and S.V.; visualization, M.M., D.B. and S.V.; supervision, M.M., D.B. and S.V.; project administration, M.M., D.B. and S.V. All authors have read and agreed to the published version of the manuscript."

Funding: This research received no external funding.

Institutional Review Board Statement: Not applicable.

Informed Consent Statement: Not applicable.

Data Availability Statement: Not applicable.

Acknowledgments: Not applicable.

Conflicts of Interest: The authors declare no conflict of interest.

References

1. Vlase, S.; Nastac, C.; Marin, M.; Mihălcică, M. A Method for the Study of the Vibration of Mechanical Bars Systems with Symmetries. *Acta Tech. Napoc. Ser. Appl. Math. Mech. Eng.* **2017**, *60*, 539–544.
2. Vlase, S. A Method of Eliminating Lagrangian Multipliers from the Equation of Motion of Interconnected Mechanical Systems. *J. Appl. Mech. Trans. ASME* **1987**, *54*, 235–237. [CrossRef]
3. Scutaru, M.L.; Vlase, S.; Marin, M.; Modrea, A. New analytical method based on dynamic response of planar mechanical elastic systems. *Bound. Value Probl.* **2020**, *2020*, 104. [CrossRef]
4. Vlase, S.; Marin, M.; Öchsner, A. Considerations of the transverse vibration of a mechanical system with two identical bars. *Proc. Inst. Mech. Eng. Part. L J. Mater. Des. Appl.* **2019**, *233*, 1318–1323. [CrossRef]
5. Marin, M.; Vlase, S.; Paun, M. Considerations on double porosity structure for micropolar bodies. *AIP Adv.* **2015**, *5*, 037113. [CrossRef]
6. Khan, A.A.; Bukhari, S.R.; Marin, M.; Ellahi, R. Effects of chemical reaction on third-grade mhd fluid flow under the influence of heat and mass transfer with variable reactive index. *Heat Transf. Res.* **2019**, *50*, 1061–1080. [CrossRef]
7. Saeed, T.; Abbas, I.; Marin, M. A GL Model on Thermo-Elastic Interaction in a Poroelastic Material Using Finite Element Method. *Symmetry* **2020**, *12*, 488. [CrossRef]
8. Zhang, L.; Bhatti, M.M.; Marin, M.; Mekheimer, K.S. Entropy Analysis on the Blood Flow through Anisotropically Tapered Arteries Filled with Magnetic Zinc-Oxide (ZnO) Nanoparticles. *Entropy* **2020**, *22*, 1070. [CrossRef]
9. Ghitescu, M.; Ghitescu, I.-M.; Vlase, S.; Borza, P. Experimental Dynamic Rigidity of an Elastic Coupling with Bolts. *Symmetry* **2021**, *13*, 989. [CrossRef]
10. Postavaru, O.; Toma, A. Symmetries for Nonconservative Field Theories on Time Scale. *Symmetry* **2021**, *13*, 552. [CrossRef]
11. Dianavinnarasi, J.; Raja, R.; Alzabut, J.; Niezabitowski, M.; Bagdasar, O. Controlling Wolbachia Transmission and Invasion Dynamics among Aedes Aegypti Population via Impulsive Control Strategy. *Symmetry* **2021**, *13*, 434. [CrossRef]
12. Ghiţescu, I.-M.; Scutaru, M.L.; Ghiţescu, M.; Borza, P.N.; Marin, M. New Command Mechanism of Flaps and Wings of a Light Sport Aircraft. *Symmetry* **2021**, *13*, 221. [CrossRef]
13. Ghiţescu, M.; Ghiţescu, I.-M.; Borza, P.; Vlase, S. A New Optimized Solution for A Flexible Coupling with Bolts Used in the Mechanical Transmissions. *Symmetry* **2021**, *13*, 171. [CrossRef]
14. Itu, C.; Bratu, P.; Borza, P.N.; Vlase, S.; Lixandroiu, D. Design and Analysis of Inertial Platform Insulation of the ELI-NP Project of Laser and Gamma Beam Systems. *Symmetry* **2020**, *12*, 1972. [CrossRef]
15. Bratu, P. Multibody System with Elastic Connections for Dynamic Modeling of Compactor Vibratory Rollers. *Symmetry* **2020**, *12*, 1617. [CrossRef]
16. Wang, C.; Chen, J.; Jia, S.; Chen, H. Parameterized Design and Dynamic Analysis of a Reusable Launch Vehicle Landing System with Semi-Active Control. *Symmetry* **2020**, *12*, 1572. [CrossRef]
17. Gerocs, A.; Gillich, G.-R.; Nedelcu, D.; Korka, Z.-I. A Multibody Inertial Propulsion Drive with Symmetrically Placed Balls Rotating on Eccentric Trajectories. *Symmetry* **2020**, *12*, 1422. [CrossRef]
18. Anghelache, G.D.M.; Debeleac, C.; Vlase, S. Experimental Assessments on the Evaluation of Wire Rope Characteristics as Helical Symmetrical Multi-body Ensembles. *Symmetry* **2020**, *12*, 1231. [CrossRef]
19. Mitu, G.L.; Chircan, E.; Scutaru, M.L.; Vlase, S. Kane's Formalism Used to the Vibration Analysis of a Wind Water Pump. *Symmetry* **2020**, *12*, 1030. [CrossRef]
20. Hobiny, A.; Alzahrani, F.; Abbas, I.; Marin, M. The Effect of Fractional Time Derivative of Bioheat Model in Skin Tissue Induced to Laser Irradiation. *Symmetry* **2020**, *12*, 602. [CrossRef]

21. El-Deeb, A.A.; Makharesh, S.D.; Baleanu, D. Dynamic Hilbert-Type Inequalities with Fenchel-Legendre Transform. *Symmetry* **2020**, *12*, 582. [CrossRef]
22. Vlase, S.; Negrean, I.; Marin, M.; Scutaru, M.L. Energy of Accelerations Used to Obtain the Motion Equations of a Three-Dimensional Finite Element. *Symmetry* **2020**, *12*, 321. [CrossRef]
23. Mahmoudi, M.R.; Nasirzadeh, R.; Baleanu, D.; Pho, K.-H. The Properties of a Decile-Based Statistic to Measure Symmetry and Asymmetry. *Symmetry* **2020**, *12*, 296. [CrossRef]

symmetry

MDPI

Article

Experimental Dynamic Rigidity of an Elastic Coupling with Bolts

Marilena Ghitescu [1], Ion-Marius Ghitescu [1,*], Sorin Vlase [1,2,*] and Paul Nicolae Borza [3]

1 Department of Mechanical Engineering, Transilvania University of Brasov, B-dul Eroilor 29, 500036 Brasov, Romania; marilenaradu@unitbv.ro
2 Romanian Academy of Technical Sciences, 700506 Bucharest, Romania
3 Department of Computers and Electronics, Transilvania University of Brasov, B-dul Eroilor 29, 500036 Brasov, Romania; borzapn@unitbv.ro
* Correspondence: marius.ghitescu@unitbv.ro (I.-M.G.); svlase@unitbv.ro (S.V.)

Abstract: The paper presents an elastic coupling with bolts and intermediary non-metallic elements, which allows for radial and axial deviation and can absorb shocks and torsional vibrations. The designed bolts have a particular shape of a circular area of a length equal to the width of the non-metallic element, a cylindrical area larger than the diameter of the cylindrical groove where the non-metallic elements are mounted, and a cylindrical area smaller than the threaded area to avoid stress concentrators and bolt breakage. The coupling represents a symmetrical piece, having two planes of symmetry. Therefore, the study of such a mechanical system can be considerably simplified considering the design and description of the repeating elements. The novelty of this coupling consists in the existence of an intermediate disc between two half-couplings (driving and driven half-coupling). The non-metallic elements with different shapes are made of different types of rubber, mounted on cylindrical bolts fixed by the driving half-coupling, transmitting the motion in both directions.

Keywords: elastic coupling; non-metallic elements; bolt; dynamic rigidity; non-collinearly shafts

Citation: Ghitescu, M.; Ghitescu, I.-M.; Vlase, S.; Borza, P.N. Experimental Dynamic Rigidity of an Elastic Coupling with Bolts. *Symmetry* **2021**, *13*, 989. https://doi.org/10.3390/sym13060989

Academic Editor: Victor A. Eremeyev

Received: 28 April 2021
Accepted: 31 May 2021
Published: 2 June 2021

1. Introduction

In the field of engingeering, an elastic coupling can make a significant contribution when two machines work together [1–8]. For this reason, a particular study must be made for every practical application.

For elastic couplings with bolts, a large variety of solutions and materials are used depending on the demands and practical role. The solutions differ in dimension, design, shape of different elements, material, etc. Two of them are standardized: version N—normal; and version B—with spacer bush [9]. In these versions, the coupling is represented by a symmetrical device with two planes of symmetry. So, in this paper, the study of such a mechanical system can be considerably simplified considering the design and description of the repeating elements. In the same time, the manufacturing of a system with repeating elements is easier.

A wide variety of flexible couplings with rubber pins exists, in which the two half-couplings are different from each other [10,11].

There are many flexible bolt coupling solutions. They stand out for their simplicity, the two semi-couplings that constitute the coupling being identical [12]. The torque is transmitted in this case through the rubber sleeves covering the bolts.

A solution with washers or rubber bushes and studs is presented in [13,14]. In literature [15], it is studied a coupling which consists of two hubs: an elastic rubber element and bolts with self-locking nuts [13]. This type of coupling has been used for different machines as welding sets or compressors, as generally machineries have large driven inertias.

Two semi-couplings with bolts and rubber elements are present in Rupex coupling [16–19].

To connect shafts with a small parallel, angular, or axial misalignment, a coupling with a barrel element is used. These types are used for the liaison between an electric machine with working machines [20,21].

A solution for a coupling with holes on the circumference where cylindrical bolts are introduced is presented in [22]. The motion is transmitted from one engine to a shaft by bolts and rubber.

The marine coupling comprises a coupling body, with connecting pins, sleeve, screws, and rubber [23]. The maritime coupling is characterized by the materials used for the components. This coupling can be easily removed.

The quick-cushion rubber sleeve coupling for quick replacement includes half of the right-hand coupling assembly, a pin, an elastic sleeve, and a nut. The sleeve can be replaced easily and does not require co-axiality correction [24,25].

A possibility for the development of the paper is the modal analysis of such a system, in order to identify the vibration behavior [26–31]. In the coupling solutions existing in this moment, the non-metallic element is mounted in one of the semi-couplings in some bores specially processed for it, which implies a higher manufacturing/execution cost of that half-coupling, and in these conditions the non-metallic element is required at crushing, the torque capable of being transmitted by the coupling being determined by the condition of resistance to crushing of the non-metallic element. The maximum angular deviation that can be taken over by these couplings is $\Delta\alpha \leq 1^0$ if the shafts are collinear.

Elastic couplings are used to damp shocks caused by loads and to prevent dangerous vibrations. In addition, the elastic couplings allow some compensation for the inaccuracies of the mutual positioning of the shafts.

The non-metallic elastic elements used in the construction of couplings have very varied construction forms. In order to achieve couplings that satisfy, to the greatest extent, the conditions imposed by the operation, shapes are used that can modify the elasticity and damping properties.

In the case of dynamic loads, the elastic couplings with non-metallic elements accumulate and partially disperse the energy. By inserting the elastic couplings in the kinematic chains, the resonance range can be exceeded and the possibility of resonant vibrations is avoided.

The dampening capacity of the coupling refers to its property to disperse and to transform energy into deformation in heat. Damping occurs when there is a difference between the elastic load characteristic at charging and discharging the coupling.

The properties of rubber are: high elasticity, high damping capacity, constructive simplicity, and low price. The non-metallic intermediate elements lead to the electrical insulation of the connected shafts, but compared to the metallic intermediate elements, they give the coupling a lower durability, the transmitted torque being limited to low-medium values.

From the study carried out in this domain of elastic couplings using non-metallic elements, we found that there are elastic couplings with bolts having simple functional and designing principles.

Such couplings are produced by specialized companies, with international reputation, in a wide range of types and sizes. The studies made in this field are not numerous, which is why the use of a new type of coupling requires a careful analysis of its characteristics. The works published in the field are relatively few and books of Machine Design (courses or monographs) present only general notions regarding the design and manufacturing [1–8].

The paper proposes a coupling that has the following advantages:

- easy assembly and disassembly of non-metallic elements; easy assembly and disassembly of bolts;
- easy assembly and disassembly of intermediary disk;
- new constructive shapes of non-metallic elements which allow for non-metallic elements to relax, to be solicited and traction, and the capable moment to be transmitted

from the condition of resistance of the non-metallic element besides crushing and shearing and torsion;

- by mounting the non-metallic element between the plates and not in bores processed in semi-couplings, it allows the non-metallic element to relax, this being required besides crushing, shearing, and traction; transmits the torque in both directions regardless of the chosen direction of rotation, as it does not become rigid;
- different elastic characteristics can be obtained, depending on the constructive shape and the material of the non-metallic elastic element; ensures the compensation of radial and angular deviations;
- the designed coupling has a simple construction, small overall dimensions and a low cost, compared to the classic ones with non-metallic elements and bolts.

2. Elastic Coupling Proposed Solution

The elastic coupling with cylindrical bolts and rubber intermediate elements presented in the paper is original, and the advantages of using such a coupling are presented below. The designed coupling was modeled 3D with the Autodesk Mechanical soft (Figure 1).

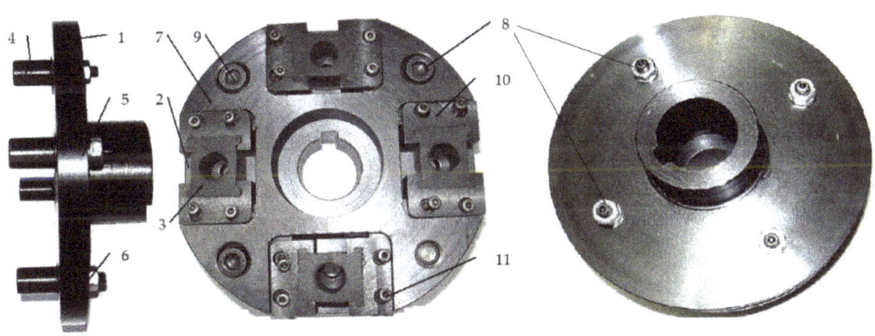

Figure 1. Proposed elastic coupling with bolts and elastic intermediary elements. 1—driver half-coupling; 2—driven half-coupling; 3—elastic intermediary pieces; 4—bolts; 5—grower washers; 6—nuts; 7—intermediary disk; 8- screws; 9—pin; 10—metal supports; 11—screws.

The driver half-coupling 1 transmits the torque from the driver machine (electric motor) to the intermediary elastic elements 3, bolts 4, fixed rigidly on the input shaft (half-coupling 1), and further through an intermediary disc 7 to the output shaft (half-coupling 2). The exterior diameter of the coupling is De = 176 mm and the axial gauge is L = 129 mm. The total weight of the coupling is m_t = 3 kg. The total moment of inertia is 0.0213 kg m^2.

In Figure 2 the constructive type of bolt is presented. The bolts 4 are fixed to the half-coupling 1.

Figure 2. Constructive shape of bolts.

The cylindrical areas of the bolt allows the centering of the bolt in the half-coupling 1 and represents a shoulder that prevents the axial displacement of the elastic element.

The cylindrical areas of the bolt, with a diameter larger than the diameter of the threaded area, aims to reduce the concentrating stress and breaking of the bolt.

The cylindrical area of the bolt with a diameter larger than the diameter of the threaded area has the role of reducing the stress at the transitions from the largest diameter to the diameter of the threaded area.

In Figure 3 the CAD model of plates 10 is presented, which are fixed to the intermediary disk 7 with two screws 11. Figure 4 presents different shapes of non-metallic element and their dimensions. The elasticity of the non-metallic element depends largely on its hardness (minimum 65° Sh).

Figure 3. Metallic plates 10, elastic intermediary piece 3, screws 11.

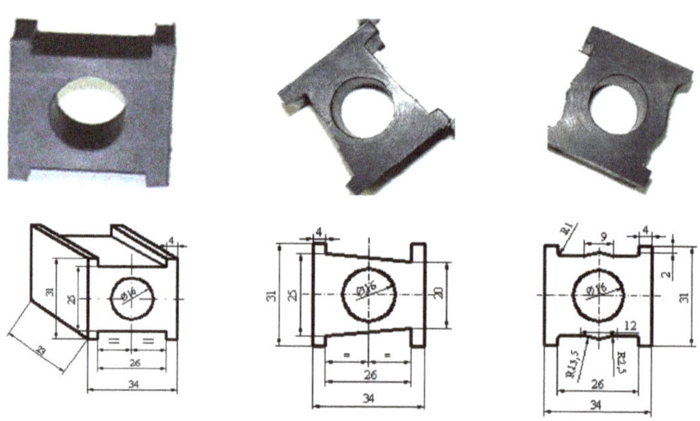

a. First version   b. Second version   c. Third version

Figure 4. Shapes of the non-metallic element.

Semi-couplings 1 and 2, the bolts 4, intermediary disc 7 and the metallic elements 10 were realized from improved OLC 45 steel. The mechanical characteristics of the C 45 steel are: density = 8.31 g/cm^3, Young's Modulus = 200 GPa, Poisssons's Ratio = 0.287.

The mold used to obtain the three shapes of non-metallic elements is presented in Figure 5. The mold consists of the upper cover 1, the intermediate ring 2 (existing in three design versions: I, II, III, corresponding to the three shapes of the non-metallic elastic element) and the bolt 3, inserted by pressing in the lower cover 4.

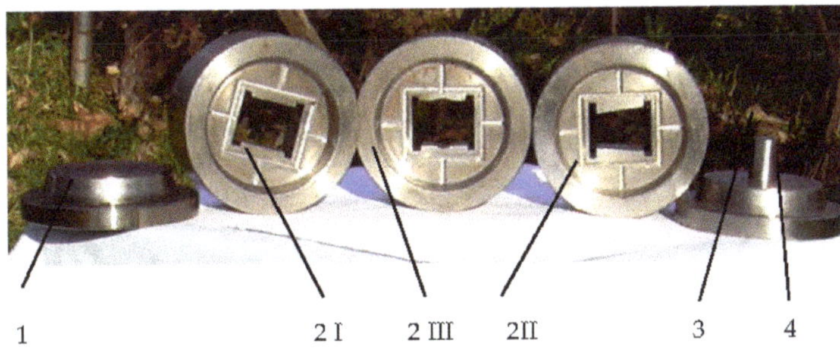

Figure 5. Molds used to obtain the three versions of the non-metallic elements.

The experimental tests performed on the designed coupling were performance tests, as well as tests that followed the operation of the coupling.

Studied functional parameters are the torque at the entry shaft and at the output shaft.

Preliminary activities were required to carry out the experimental tests in good conditions, which consisted of:

- balancing the ends of the entrance and exit shafts of the stand;
- balancing the sensor applied on the coupling to determine the angular deformation of the coupling in the dynamic mode.

An activity prior to mounting the coupling on the stand considered the following stages:

- balancing the subassembly formed by the intermediate disc and plates;
- balancing the subassembly consisting of the intermediate disc, plates and the driven half-coupling;
- balancing the assembly consisting of the driving half-coupling, the intermediate disc, the plates, and the driven half-coupling.

The purposes of the experiments are:

➢ measuring the torque at the input and at the output shaft;
➢ measuring the relative deformation that appears between the two half-couplings;
➢ obtaining the dynamic rigidity of the elastic coupling.

Figures 6 and 7 present the experimental stand to measure the torque in the elastic coupling.

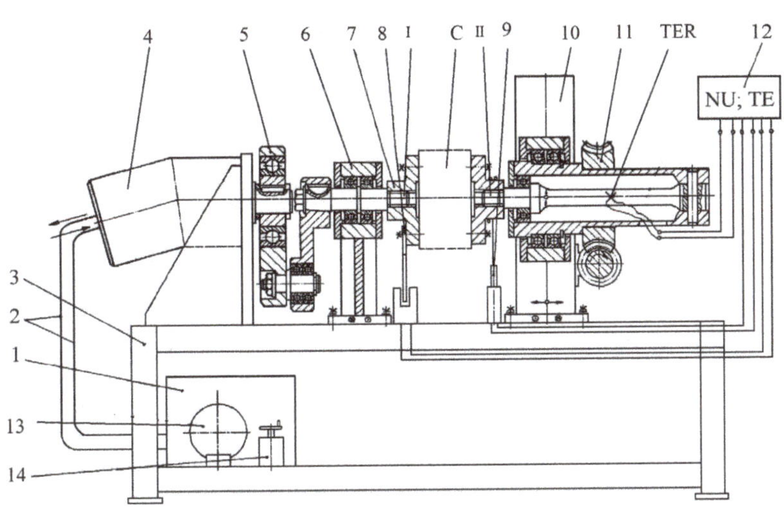

Figure 6. Componence of experimental stand. The number are explained in the text bellow.

Figure 7. Componence of experimental stand-detail.

The installation for the dynamic testing of the prototype (Figure 6) consists of the drive group 1, consisting of the electric motor and pump, supply pipes 2, frame 3, hydraulic motor 4, quadrilateral mechanism 5, support bearing 6, support flange 7, transducer pulse 8, the elastic test coupling C, the angle transducer 9, the adjustable support 10, the pre-tensioning system 11, the output shaft II, the measuring and recording system 12, the pressure gauge 13, the hydraulic regulator 14, and the TER resistive electrical transducers.

The dimensions of the existing stand in the endowment of the department at the moment of performing the test were: width $l = 1$ m, length $L = 1.5$ m, and height $H = 1.5$ m.

The measuring system 12 consists of a universal pulse NO counter, which counts the pulses detected by the pulse transducer 8 and the TE electronic tensometer, which indicates the value of the specific deformation. The power circuit is hydraulically driven from the drive group 1, and by means of the supply pipes 2 the hydraulic motor 4 is supplied. The hydraulic motor drives, by means of the quadrilateral mechanism 5, the inlet drive shaft I supported by the support bearing 6. Support flange 7 is fixed by means of grooves to the end of the input drive shaft I. The elastic test coupling C (half-coupling I) is attached to one end of the flange 7, its other end being fixed by grooves to the flange of the output shaft II. By means of the pre-tensioning system 11, the output shaft II is fixed on the adjustable support 10, which can move in two directions orthogonal to the frame of the frame 3, thus performing the coupling test.

To perform the dynamic coupling test, the following steps are performed: check the position and correct operation of the prototype C (Figure 6) and of the angular transducer 9; before starting the test, move the support bearing 6 to the desired position to achieve the amplitude of the variable moment; the power circuit is rotated until the quadrilateral mechanism 5 reaches neutral, thus being blocked. By actuating the pretensioning system 11, the prototype is twisted, the static torque being created in the power circuit. This moment produces a deformation, which is measured by means of the output shaft II and the TER electroresistive transducers glued to the shaft; when the desired stress cycle is obtained, the action on the prestressing system 11 ceases. The stress cycle can also be identified by means of the indicator with which the test installation is provided; by actuating the hydraulic regulator 14 (Figure 6), the pressure in the hydraulic circuit is regulated, thus obtaining the desired test frequency. The manometer 13 indicates the variation of the pressure in the hydraulic circuit, and the pulse transducer 8 together with the universal numerator in the measuring and recording system 12 signals the value of the frequency; depending on the deformation value, from the torsion sensor calibration diagram, the values of the maximum torque and the amplitude of the oscillating moment are estimated. Using the angular transducer 9 and the measuring and recording system 12, the elastic deformation of the tested coupling is measured and the rigidity of the tested coupling is calculated.

Figures 8 and 9 present the experimental stand to measure the torque in the elastic coupling.

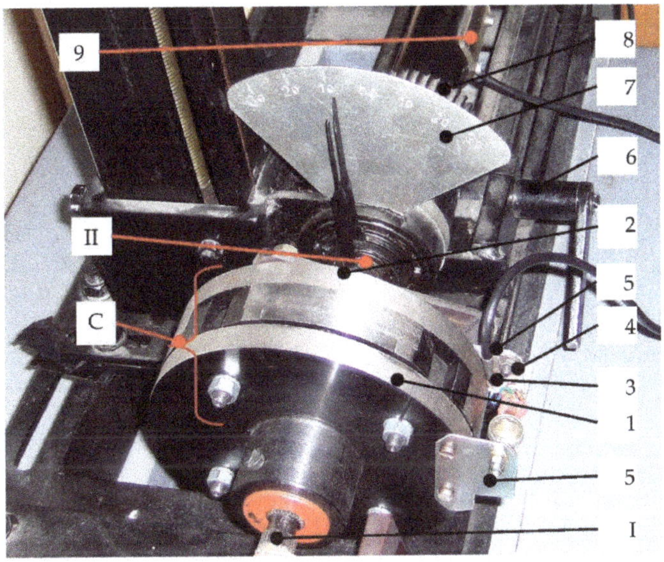

Figure 8. Experimental stand for the study of elastic coupling.

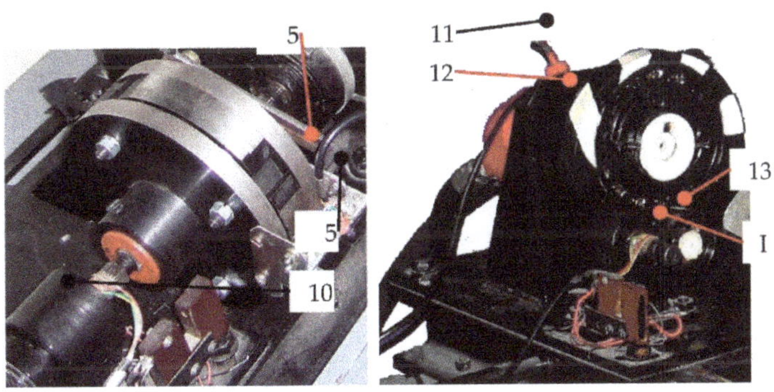

Figure 9. Experimental stand for the study of elastic coupling details.

To determine the desired parameters, the installation is equipped with transducers (pulse transducer, angle transducer), measuring and control devices, devices for visualizing the studied phenomenon.

In order to determine the angular deformations of the coupling in a dynamic regime, it was necessary to use a deformation sensor, which was applied on the driven half-coupling 2. The deformation sensor 3 was glued on a metal plate 4 made of spring steel, fixed to a metal plate 5 (bent in the shape of the letter "L"), by means of two screws and two nuts (Figure 8).

- Figure 8 shows the coupling C to be tested mounted on the stand, the input (I) and output shafts (II), the conductive and driven half-coupling 2, the deformation measuring sensor 3 fixed on a metal plate 4 in the shape of the letter L which is attached to the half-coupling 2, where: 4—metal lamella, 5—metal plate fixed to the half-coupling 2, 6—crank, 7—angular transducer, 8—prestressing system, 9—TER represents resistive electrical transducers, 10—support bearing, 11—hydraulic motor, and 12 -, 13 -, and 14—the hydraulic regulator of the stand.

- The driving half-coupling is mounted on the input shaft I of the test stage, and the driven half-coupling on the output shaft II of the stand.
- The electronic equipment necessary for the dynamic tests of the tested coupling consists of two tensometric bridges and an oscilloscope. In Figure 9, item 11 represents the tensometric bridges that measure the variation of the relative rotation angle between the half-couplings and the torque at the output shaft and 13 is the oscillograph that records the torque at the output shaft as a function of the relative rotation angle between the half-couplings.

The purposes of the experiments are:

➢ measuring the torque at the input and at the output shaft;
➢ measuring the relative deformation that appears between the two half-couplings;
➢ obtaining the dynamic rigidity of the elastic coupling.

The coupling tests were performed in dynamic mode, taking into account two distinct situations: (1) the two input shaft and output shaft mounted collinearly; (2) the two input shaft-output shafts radially offset by 0.4 mm. In these situations, the variation of the torque at the output shaft was analyzed according to the angular deformation between the two half-couplings, a deformation that occurred as a result of the loading and unloading of the coupling.

First, the coupling was tested in a dynamic regime without radial misalignment of the input and output shafts. The maximum relative rotation angle between the two half-couplings being of maximum 2.2° at a maximum moment at the output shaft of 44 Nm. The second dynamic determination was performed with the input and output shafts radial offset by 0.5 mm. The radial displacement between the input and output shafts was obtained by the radial displacement of the output shaft.

By actuating the crank, the dynamic loading and unloading of the prototype was performed in case of radial misalignment of the two input shafts, the oscilloscope recording the variation of the angular deformation and the torque at the output shaft II.

The data obtained after measurement are presented in Table 1. The dynamic rigidity of the tested couplings varies depending on the relative value of the angle between the 2 half-couplings. In the experiment a radial deviation of $\Delta r = 0.4$ mm is considered.

Table 1. Dynamic rigidity of tested prototype with radial deviation $\Delta r = 0.4$ mm, input-output shafts, in dynamic regime.

Zone	**Rotation Relative Angle φ between Semi-Couplings**	**Torsion Moment to Entry-Out-Put Shaft**	**Dynamic Rigidity**	**Medium Dynamic Rigidity,**
-	**[Degree]**	**[Nmm]**	**[Nmm/Degree]**	
Radial deviation $\Delta r = 0.4$ mm by input-out-put shafts, in dynamic regime				
I	0.185184	5600	30,240.27	34,585
	0.555551	21,730	39,114.35	
	0.740734	24,500	33,075.30	
	0.740734	24,500	33,075.30	
II	0.246000	6700	27,235.77	31,979
	0.462959	19,200	41,472.37	
	0.740734	24,500	33,075.30	
	0.740734	24,500	33,075.30	
	0.581000	19,090	32,857.14	
	0.442959	10,700	24,155.75	

The dynamic rigidity of elastic coupling is defined as the ratio of the torque obtained in dynamic regime to the relative rotation angle between the semi-couplings. The medium dynamic rigidity of coupling is obtained with relation:

$$k_{dynamic} = \frac{\sum_{i=1}^{n} k_{dynamic\ i}}{n}, \quad (1)$$

where n represents the number of dates on graphic and $k_{dynamic\ i}$ is the partial dynamic rigidity same of each dates

After processing the experimental data obtained with a radial misalignment by 0.4 mm between the input and output shaft, the torque diagram is obtained. The variation of the torque and the hysteresis of the elastic non-metallic element can be seen in Figure 10.

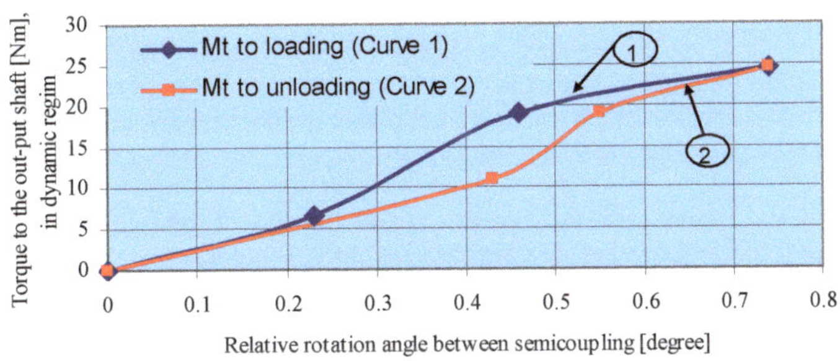

Figure 10. The dynamic variation of torque to the output shaft depending on the relative rotation angle between the two half-couplings (radial deviation Δr = 0.4 mm).

The coupling was also tested in dynamic regime without radial misalignment of the input and output shafts, the maximum relative rotation angle between the two half-couplings being of maximum 2.2° at a maximum moment at the output shaft of 44 Nm.

The rigidity of the coupling is represented by the ratio between the torque and the relative angular deformation between the half-couplings.

The damping causes the moment-deformation curve to change into a hysteresis curve. The most important factors that influence the dynamic rigidity are: the operating temperature of the coupling, the average torque, the frequency, and the amplitude of the torque. Heating of the elastic elements can occur due to the ambient temperature or due to internal friction under the influence of the torque. Due to the increase of the temperature of the elastic elements, the relative angular deformations of the half-couplings increases and both the static and the dynamic rigidity decreases. As the temperature increases, the modulus of elasticity decreases. The rubber chosen to manufacture the non-metallic elements is resistant to oils or diesel fuel.

The nonlinear characteristics of nonmetallic elements also depend on their geometry.

Table 2 shows the dynamic rigidity of the tested coupling calculated when the input and output shafts are parallel, with a radial deviation of Δr = 0.4 mm.

Table 2. Medium rigidity of tested coupling with radial deviation Δr = 0.4 mm by input-output shafts, in dynamic regime.

Medium Rigidity of Coupling [N·mm/Degree] in Dynamic Regime		
Input-Output Shafts: *Collinear*	**With *Radial Deviation* Δr = 0.4 mm by Input-Output Shafts**	
18,243.25	Zone I:	34,585
	Zone II:	31,979
	Medium (Average):	33,282

3. Discussion

Before testing the elastic coupling in dynamic regime, the input and output shafts of the stand were balanced, the coupling was balanced, and a sensor was applied to the coupling. During the experiments, progressive loading and unloading was performed. Deformations and stress in the non-metallic material depend on the loading torque.

The rigidity depends as well on the torque M_t and on the relative rotation angle between the half-couplings. After testing the coupling in dynamic mode, the elastic characteristic of the coupling was obtained.

The shape of the elastic element and the material from which it is obtained, determine the elastic characteristic. The maximum angular deformation between the two half-couplings is 0.7407340 degrees when the input and output shafts have a deviation between axes by 0.4 mm. The maximum relative angular deformation between the two half-couplings is $\Delta\varphi = 2.2°$ from when no radial misalignment is introduced.

Value of medium dynamic rigidity of coupling, on the first zone of testing is 34,585 Nmm/degree. The value of medium dynamic rigidity of coupling, on the second zone of testing is 31,979 Nmm/degree. The value of mean dynamic rigidity of coupling is 33,282 N.mm/degree. If the two shafts are collinearly, the medium dynamic rigidity of coupling is 18,243.25 Nmm/degree.

From the comparison of the results obtained in the dynamic test without radial offset and with radial offset, it was found that the dynamic stiffness of the coupling is higher in the dynamic regime. From the comparison of the results obtained in the dynamic test without radial misalignment (input-output shafts being collinear) with the results obtained in the dynamic test with radial misalignment (input-output shafts being non-collinear) it is found that the maximum relative angular deformation between two half-couplings is $\Delta\varphi = 2.2°$ from when no radial misalignment is introduced and $\Delta\varphi = 0.74°$ from when the input-output shafts are radial offset by $\Delta r = 0.4$ mm.

The damping capacity of torsional shocks for an elastic coupling with bolts and non-metallic intermediate elements help the system to convert a part of energy into heat and the rest is transformed into deformation energy.

4. Conclusions

For this prototype, depending on the design and material of the elastic element, regardless of whether the input-output shafts are non-collinear, can be obtained different elastic characteristics and different partial and average dynamic.

The experimental dynamic stiffness obtained in the case of radial misalignment between input-output shafts is higher as in the case when input-output shafts are collinear. During the experimental testing of this coupling, a silent operation of the prototype is detected.

It was found that the dynamic average stiffness, if the input-output shafts are a radially deviation of 0.4 mm, is lower than the static average stiffness, if the input-output shafts are collinear.

Therefore, it is found that in dynamic regime without radial misalignment the tested design version takes a maximum angular deformation of relative rotation between the two half-couplings greater than the maximum angular deformation of relative rotation between the two semi-couplings in dynamic regime with radial misalignment.

Figure 10 shows the nonlinear elastic characteristic of the coupling at loading and unloading with load, the hysteresis of the coupling, as well as the fact that the coupling has variable rigidity.

The designed coupling takes over larger angular deviations in the dynamic regime without misalignment than the classic couplings in the current study for the constructive dimension at which it was designed.

During the dynamic experimental testing of prototype, the following were found: different elastic characteristics can be obtained, depending on the constructive shape and the material of the non-metallic element; and no vibrations or radial beats occurred, which confirms a good design and a good execution.

Author Contributions: Conceptualization, M.G., I.-M.G., S.V., P.N.B.; methodology, M.G., I.-M.G., S.V.; software, M.G., I.-M.G. and P.N.B.; validation, M.G., I.-M.G., and S.V.; formal analysis, M.G., I.-M.G., and P.N.B.; investigation M.G.; resources, M.G., S.V. and P.N.B.; data curation, I.-M.G., M.G. and P.N.B.; writing—original draft preparation, M.G.; writing—review and editing, M.G., I.-M.G., S.V.; visualization, M.G. and S.V.; supervision, I.-M.G., M.G., S.V. and P.N.B. All authors have read and agreed to the published version of the manuscript.

Funding: This research received no external funding.

Institutional Review Board Statement: Not applicable.

Informed Consent Statement: Not applicable.

Data Availability Statement: Not applicable.

Acknowledgments: We want to thank the reviewers who have read the manuscript carefully and have proposed pertinent corrections that have led to an improvement in our manuscript.

Conflicts of Interest: The authors declare no conflict of interest.

References

1. Budynas, R.; Nisbett, K. *Mechanical Engineering Design*, 10th ed.; McGraw-Hill: New York, NY, USA, 2014.
2. Dieter, G.; Schmidt, L. *Engineering Design*, 5th ed.; McGraw-Hill Education: New York, NY, USA, 2012.
3. Mott, R.; Vavrek, E.; Wang, J. *Machine Elements in Mechanical Design*, 6th ed.; Pearson: London, UK, 2017.
4. Mzsyka, D. *Machines & Mechanisms: Applied Kinematic Analysis*, 4th ed.; Pearson: London, UK, 2010.
5. Oberg, E. *Machinery's Handbook Toolbox*, 31st ed.; Industrial Press Inc.: Norwalk, CT, USA, 2020.
6. Parmlez, R. *Mechanical Components*, 1st ed.; McGraw-Hill: New York, NY, USA, 2000.
7. Norton, R. *Design of Machinery*, 5th ed.; McGraw-Hill: New York, NY, USA, 2011.
8. Spotts, M.; Shoup, T.; Hornberger, L. *Design of Machine Elements*, 8th ed.; Pearson: New Delhi, India, 2003.
9. Standard STAS 5982. Flexible Coupling with Bolts. Available online: https://www.etansari-mecanice-pompe.ro/cat/cuplaje-elastice-mecanice-permanente/cuplaje-mobile-elastice-cu-bolturi-element-elastic-din-cauciuc/ (accessed on 20 October 2020).
10. Flexible Couplings with Rubber Studs. Available online: http://www.velhightech.com/Documents/ME8593%20Design%20of%20Machine%20Elements.pdf (accessed on 20 October 2020).
11. A Flexible Coupling with Stifts and Washers or Rubber Bushes. Available online: https://www.kdkce.edu.in/upload//Machine%20elements,%20Power%20Transmission%20Devices.pdf (accessed on 20 October 2020).
12. Guardian-Couplings. Available online: https://www.guardiancouplings.com/-/media/Files/Literatur/Brand/guardian-couplings/related/brochures/p-7733-gc.ashx (accessed on 20 October 2020).
13. RUPEX®Couplings. Available online: https://www.flender.com/en/Products/Couplings/RUPEX-Pin-and-Bush-Coupling/p/ATN03101 (accessed on 26 October 2020).
14. Module 5 Couplings. A Typical Flexible Coupling with Rubber Bushings. Available online: https://nptel.ac.in/content/storage2/courses/112105125/pdf/module-5%20lesson-2.pdf (accessed on 20 October 2020).
15. Flexible-Coupling. Available online: https://www.comintec.com/en/project/flexible-coupling (accessed on 20 October 2020).
16. Construction-Installation-and-Maintenance-of-Flexible-Couplings.pdf. Available online: https://practicalmaintenance.net/wp-content/uploads/Construction-Installation-and-Maintenance-of-Flexible-Couplings.pdf (accessed on 26 October 2020).
17. Rupex-Claw-Couplings. Available online: https://www.geartech.no/en/products/transmission/shaft-couplings/rupex-claw-couplings/ (accessed on 26 October 2020).
18. A Flange Coupling and Non-Metallic Barrel elements. Available online: https://www.researchgate.net/publication/331154663_Design_and_Analysis_of_Bushed_Pin_Flexible_Co.upling?enrichId=rgreq-10a027bed8d6a339a6fd0b9e3c43d0b4-XXX&enrichSource=Y292ZXJQYWdlOzMzMTE1NDY2MztBUzo3MjY5OTMwMTY0Nzk3NDRAMTU1MDM0MDA1MDY4Ng%3D%3D&el=1_x_2&_esc=publicationCoverPdf (accessed on 25 October 2020).
19. Patent SU 1262417 A1. Elastic Coupling with Rubber-Bushed Studs. Available online: https://worldwide.espacenet.com/publicationDetails/biblio?II=15&ND=3&adjacent=true&locale=en_EP&FT=D&date=19861007&CC=SU&NR=1262147A1&KC=A1 (accessed on 10 November 2020).
20. Patent CN 103925303/16.07.2014. Rubber-Cushioned Sleeve Bearing Coupling Fast to Replace. Patent CN 106321676A, 11 January 2017. Elastic Coupling. Available online: https://worldwide.espacenet.com/patent/search/family/057733218/publication/CN106321676A?q=elastic%20coupling%20rubber%20bolt (accessed on 11 November 2020).
21. Timur, C.K. Design and Analysis of Bushed Pin Flexible Coupling. *IJARESM* **2017**, *5*, 38–44.
22. Prodănescu, G.S.; Iliuc, I. Elastic Coupling, with Elastic, Non-Metallic, Radial, Discrete Elements. Patent RO 116430 (B1), 30 January 2001.

23. Hangzhou, X.K. Transmission Machinery Co Ltd. Maritime Coupling. Utility Model. CN 203847581U, 24 September 2014.
24. Herwarth, R. Elastic Coupling. U.S. Patent 2924082A, 9 February 1960.
25. Vlase, S. A Method of Eliminating Lagrangian-Multipliers from the Equation of Motion of Interconnected Mechanical Systems. *J. Appl. Mech. Trans. ASME* **1987**, *54*, 235–237. [CrossRef]
26. Radu, M. Theoretical and Experimental Studies as Concerns Couplings with Nonmetallic Elastic Elements. Ph.D. Thesis, Transylvania University of Brasov, Brașov, Romania, 2005.
27. Autodesk Mechanical Desktop 6 Power Pack. Available online: https://usermanual.wiki (accessed on 26 October 2020).
28. Martini, A.; Troncossi, M.; Vincenzi, N. Structural and elastodynamic analysis of rotary transfer machines by Finite Element model. *J. Serbian Soc. Comput. Mech.* **2017**, *11*, 1–16. [CrossRef]
29. Manzato, S.; Peeters, B.; Osgood, R.; Luczak, M. Wind turbine model validation by full-scale vibration test. In Proceedings of the European Wind Energy Conference (EWEC), Warsaw, Poland, 20–23 April 2010.
30. Troncossi, M.; Taddia, S.; Rivola, A.; Martini, A. Experimental Characterization of a High-Damping Viscoelastic Material Enclosed in Carbon Fiber Reinforced Polymer Components. *Appl. Sci.* **2020**, *10*, 6193. [CrossRef]
31. Carden, E.P.; Lindblad, M. Operational modal analysis of torsional modes in rotating machinery. *J. Eng. Gas Turbines Power* **2015**, *137*, 022501. [CrossRef]

symmetry MDPI

Article

Symmetries for Nonconservative Field Theories on Time Scale

Octavian Postavaru *,† and Antonela Toma †

Center for Research and Training in Innovative Techniques of Applied Mathematics in Engineering, University Politehnica of Bucharest, 060042 Bucharest, Romania; antonela.toma@upb.ro
* Correspondence: opostavaru@linuxmail.org; Tel.: +40-0770-241-912
† These authors contributed equally to this work.

Abstract: Symmetries and their associated conserved quantities are of great importance in the study of dynamic systems. In this paper, we describe nonconservative field theories on time scales—a model that brings together, in a single theory, discrete and continuous cases. After defining Hamilton's principle for nonconservative field theories on time scales, we obtain the associated Lagrange equations. Next, based on the Hamilton's action invariance for nonconservative field theories on time scales under the action of some infinitesimal transformations, we establish symmetric and quasi-symmetric Noether transformations, as well as generalized quasi-symmetric Noether transformations. Once the Noether symmetry selection criteria are defined, the conserved quantities for the nonconservative field theories on time scales are identified. We conclude with two examples to illustrate the applicability of the theory.

Keywords: time scales; Noether theory; conserved quantity

Citation: Postavaru, O.; Toma, A. Symmetries for Nonconservative Field Theories on Time Scale. *Symmetry* **2021**, *13*, 552. https://doi.org/10.3390/sym13040552

Academic Editors: Marin Marin and Dumitru Baleanu

Received: 22 February 2021
Accepted: 24 March 2021
Published: 26 March 2021

1. Introduction

Noether's theorem, considered the most beautiful theorem in mathematical physics, attributes a conservation law to each symmetry [1]. In fact, Noether's theorem states that if the Lagrangian of a physical system is not affected by a continuous transformation in its proper coordinate system, then there is a conservation law, more precisely a quantity that remains constant. There is also an extension of Noether's theorem, called Noether's second theorem, which states that a discretization that maintains a variational symmetry creates a proper conservation law.

In general, the studied systems are dissipative, regardless of whether they have applicability in physics or engineering. This field of research is of particular interest due to the few existing contributions in the literature. A consistent formalism for describing the interaction of certain systems with the environment is described in [2]. This problem is a particular case of a class of problems called the inverse problem of mechanics and occurs when a Lagrangian is constructed from the equations of motion of a mechanical system. Since the end of the last century, this topic has long been studied by many mathematicians and theoretical physicists [3–6], and the interest of physicists in this subject has recently increased with the quantification of dissipative systems [7–10].

The study of real-time dissipative processes, such as radiative processes, is not possible without a technique that is capable of introducing nonconservative interactions into action. In [11], we find a series of effective field theory techniques that can be seen as modern theoretical tools to find the existence of hierarchies of scale for a physical problem. Power-counting results are presented for several situations of practical interest, and several applications of quantum electrodynamics are also illustrated. In the course notes [12], some of the basic ideas of effective field theories are introduced, such as relevant and irrelevant operators and scaling, renormalization in effective field theories, the decoupling of heavy particles, power counting and naive dimensional analysis. A number of introductions to some of the latest techniques and applications in the field can be found in [13].

To unify continuous analysis with discrete analysis, Hilger [14] introduced calculation on time scales. The paper presented in [15] developed the concept of the discrete calculus of variations and showed how it can be applied to the optimization problems of discrete systems. In [16], the authors investigate the invariant properties of the discrete Lagrangian for conservative systems, the discrete analog of the calculus of variations and the Noether theorem, and another work [17] extends this theory to nonconservative systems. Noether conservation symmetries and laws for nonconservative discrete systems with irregular lattices were developed in [18], first by finding the discrete analog of Noether identities and then by introducing the generalized quasi-external equations and their properties. In [19,20], the Noether symmetries and the conservation laws of nonconservative and nonholonomic mechanical systems on time scales were analyzed—a theory that unifies the two cases of continuous and discrete theories. For the first time, a relationship between isochronous variations and delta derivatives was established, as well as a link between isochronous variations and total variations on time scale. Q-versions of some basic concepts of continuous variational calculus such as the Euler–Lagrange equation and its applications to the isoperimetric Lagrange and optimal control problems were introduced in [21].

In the present paper, we extend the formalism of the nonconservative field theory presented in [22] to time scales. The field theory on discrete time is very different from classical dynamics except for the fact that it is a multi-freedom system, because every point of discrete time has a dynamical freedom. Apart from this fundamental character, the field theory has engineering applications in gravitation, electrostatics, magnetism, electric current flow, conductive heat transfer, fluid flow and seepage [23].

The paper is structured as follows: In Section 2, we examine delta differentiation and integration on time scales and review some properties that we use later. In Section 3, we derive the Lagrange equation for nonconservative field theories on time scales. Further, in Section 4 we discuss the criteria of symmetric and quasi-symmetric Noether transformations for two types of transformations, and then in Section 5, we define and identify the quantities that are conserved. We end the paper with some illustrative examples.

2. Preliminaries and Notations

In this section, we give some basic knowledge about the calculus on time scales introduced and developed by Hilger [14]. A time scale $\mathbb{T}$ is an arbitrary nonempty closed subset of the real numbers. From the multitude of examples, we would like to mention the following: the real numbers $\mathbb{R}$, integers $\mathbb{Z}$, natural numbers $\mathbb{N}$, a sequence of points of $\mathbb{R}$ with a varying step size $\mathbb{S}$, the Cantor set $\mathbb{D}$ and sequence of closed intervals $\mathbb{P}$. In general, $\mathbb{P}$ is understood as a time scale which underpins differential equations with pulses. The theory presented in this section is valid for any type of time scale.

Definition 1. *The mapping $\sigma : \mathbb{T} \to \mathbb{T}$ defined by $\sigma(\tau) \equiv \inf\{s \in \mathbb{T} | \, s > \tau\}$ is called the jump operator. Accordingly, we may define the backward jump operator $\rho : \mathbb{T} \to \mathbb{T}$ to be the mapping $\rho(\tau) \equiv \sup\{s \in \mathbb{T} | \, s < \tau\}$. Via these two operators, we may classify the points $\tau \in \mathbb{T}$ in terms of their right and left neighborhood as follows: τ is called right-dense, right-scattered, left-dense or left-scattered if $\sigma(\tau) = \tau$, $\sigma(\tau) > \tau$, $\rho(\tau) = \tau$ or $\rho(\tau) < \tau$, respectively.*

Definition 2. *If $\mathbb{T}$ has a left-scattered maximum M, then we define $\mathbb{T}^k = \mathbb{T} - \{M\}$; thus, this k-operator cuts off an eventually existing isolated maximum of $\mathbb{T}$. The graininess function $\mu : \mathbb{T}^k \to [0, \infty)$ is defined by $\mu(\tau) = \sigma(\tau) - \tau$.*

Remark 1. *If $\mathbb{T} = \mathbb{R}$, then $\sigma(\tau) = \rho(\tau) = \tau$. If $\mathbb{T} = \mathbb{Z}$, $\sigma(\tau) = \tau + 1$, $\rho(\tau) = \tau - 1$, and $\mu(\tau) = 1$.*

Definition 3. *A function $f : \mathbb{T} \to \mathbb{R}$ is called differentiable at $\tau \in \mathbb{T}$, with derivative $f^{\Delta}(\tau) \in \mathbb{T}$, if for each $\epsilon > 0$ there is a neighborhood U of τ such that*

$$|f(\sigma(\tau)) - f(s) - f^{\Delta}(\tau)(\sigma(\tau) - s)| \leq \epsilon|\sigma(\tau) - s|, \quad \text{for all} \quad s \in U.$$

The derivative can also be defined in terms of a limit as follows:

$$f^{\Delta}(\tau) \equiv \frac{\Delta f}{\Delta \tau} = \lim_{s \to \tau} \frac{f(\sigma(\tau)) - f(s)}{\sigma(\tau) - s}.$$

If $f, g : \mathbb{T} \to \mathbb{R}$ are differentiable in τ, then one has the following properties:

$$\begin{aligned} f(\sigma(t)) = f^{\sigma}(\tau) &= f(\tau) + \mu(\tau) f^{\Delta}(\tau), \\ (af + bg)^{\Delta}(\tau) &= af^{\Delta}(\tau) + bg^{\Delta}(\tau), \\ (fg)^{\Delta}(\tau) &= f^{\Delta}(\tau) g^{\sigma}(\tau) + f(\tau) g^{\Delta}(\tau) = f^{\Delta}(\tau) g(\tau) + f^{\sigma}(\tau) g^{\Delta}(\tau). \end{aligned} \tag{1}$$

Remark 2. *If $\mathbb{T} = \mathbb{R}$, then $f^{\Delta}(\tau) = f'(\tau)$. If $\mathbb{T} = \mathbb{Z}$, then $f^{\Delta}(\tau) = f(\tau + 1) - f(\tau)$.*

Definition 4. *A function $f : \mathbb{T} \to \mathbb{R}$ is called rd-continuous if it is continuous at right-dense points and its left-sided limits exist (finite) at left-dense points. The set of rd-continuous functions is denoted by C_{rd}, and the set of differentiable functions with rd-continuous derivative is denoted by C^1_{rd}.*

Remark 3. *Every rd-continuous functions have an antiderivative. In particular, if $\tau_1, \tau_2 \in \mathbb{T}$, then the antiderivative function F is defined as*

$$F(\tau) \equiv \int_{\tau_1}^{\tau_2} f(\tau)\, \Delta\tau,$$

with $\Delta\tau$ the measure on $\mathbb{T}$ [14].

Theorem 1. *For a strictly increasing function $\nu : \mathbb{T} \to \mathbb{R}$, we have the following properties:*
(1) (Chain rule) Let $\omega : \mathbb{T}^* \to \mathbb{R}$, where $\mathbb{T}^* \equiv \nu(\mathbb{T})$. If $\nu^{\Delta}(\tau)$ and $\omega^{\Delta^*}(\nu(\tau))$ exists, then

$$(\omega \circ \nu)^{\Delta} = \left(\omega^{\Delta^*} \circ \nu\right) \nu^{\Delta}, \tag{2}$$

(2) (Derivative of the inverse) For $\nu^{\Delta}(\tau) \neq 0$, we have the following relation:

$$\left(\nu^{-1}\right)^{\Delta^*}(\nu(\tau)) = \frac{1}{\nu^{\Delta}(\tau)},$$

(3) (Substitution) If $f : \mathbb{T} \to \mathbb{R}$ is a C_{rd} function and ν is a C^1_{rd} function, then

$$\int_a^b f(\nu(\tau)) \nu^{\Delta}(\tau) \Delta\tau = \int_{\nu(a)}^{\nu(b)} \left(f \circ \nu^{-1}\right)(s) \Delta^* s, \tag{3}$$

with $a, b \in \mathbb{T}$.

Lemma 1. *(Dubois–Reymond) Let $g : [a, b] \to \mathbb{R}^n$, and $g \in C_{\text{rd}}$, then*

$$\int_a^b g(\tau)\, \eta^{\Delta}(\tau) \Delta\tau = 0,$$

for all $\eta \in C^1_{rd}$ *with* $\eta(a) = \eta(b) = 0$, *if and only if* $g(\tau) = c$, *with* $c \in \mathbb{R}^n$.

3. Lagrange Equation for Nonconservative Field Theories on Proper Time Scales

We consider N fields $\Phi^I(x^\mu)$, with $I = \overline{1,N}$ and $x^\mu = (x^0, x^1, x^2, x^3) \in \{T \times V\}$ with the time segment $T = [t_1, t_2]$ with $t \in \mathbb{T}$ and space volume V. To capture the nonconservative (e.q., dissipative) effects of a field that is subject to nonconservative interactions, we must double its degrees of freedom $\Phi^I \to (\Phi^I_1, \Phi^I_2)$. Thus, we can explain the correct causal evolution of the dynamics of the open system.

We denote the nonconservative Lagrange density with Ω, and it is generally an arbitrary function of two variables, their derivatives and their space-time coordinates x_μ:

$$\Omega\left[\Phi^I_a\right] = \Omega\left(x^\mu, \left(\Phi^I_a\right)^\sigma, \left(\Phi^I_a\right)^\Delta, \partial_i\left(\Phi^I_a\right)^\sigma\right), \tag{4}$$

with $i = \overline{1,3}$ and $(\Phi^I_a)^\sigma(x^\mu) = \Phi^I_a(\sigma(t), x^1, x^2, x^3)$. Additionally, Φ^I_a satisfies Definition 1 in the sense that

$$\frac{\Delta}{\Delta t}\Phi^I_a(t, x^1, x^2, x^3) = \lim_{s \to t} \frac{\Phi^I_a(\sigma(t), x^1, x^2, x^3) - \Phi^I_a(s, x^1, x^2, x^3)}{\sigma(t) - s}.$$

In the case of nonconservative field theories, we state Hamilton's principle on a time scale as

$$\int_{t_1}^{t_2} \Delta t \int_V d^3x\, \delta\Omega\left[\Phi^I_a\right] = 0, \tag{5}$$

where the isochronous variations are represented by δ.

Proposition 1. *For isochronous variations, we have the following relations:*

$$\delta\left(\Phi^I_a\right)^\Delta = \left(\delta\Phi^I_a\right)^\Delta, \quad \text{and} \quad \delta\left(\Phi^I_a\right)^\sigma = \left(\delta\Phi^I_a\right)^\sigma. \tag{6}$$

Proof. For the first relationship, we have

$$\begin{aligned}\left(\delta\Phi^I_a(t)\right)^\Delta &= \frac{\Delta}{\Delta t}\left(\delta\Phi^I_a\right) = \frac{\Delta}{\Delta t}\left(\frac{\partial\Phi^I_a}{\partial x^\mu}\delta x^\mu\right) = \frac{\Delta}{\Delta t}\left(\frac{\partial\Phi^I_a}{\partial x^i}\delta x^i\right) \\ &= \frac{\partial}{\partial x^i}\left(\frac{\Delta\Phi^I_a}{\Delta t}\right)\delta x^i = \frac{\partial(\Phi^I_a)^\Delta}{\partial x^i}\delta x^i = \delta\left(\Phi^I_a\right)^\Delta,\end{aligned}$$

where $\mu = \overline{0,3}$, $i = \overline{1,3}$ and we used Einstein's summation convention.

For the second relationship, we have

$$\left(\delta\Phi^I_a\right)^\sigma = \left(\frac{\partial\Phi^I_a}{\partial x^i}\delta x^i\right)^\sigma = \frac{\partial(\Phi^I_a)^\sigma}{\partial x^i}\delta x^i = \delta\left(\Phi^I_a\right)^\sigma.$$

□

Proposition 2. *The relationship between the total variation and the isochronous variation is given by the formula*

$$\Delta\Phi^I_a = \left(\Phi^I_a\right)^\Delta \Delta t + \delta\Phi^I_a. \tag{7}$$

Proof. The total variation of Φ_a^I is written as follows:

$$\Delta\Phi_a^I = \frac{\partial \Phi_a^I}{\partial t}\delta t + \frac{\partial \Phi_a^I}{\partial x^i}\delta x^i \,, \tag{8}$$

and the isochronous one is as follows:

$$\delta\Phi_a^I = \frac{\partial \Phi_a^I}{\partial x^i}\delta x^i \,. \tag{9}$$

Next, we consider $t = t(\alpha)$, where α is a generalized parameter that has convenient interpretations for the classical and relativistic case. We can write

$$\Delta t = \frac{\partial t}{\partial \alpha}\delta\alpha \,.$$

From Equations (8) and (9), we get

$$\Delta\Phi_a^I = \frac{\partial \Phi_a^I}{\partial t}\frac{\partial t}{\partial \alpha}\delta\alpha + \delta\Phi_a^I \,,$$

and the demonstration is done. □

We can write Hamilton's principle (5) in the following form:

$$\int_{t_1}^{t_2} \Delta t \int_V d^3x \left(\frac{\partial \Omega}{\partial x^\mu}\delta x^\mu + \frac{\partial \Omega}{\partial (\Phi_a^I)^\sigma}\delta\left(\Phi_a^I\right)^\sigma + \frac{\partial \Omega}{\partial (\Phi_a^I)^\Delta}\delta\left(\Phi_a^I\right)^\Delta \right.$$
$$\left. + \frac{\partial \Omega}{\partial \partial_i (\Phi_a^I)^\sigma}\delta\partial_i\left(\Phi_a^I\right)^\sigma \right) = 0 \,, \tag{10}$$

where the I indices satisfy the Einstein summation convention.

Next, we use the following relation:

$$\int d^3x \frac{\partial \Omega}{\partial(\partial_i\Phi)}\delta(\partial_i\Phi) = \int d^3x \left[-\partial_i\left(\frac{\partial \Omega}{\partial(\partial_i\Phi)}\right)\delta\Phi + \partial_i\left(\frac{\partial \Omega}{\partial(\partial_i\Phi)}\delta\Phi\right)\right] . \tag{11}$$

Because the edges of the volume are considered fixed, the second term of this expression does not contribute to the result. The first term in Equation (10) is a constant that does not influence the final result. Thus, Equation (10) together with Equation (11) reduce the Hamilton problem to

$$\int_{t_1}^{t_2} \Delta t \int_V d^3x \left(\frac{\partial \Omega}{\partial (\Phi_a^I)^\sigma}\delta\left(\Phi_a^I\right)^\sigma + \frac{\partial \Omega}{\partial (\Phi_a^I)^\Delta}\delta\left(\Phi_a^I\right)^\Delta \right.$$
$$\left. - \partial_i\left(\frac{\partial \Omega}{\partial \partial_i (\Phi_a^I)^\sigma}\right)\delta\left(\Phi_a^I\right)^\sigma \right) = 0 \,,$$

or

$$\int_{t_1}^{t_2} \Delta t \int_V d^3x \left[\left(\frac{\partial \Omega}{\partial (\Phi_a^I)^\sigma} - \partial_i\left(\frac{\partial \Omega}{\partial \partial_i (\Phi_a^I)^\sigma}\right)\right)\delta\left(\Phi_a^I\right)^\sigma \right.$$
$$\left. + \frac{\partial \Omega}{\partial (\Phi_a^I)^\Delta}\delta\left(\Phi_a^I\right)^\Delta \right] = 0 \,. \tag{12}$$

Proposition 3. *To transform Φ^σ in Φ^Δ, we use the following relation:*

$$\int_{t_1}^{t_2} f(t)(\delta\Phi(t))^\sigma \Delta t = -\int_{t_1}^{t_2}\int_{t_1}^{t} f(\tau)\Delta\tau(\delta\Phi(t))^\Delta \Delta t \,. \tag{13}$$

Proof. We use Equation (1) to achieve

$$\begin{aligned}\left[\int_{t_1}^{t} f(\tau)\Delta\tau\,\delta\Phi(t)\right]^{\Delta} &= \frac{\Delta}{\Delta t}\left[\int_{t_1}^{t} f(\tau)\Delta\tau\,\delta\Phi(t)\right] \\ &= \left[\int_{t_1}^{t} f(\tau)\Delta\tau\right]^{\Delta}(\delta\Phi(t))^{\sigma} + \int_{t_1}^{t} f(\tau)\Delta\tau\,(\delta\Phi(t))^{\Delta},\end{aligned}$$

or

$$\left[\int_{t_1}^{t} f(\tau)\Delta\tau\,\delta\Phi(t)\right]^{\Delta} = f(t)(\delta\Phi(t))^{\sigma} + \int_{t_1}^{t} f(\tau)\Delta\tau\,(\delta\Phi(t))^{\Delta}.$$

Integrating on both sides over Δt and applying Lemma 1, on the left side of the equality, we get zero. In Lemma 1, $g(t) = 1$ and

$$\eta(t) = \int_{t_1}^{t} f(\tau)\Delta\tau\,\delta\Phi(t) = (F(t) - F(t_1))\,\delta\Phi(t).$$

The four-volume is fixed at the boundary; therefore, $\eta(t_1) = \eta(t_2) = 0$. □

Theorem 2. *Lagrange equations for nonconservative theories on time scale are written*

$$\frac{\partial\Omega}{\partial(\Phi_a^I)^{\sigma}} - \partial_i\left(\frac{\partial\Omega}{\partial\partial_i(\Phi_a^I)^{\sigma}}\right) - \frac{\Delta}{\Delta t}\frac{\partial\Omega}{\partial(\Phi_a^I)^{\Delta}} = 0.$$

Proof. Combining Proposition 3 with Equation (12), we obtain

$$\begin{aligned}\int_{t_1}^{t_2}\Delta t\int_V d^3x\Bigg[-\int_{t_1}^{t}\left(\frac{\partial\Omega}{\partial(\Phi_a^I)^{\sigma}} - \partial_i\left(\frac{\partial\Omega}{\partial\partial_i(\Phi_a^I)^{\sigma}}\right)\right)\Delta\tau(\delta\Phi(\tau))^{\Delta} \\ + \frac{\partial\Omega}{\partial(\Phi_a^I)^{\Delta}}\delta\left(\Phi_a^I\right)^{\Delta}\Bigg] = 0.\end{aligned}$$

Because $\delta(\Phi_a^I)$ are independent of each other, the following expression is obtained with the help of Lemma 1:

$$-\int_{t_1}^{t}\left(\frac{\partial\Omega}{\partial(\Phi_a^I)^{\sigma}} - \partial_i\left(\frac{\partial\Omega}{\partial\partial_i(\Phi_a^I)^{\sigma}}\right)\right)\Delta\tau + \frac{\partial\Omega}{\partial(\Phi_a^I)^{\Delta}} = \text{const.}$$

Taking the delta derivatives of the above formula, the demonstration is finished. □

We define the current densities on time scale as follows:

$$\Pi_{Ia}^{0} = \frac{\partial\Omega}{\partial(\Phi_a^I)^{\Delta}}, \quad \Pi_{Ia}^{i} = \frac{\partial\Omega}{\partial\partial_i(\Phi_a^I)^{\sigma}}.$$

For historical reasons, it is advisable to separate the Lagrange density in conservative and nonconservative terms; i.e.,

$$\begin{aligned}\Omega\left[\Phi_1^I,\Phi_2^I\right] &= \mathcal{L}\left(x^{\mu},\left(\Phi_1^I\right)^{\sigma},\left(\Phi_2^I\right)^{\sigma},\left(\Phi_1^I\right)^{\Delta},\left(\Phi_2^I\right)^{\Delta},\partial_i\left(\Phi_1^I\right)^{\sigma},\partial_i\left(\Phi_2^I\right)^{\sigma}\right) \\ &\quad+\mathcal{K}\left(x^{\mu},\left(\Phi_1^I\right)^{\sigma},\left(\Phi_2^I\right)^{\sigma},\left(\Phi_1^I\right)^{\Delta},\left(\Phi_2^I\right)^{\Delta},\partial_i\left(\Phi_1^I\right)^{\sigma},\partial_i\left(\Phi_2^I\right)^{\sigma}\right), \quad (14)\end{aligned}$$

where $\mathcal{L}$ is a sum of independent non-interacting field Lagrangians for each of the fields to be considered to be relevant within the interaction Lagrangian density $\mathcal{K}$.

A more correct parametrization is obtained using the $\pm$ base, defined as [22]

$$\Phi^I_+ = \frac{\Phi^I_1 + \Phi^I_2}{2}, \quad \Phi^I_- = \Phi^I_1 - \Phi^I_2. \tag{15}$$

The physical limit (PL) is given by the equations

$$\Phi^I_+ \to \Phi^I, \quad \Phi^I_- \to 0. \tag{16}$$

Because only the variable "-" contributes to the physical result, we get

$$\left[\frac{\partial \Omega}{\partial (\Phi^I_-)^\sigma} - \partial_i \left(\frac{\partial \Omega}{\partial \partial_i (\Phi^I_-)^\sigma}\right) - \frac{\Delta}{\Delta t} \frac{\partial \Omega}{\partial (\Phi^I_-)^\Delta}\right]_{PL} = 0,$$

or using Equation (14),

$$\begin{aligned} &\frac{\partial \mathcal{L}}{\partial (\Phi^I)^\sigma} - \partial_i \left(\frac{\partial \mathcal{L}}{\partial \partial_i (\Phi^I)^\sigma}\right) - \frac{\Delta}{\Delta t} \frac{\partial \mathcal{L}}{\partial (\Phi^I)^\Delta} \\ &= \left[\partial_i \left(\frac{\partial \mathcal{K}}{\partial \partial_i (\Phi^I_-)^\sigma}\right) + \frac{\Delta}{\Delta t} \frac{\partial \mathcal{K}}{\partial (\Phi^I_-)^\Delta} - \frac{\partial \mathcal{K}}{\partial (\Phi^I_-)^\sigma}\right]_{PL} = \mathcal{Q}_I. \end{aligned}$$

If $\mathcal{Q}_I = 0$, the Lagrange equations for the nonconservative field theory turn into the usual differential equations of motion on time scales.

4. Noether Symmetries on Time Scales

In this section, we discuss some criteria for symmetric Noether transformations for nonconservative field theories on time scales. There is no concept of a complete classification of the conserved currents. In the following, all Φ functions will be read as Φ^I_a.

On time scale, the action has the following definition:

$$S[\Phi] = \int_{t_1}^{t_2} \Delta t \int_V d^3x \, \Omega[\Phi]. \tag{17}$$

Definition 5. *We say that the Hamilton action (17) is invariant to the following infinitesimal transformations, which depend on a parameter r*

$$t^* = t, \quad (\Phi(x^\mu))^* = \Phi(x^\mu) + \epsilon_\alpha \xi^\alpha(x^\mu), \tag{18}$$

with infinitesimal parameters ϵ_α $(\alpha = \overline{1,r})$ *and with the generators of the infinitesimal transformation* $\xi^\alpha \equiv \xi^{I\alpha}_a$, *if and only if*

$$\begin{aligned} &\int_{t_1}^{t_2} \Delta t \int_V d^3x \, \Omega\left(x^\mu, \Phi^\sigma, \Phi^\Delta, \partial_i \Phi^\sigma\right) \\ &= \int_{t_1}^{t_2} \Delta t \int_V d^3x \, \Omega\left(x^{*\mu}, \Phi^{*\sigma}, \Phi^{*\Delta}, \partial_i \Phi^{*\sigma}\right). \end{aligned}$$

This transformation is called a symmetrical Noether transformation.

Theorem 3. *Noether symmetric transformations corresponding to infinitesimal transformations (18) are subject to the following conditions:*

$$\frac{\partial \Omega}{\partial \Phi^\sigma} (\xi^\alpha)^\sigma + \frac{\partial \Omega}{\partial \Phi^\Delta} (\xi^\alpha)^\Delta + \frac{\partial \Omega}{\partial \partial_i \Phi^\sigma} (\partial_i \xi^\alpha)^\sigma = 0.$$

Proof. Equation (18) together with Equation (19) give

$$\begin{aligned}&\int_{t_1}^{t_2}\Delta t\int_V d^3x\,\Omega\Big(x^\mu,\Phi^\sigma,\Phi^\Delta,\partial_i\Phi^\sigma\Big)\\&=\int_{t_1}^{t_2}\Delta t\int_V d^3x\Omega\Big(x^{*\mu},\Phi^\sigma+\epsilon_\alpha(\xi^\alpha)^\sigma,\Phi^\Delta+\epsilon_\alpha(\xi^\alpha)^\Delta,\partial_i\Phi^\sigma+\epsilon_\alpha(\partial_i\xi^\alpha)^\sigma\Big).\end{aligned}$$

Deriving both sides of the equals at ϵ_α and doing $\epsilon_\alpha=0$, we obtained the desired result. □

Definition 6. *We say that the Hamiltonian action (17) is invariant to infinitesimal transformations (18) if and only if*

$$\begin{aligned}&\int_{t_1}^{t_2}\Delta t\int_V d^3x\,\Omega\Big(x^\mu,\Phi^\sigma,\Phi^\Delta,\partial_i\Phi^\sigma\Big)\\&=\int_{t_1}^{t_2}\Delta t\int_V d^3x\left[\Omega\Big(x^{*\mu},\Phi^{*\sigma},\Phi^{*\Delta},\partial_i\Phi^{*\sigma}\Big)+\frac{\Delta}{\Delta t}(\tilde{\Delta}G)\right],\end{aligned}$$

where $\tilde{\Delta}G=\epsilon_\alpha G^\alpha$, $G^\alpha=G^\alpha\left(x^\mu,\Phi^\sigma,\Phi^\Delta,\partial_i\Phi^\sigma\right)$. *This transformation is called a quasi-symmetric Noether transformation.*

Theorem 4. *Noether quasi-symmetric transformations corresponding to infinitesimal transformations (18) are subject to the following conditions:*

$$\frac{\partial\Omega}{\partial\Phi^\sigma}(\xi^\alpha)^\sigma+\frac{\partial\Omega}{\partial\Phi^\Delta}(\xi^\alpha)^\Delta+\frac{\partial\Omega}{\partial\partial_i\Phi^\sigma}(\partial_i\xi^\alpha)^\sigma=-\frac{\Delta}{\Delta t}G^\alpha.$$

Proof. Demonstration similar to that of the Theorem 3. □

Next, we consider the following sets of transformations:

$$\begin{aligned}t^*&=T(x^\mu,\epsilon^\alpha)=t+\epsilon_\alpha\xi_0^\alpha(x^\mu),\\(\Phi(x^\mu))^*&=P_k(x^\mu,\epsilon^\alpha)=\Phi(x^\mu)+\epsilon_\alpha\xi^\alpha(x^\mu),\end{aligned}\tag{19}$$

$k=\overline{1,a\,N}$, and $\xi_0^\alpha\equiv\xi_0^{I\alpha}$.

Consider the function $t\in[t_1,t_2]\to\beta(t)=T(x^\mu,\epsilon^\alpha)\in\mathbb{R}$ to be a strictly increasing of class $C^1_{\rm rd}$ whose image is a new time scale. We denote the jump operator and delta derivatives with σ^* and Δ^*. Between σ and σ^*, we have the following relation:

$$\sigma^*\circ\beta=\beta\circ\sigma.\tag{20}$$

Definition 7. *We say that the Hamilton action (17) is invariant to the infinitesimal transformations (19) if and only if*

$$\begin{aligned}&\int_{t_1}^{t_2}\Delta t\int_V d^3x\,\Omega\Big(x^\mu,\Phi^\sigma,\Phi^\Delta,\partial_i\Phi^\sigma\Big)\\&=\int_{\beta(t_1)}^{\beta(t_2)}\Delta^*t^*\int_V d^3x\,\Omega\Big(x^{*\mu},\Phi^{*\sigma^*},\Phi^{*\Delta^*},\partial_i\Phi^{*\sigma^*}\Big)+\int_{t_1}^{t_2}\Delta t\int_V d^3x\,\frac{\Delta}{\Delta t}(\tilde{\Delta}G),\end{aligned}$$

where $\tilde{\Delta}G=\epsilon_\alpha G^\alpha$, $G^\alpha=G^\alpha\left(x^\mu,\Phi^\sigma,\Phi^\Delta,\partial_i\Phi^\sigma\right)$. *This transformation is called a generalized quasi-symmetric Noether transformation.*

Theorem 5. *Noether generalized quasi-symmetric transformations corresponding to infinitesimal transformations (18) are subject to the following conditions:*

$$\frac{\partial\Omega}{\partial t}\xi_0^{I\alpha} + \Omega\left(\xi_0^\alpha\right)^\Delta + \frac{\partial\Omega}{\partial\Phi^\sigma}(\xi^\alpha)^\sigma + \frac{\partial\Omega}{\partial\Phi^\Delta}\left((\xi^\alpha)^\Delta - \Phi^\Delta(\xi_0^\alpha)^\Delta\right) + \frac{\partial\Omega}{\partial\partial_i\Phi^\sigma}(\partial_i\xi^\alpha)^\sigma = -\frac{\Delta}{\Delta t}G^\alpha .$$

Proof.

$$\begin{aligned} &\int_{\beta(t_1)}^{\beta(t_2)} \Delta^* t^* \int_V d^3x\,\Omega\left(x^{*\mu}, \Phi^{*\sigma^*}, \Phi^{*\Delta^*}, \partial_i\Phi^{*\sigma^*}\right) \\ = \quad & \int_{t_1}^{t_2} \Delta t \beta^\Delta(t) \int_V d^3x\,\Omega\left(\beta(t), x^{*i}, \Phi^* \circ \sigma^* \circ \beta, \frac{(\Phi^* \circ \beta)^\Delta}{\beta^\Delta(t)}, \partial_i\Phi^* \circ \sigma^* \circ \beta\right), \end{aligned} \tag{21}$$

where $(\Phi \circ f)(t) = \Phi(f(t), x^i)$, $i = \overline{1,3}$. To obtain the formula (21), we used the following property:

$$\left(\Phi^{*\Delta^*} \circ f\right)(t) = \frac{(\Phi^* \circ f)^\Delta}{f^\Delta(t)}(t) .$$

Using Equations (20), Equation (21) becomes

$$\int_{t_1}^{t_2} \Delta t \beta^\Delta(t) \int_V d^3x\,\Omega\left(\beta(t), x^{*i}, \Phi^* \circ \beta \circ \sigma, \frac{(\Phi^* \circ \beta)^\Delta}{\beta^\Delta(t)}, \partial_i\Phi^* \circ \beta \circ \sigma\right).$$

This equation together with the transformations (20) gives

$$\int_{t_1}^{t_2} \Delta t\, T^\Delta \int_V d^3x\,\Omega\left(T, x^{*i}, P_k^\sigma, \frac{P_k^\Delta}{T^\Delta}, \partial_i P_k^\sigma\right).$$

Differentiating this formula at ϵ_α and performing $\epsilon_\alpha = 0$ together with Equations (2) and (3), we get

$$\begin{aligned} &\int_{t_1}^{t_2} \Delta t \int_V d^3x\left[\frac{\partial\Omega}{\partial t}\xi_0^\alpha + \Omega\left(\xi_0^\alpha\right)^\Delta + \frac{\partial\Omega}{\partial\Phi^\sigma}(\xi^\alpha)^\sigma\right. \\ &\left.+\frac{\partial\Omega}{\partial\Phi^\Delta}\left((\xi^\alpha)^\Delta - \Phi^\Delta(\xi_0^\alpha)^\Delta\right) + \frac{\partial\Omega}{\partial\partial_i\Phi^\sigma}(\partial_i\xi^\alpha)^\sigma + \frac{\Delta}{\Delta t}G^\alpha\right] = 0 . \end{aligned}$$

Given the arbitrariness of the integration interval, we obtain the desired result. □

5. Noether's Theorems on a Time Scale

In this section, we define conserved quantities in nonconservative field theories on time scales.

Definition 8. *We say a function of type*

$$I\left(x^\mu, \Phi^\sigma, \Phi^\Delta, \partial_i\Phi^\sigma\right),$$

is a conserved quantity in nonconservative field theories on time scales if and only if

$$\frac{\Delta}{\Delta t} I\left(x^\mu, \Phi^\sigma, \Phi^\Delta, \partial_i\Phi^\sigma\right) = 0 .$$

Below, we present a series of conserved quantities that are obtained from Noether symmetries.

Theorem 6. *The infinitesimal transformations given by the Theorem 3 have the conserved quantities of the form*

$$I^{\alpha} = \frac{\partial \Omega}{\partial \Phi^{\Delta}} \xi^{\alpha} = c^{\alpha} \,. \tag{22}$$

Proof. Applying $\Delta/\Delta t$ to the right side of Equation (22), and after the operator $\int_{t_1}^{t_2} \Delta t \int_V d^3x$; i.e.,

$$\int_{t_1}^{t_2} \Delta t \int_V d^3x \left[\frac{\Delta}{\Delta t} \frac{\partial \Omega}{\partial \Phi^{\Delta}} (\xi^{\alpha})^{\sigma} + \frac{\partial \Omega}{\partial \Phi^{\Delta}} (\xi^{\alpha})^{\Delta} \right].$$

Using Theorem 2, we obtain

$$\int_{t_1}^{t_2} \Delta t \int_V d^3x \left[\frac{\partial \Omega}{\partial \Phi^{\sigma}} (\xi^{\alpha})^{\sigma} - \partial_i \frac{\partial \Omega}{\partial \partial_i \Phi^{\sigma}} (\xi^{\alpha})^{\sigma} + \frac{\partial \Omega}{\partial \Phi^{\Delta}} (\xi^{\alpha})^{\Delta} \right],$$

and integrating by parts, we obtain

$$\int_{t_1}^{t_2} \Delta t \int_V d^3x \left[\frac{\partial \Omega}{\partial \Phi^{\sigma}} (\xi^{\alpha})^{\sigma} + \frac{\partial \Omega}{\partial \Phi^{\Delta}} (\xi^{\alpha})^{\Delta} + \frac{\partial \Omega}{\partial \partial_i \Phi^{\sigma}} (\partial_i \xi^{\alpha})^{\sigma} \right] = 0,$$

where we used Theorem 3. □

Theorem 7. *The infinitesimal transformations given by Theorem 4 have conserved quantities of the form*

$$I^{\alpha} = \frac{\partial \Omega}{\partial \Phi^{\Delta}} \xi^{\alpha} + G^{\alpha} = c^{\alpha} \,.$$

Proof. Similar to the proof of Theorem 6. □

Theorem 8. *Infinitesimal transformations (19) together with Theorem 5 have conserved quantities of the form*

$$I^{\alpha} = \left(\Omega - \frac{\partial \Omega}{\partial t} \mu(t) \right) \xi_0^{\alpha} + \frac{\partial \Omega}{\partial \Phi^{\Delta}} \left(\xi^{\alpha} - \Phi^{\Delta} \xi_0^{\alpha} \right) + G^{\alpha} = c^{\alpha} \,.$$

Proof. We start from the following two definitions:

$$S[\Phi] = \int_{t_1}^{t_2} \Delta t \int_V d^3x \Omega \left(x^{\mu}, \Phi^{\sigma}, \Phi^{\Delta}, \partial_i \Phi^{\sigma} \right),$$

and

$$\overline{S}[\Phi, \Psi] = \int_{t_1}^{t_2} \Delta t \int_V d^3x \overline{\Omega} \left(x^{\mu}, \Psi^{\sigma}, \Phi^{\sigma}, \Psi^{\Delta}, \Phi^{\Delta}, \partial_i \Psi^{\sigma}, \partial_i \Phi^{\sigma} \right).$$

Let

$$\overline{\Omega} \left(x^{\mu}, \Psi^{\sigma}, \Phi^{\sigma}, \Psi^{\Delta}, \Phi^{\Delta}, \partial_i \Psi^{\sigma}, \partial_i \Phi^{\sigma} \right) = \Omega \left(\Psi^{\sigma} - \mu(t) \Psi^{\Delta}, x^i, \Phi^{\sigma}, \frac{\Phi^{\Delta}}{\Psi^{\Delta}}, \partial_i \Phi^{\sigma} \right).$$

When $\Psi(t) = t$, $S[\Phi] = \overline{S}[\Phi, \Psi]$. According to Definition 7, we have

$$\begin{aligned} S\left[\Phi_a^I\right] &= \int_{t_1}^{t_2} \Delta t \int_V d^3x \frac{\Delta}{\Delta t}(\tilde{\Delta} G) \\ &+ \int_{\beta(t_1)}^{\beta(t_2)} \Delta^* t^* \int_V d^3x \Omega\left(x^{*\mu}, \Phi^{*\sigma^*}, \Phi^{*\Delta^*}, \partial_i \Phi^{*\sigma^*}\right) \end{aligned}$$

$$\begin{aligned} &= \int_{t_1}^{t_2} \Delta t \beta^\Delta(t) \int_V d^3x \Omega\left(\beta(t), x^{*i}, \Phi^* \circ \sigma^* \circ \beta, \frac{(\Phi^* \circ \beta)^\Delta}{\beta^\Delta(t)}, \partial_i \Phi^* \circ \sigma^* \circ \beta\right) \\ &+ \int_{t_1}^{t_2} \Delta t \int_V d^3x \frac{\Delta}{\Delta t}(\tilde{\Delta} G) \\ &= \overline{S}[\Phi \circ \beta, \beta(t)] + \int_{t_1}^{t_2} \Delta t \int_V d^3x \frac{\Delta}{\Delta t}(\tilde{\Delta} G). \end{aligned}$$

It follows that the action $\overline{S}[\Phi \circ \beta, \beta(t)]$ corresponds to the Noether symmetry in Definition 7. Applying Theorem 7, the Lagrange equation given in Theorem 2 has the following conserved quantities:

$$I^\alpha = \frac{\partial \overline{\Omega}}{\partial \Psi^\Delta} \xi_0^\alpha + \frac{\partial \overline{\Omega}}{\partial \Phi^\Delta} \xi^\alpha + G^\alpha = c^\alpha .$$

This equation, along with the relationships

$$\begin{aligned} \frac{\partial \overline{\Omega}}{\partial \Psi^\wedge} &= -\frac{\partial \Omega}{\partial t} \mu(t) - \frac{\partial \Omega}{\partial \Phi^\Delta} \Phi^\Delta + \Omega, \\ \frac{\partial \overline{\Omega}}{\partial \Phi^\Delta} &= \frac{\partial \Omega}{\partial \Phi^\Delta}, \end{aligned}$$

ends the demonstration. □

6. Illustrative Examples

In this section, we discuss two particular cases of this theory; namely, a continuous case and a discrete case.

6.1. Example 1

If $\mathbb{T} = \mathbb{R}$, then $\sigma(t) = t$ and $\mu(t) = 0$, and the Euler–Lagrange equations in Theorem 2 are written

$$\frac{\partial \Omega}{\partial \Phi} - \partial_\mu \left(\frac{\partial \Omega}{\partial \partial_\mu \Phi} \right) = 0 .$$

Further, the Theorem 5 turns into

$$\dot{\Omega}\, \xi_0^\alpha + \Omega\, \dot{\xi}_0^\alpha + \frac{\partial \Omega}{\partial \Phi} \xi^\alpha - \frac{\partial \Omega}{\partial \dot{\Phi}} \dot{\Phi} \dot{\xi}_0^\alpha + \frac{\partial \Omega}{\partial \partial_\mu \Phi} \partial_\mu \xi^\alpha = -\dot{G}^\alpha ,$$

and the conserved quantity in Theorem 8 becomes

$$I^\alpha = \Omega\, \xi_0^\alpha + \frac{\partial \Omega}{\partial \dot{\Phi}} (\xi^\alpha - \dot{\Phi} \xi_0^\alpha) + G^\alpha = c^\alpha .$$

6.2. Example 2

Next, we consider the discrete case $\mathbb{T} = h\mathbb{Z}$. Consequently, $\sigma(t) = t + h$ and $\mu(t) = h$, and the Ω function becomes

$$\Omega\left(x^\mu, \Phi(t+h, x^i), \frac{\Phi(t+h, x^i) - \Phi(t, x^i)}{h}, \partial_i \Phi(t+h, x^i)\right).$$

Theorems 5 and 8 are transformed accordingly, taking into account the fact that

$$\frac{\Delta}{\Delta t} f(t, x^i) = \frac{f(t+h, x^i) - f(t, x^i)}{h}, \quad \text{and} \quad f(t, x^i)^\sigma = f(t+h, x^i).$$

7. Conclusions and Outlook

In this paper, we studied the Noether theorem for nonconservative field theories on time scales. After establishing Hamilton's principle, we extracted from it the Lagrange equations on time scales. For two different types of infinitesimal transformations, we established the selection criteria of Noether symmetries and based on these, we found the conserved quantities in the nonconservative field theories on time scales. In conclusion, we illustrated two examples for the discrete case and the continuous case to show that they were particular cases of the theory presented in this paper.

Continuous cases are common in the literature and underpin, for example, quantum field theory [24]. For future investigations, we want to extend our analysis to the following physical equations: Klein–Gordon, Maxwell and Dirac, from which it would be possible to extract the principles of time scale mechanics. Continuous theory is elegant but badly defined mathematically in many places, whilst the discrete time analogues are perhaps better defined mathematically. We would like to mention the success of the lattice field theory in the Yang–Mills problem. If no empirical test could distinguish between their predictions, we would rather choose the discrete theory. Possibly, using the time scale expression, we can highlight certain intervals of interest for which we can use discrete calculation.

The study of Maxwell's equations allows us to address this theory to particular problems. We can discuss the case of a simple relativistic engine made of two current loops of arbitrary geometry, in which we shall consider the mechanical momentum and energy gained by the engine [25]. Extending this theory to the gravitational field on a time scale, we could calculate the interaction field Lagrangians for the electromagnetic and gravitational interactions [26]. Next, we can look for analogies between the Fibonacci sequence and certain spatially homogeneous and isotropic universes in Friedmann–Lemaitre–Robertson–Walker cosmology on time scales [27].

Beyond the Lagrangian and Hamiltonian dynamics of typical nonconservative field theory investigated in this work, a range of other phenomena such as spontaneous breaking symmetry can be theoretically studied using the same framework.

The standard case of the Nambu–Goldstone theorem remains unchanged on this time scale, and as a consequence, the numbers of broken generators are definitely related to the existence of the Nambu–Goldstone bosons.

Using the pioneering work of [28,29], in a subsequent work, we could calculate both relativistically and non-relativistically the appropriate dispersion relation and the appropriate counting for the Nambu–Goldstone bosons if we study systems in which the symmetry is spontaneously broken. As benchmarks, within the same scenario, in future we can use the method of operators used in [30] in non-relativistic theory, or more generally, for both relativistic and non relativistic cases, we can check if the dynamics of Nambu–Goldstone bosons are somehow governed by the quantum Yang–Baxter equations [31].

Author Contributions: Conceptualization, O.P. and A.T.; methodology, O.P. and A.T.; software, O.P. and A.T.; validation, O.P. and A.T.; formal analysis, O.P. and A.T.; investigation, O.P. and A.T.; resources, O.P. and A.T.; data curation, O.P. and A.T.; writing—original draft preparation, O.P. and A.T.; writing—review and editing, O.P.; visualization, O.P. and A.T. All authors have read and agreed to the published version of the manuscript.

Funding: This research received no external funding.

Institutional Review Board Statement: Not applicable.

Informed Consent Statement: Not applicable.

Data Availability Statement: Not applicable.

Conflicts of Interest: The authors declare no conflict of interest.

References

1. Noether, E. Invariante Variationsprobleme. *Math. Phys. Kl.* **1918**, *1918*, 235–257.
2. Sebens, C.T.; Carroll, S.M. Self-Locating Uncertainty and the Origin of Probability in Everettian Quantum Mechanics. *arXiv* **2014**, arXiv:1405.7577.
3. Darboux, G. *Lecons Sur la Theorie General des Surfaces et les Applications Geometriques du Calcul Infinitesimal*; Gauthier-Villars: Paris, France, 1984.
4. Douglas, J. Solution of the inverse problem of the calculus of variations. *Trans. Am. Math. Soc.* **1941**, *50*, 71–128. [CrossRef]
5. Havas, P. The range of application of the lagrange formalism–I. *Il Nuovo Cimento* **1957**, *5*, 363–388. [CrossRef]
6. Santilli, R.M. *Foundations of Theoretical Physics I*; Springer: New York, NY, USA, 1978.
7. Dekker, H. *Classical and Quantum Mechanics of the Damped Harmonic Oscillator*; North-Holland: Amsterdam, The Netherlands, 1981.
8. Glauber, R.; Man'ko, V.I. Damping and fluctuations in coupled quantum oscillator systems. *Zh. Eksp. Teor. Fiz.* **1984**, *87*, 790–804.
9. Dodonov, V.V.; Man'ko, V.I.; Skarzhinsky. Inverse variational problems and ambiguities of quantization for a particle in a magnetic field. *Hadron. J.* **1983**, *6*, 159–179.
10. Okubo, S. Canonical quantization of some dissipative systems and nonuniqueness of Lagrangians. *Phys. Rev. A* **1981**, *23*, 2776–1784. [CrossRef]
11. Burgess, C.P. An Introduction to Effective Field Theory. *Ann. Rev. Nucl. Part. Sci.* **2007**, *57*, 329–362. [CrossRef]
12. Manohar, A.V. Effective field theories. *Lect. Notes Phys.* **1997**, *479*, 311–362.
13. Rothstein, I.Z. Effective field theories. In *Theoretical Advanced Study Institute in Elementary Particle Physics (TASI 2002): Particle Physics and Cosmology: The Quest for Physics Beyond the Standard Model(s)*; World Scientific: Singapore, 2002; pp. 101–189.
14. Hilger, S. *Ein Maßkettenkalkül mit Anwendung auf Zentrumsmannigfaltigkeiten*; Universität Würzburg: Würzburg, Germany, 1988.
15. Cadzow, J.A. Discrete calculus of variations. *Int. J. Control.* **1970**, *11*, 393–407. [CrossRef]
16. Logan, J.D. First integrals in the discrete variational calculus. *Aequationes Math.* **1973**, *9*, 210–220. [CrossRef]
17. Chen, L.Q.; Zhang, H.B.; Liu, R.W. First integrals of the discrete nonconservative and nonholonomic systems. *Chin. Phys.* **2005**, *14*, 238–243.
18. Fu, J.L.; Chen, L.Q.; Chen, B.Y. Noether-type theorem for discrete nonconservative dynamical systems with nonregular lattices. *Sci. China-Phys. Mech. Astron.* **2010**, *53*, 545–554. [CrossRef]
19. Fu, J.L.; Chen, L.Q.; Chen, B.Y. Noether-type theory for discrete mechanico-electrical dynamical systems with nonregular lattices. *Sci. China-Phys. Mech. Astron.* **2010**, *53*, 1687–1698. [CrossRef]
20. Cai, P.P.; Fu, J.L.; Guo, Y.X. Noether symmetries of the nonconservative and nonholonomic systems on time scales. *Sci. China-Phys. Mech. Astron.* **2013**, *56*, 1017–1028. [CrossRef]
21. Bangerezako, G. Variational q-calculus. *J. Math. Anal. Appl.* **2004**, *289*, 650–665. [CrossRef]
22. Galley, C.R. Classical Mechanics of Nonconservative Systems. *Phys. Rev. Lett.* **2013**, *110*, 174301. [CrossRef]
23. Fuller, A.J.B. *Engineering Field Theory*; Pergamon: Oxford, UK, 1973.
24. Peskin, M.E.; Schroeder, D.V. *An Introduction to Quantum Field Theory*; Perseus Books: Reading, MA, USA, 1995.
25. Rajput, S.; Yahalom, A.; Qin, H. Lorentz Symmetry Group, Retardation and Energy Transformations in a Relativistic Engine. *Symmetry* **2021**, *13*, 420. [CrossRef]
26. Gueorguiev, V.G.; Maeder, A. Geometric Justification of the Fundamental Interaction Fields for the Classical Long-Range Forces. *Symmetry* **2021**, *13*, 379. [CrossRef]
27. Faraoni, V.; Atieh, F. Generalized Fibonacci Numbers, Cosmological Analogies, and an Invariant. *Symmetry* **2021**, *13*, 200. [CrossRef]
28. Goldstone, J. Field Theories with Superconductor Solutions. *Nuovo Cim.* **1961**, *19*, 154. [CrossRef]
29. Nambu, Y.; Jona-Lasinio, G. Dynamical Model of Elementary Particles Based on an Analogy with Superconductivity. I. *Phys. Rev.* **1961**, *122*, 345. [CrossRef]
30. Arraut, I. The Nambu Goldstone theorem in nonrelativistic systems. *Int. J. Mod. Phys. A* **2017**, *32*, 1750127. [CrossRef]
31. Arraut, I. The Quantum Yang-Baxter Conditions: The Fundamental Relations behind the Nambu-Goldstone Theorem. *Symmetry* **2019**, *11*, 803. [CrossRef]

symmetry

MDPI

Article

Controlling Wolbachia Transmission and Invasion Dynamics among Aedes Aegypti Population via Impulsive Control Strategy

Joseph Dianavinnarasi [1], Ramachandran Raja [2], Jehad Alzabut [3,*], Michał Niezabitowski [4] and Ovidiu Bagdasar [5]

1 Department of Mathematics, Alagappa University, Karaikudi 630 004, India; josephdiana4866@gmail.com
2 Ramanujan Centre for Higher Mathematics, Alagappa University, Karaikudi 630 004, India; rajar@alagappauniversity.ac.in
3 Department of Mathematics and General Sciences, Prince Sultan University, Riyadh 12435, Saudi Arabia
4 Department of Automatic Control and Robotics, Faculty of Automatic Control, Electronics and Computer Science, Silesian University of Technology, Akademicka 16, 44-100 Gliwice, Poland; michal.niezabitowski@polsl.pl
5 Department of Electronics, Computing and Mathematics, University of Derby, Derby DE22 1GB, UK; o.bagdasar@derby.ac.uk
* Correspondence: jalzabut@psu.edu.sa

Abstract: This work is devoted to analyzing an impulsive control synthesis to maintain the self-sustainability of Wolbachia among Aedes Aegypti mosquitoes. The present paper provides a fractional order Wolbachia invasive model. Through fixed point theory, this work derives the existence and uniqueness results for the proposed model. Also, we performed a global Mittag-Leffler stability analysis via Linear Matrix Inequality theory and Lyapunov theory. As a result of this controller synthesis, the sustainability of Wolbachia is preserved and non-Wolbachia mosquitoes are eradicated. Finally, a numerical simulation is established for the published data to analyze the nature of the proposed Wolbachia invasive model.

Keywords: sustainability; mosquito borne diseases; Aedes Aegypti; Wolbachia invasion; impulsive control

Citation: Dianavinnarasi, J.; Raja, R.; Alzabut, J.; Niezabitowski, M.; Bagdasar, O. Controlling Wolbachia Transmission and Invasion Dynamics among Aedes Aegypti Population via Impulsive Control Strategy. *Symmetry* **2021**, *13*, 434. https://doi.org/10.3390/sym13030434

Academic Editors: Marin Marin and Jan Awrejcewicz

Received: 15 February 2021
Accepted: 3 March 2021
Published: 8 March 2021

1. Introduction

In the 19th century, fractional calculus (FC) theory has been built by some famous mathematicians like Grunwald, Letnikov, Riemann, Liouville, Euler and Caputo [1–3]. Fractional order derivatives are the generalization of integer order derivatives. FC is unavoidable due to its extensive applications in the study of real-world problems. The main advantage of FC is that it can provide a path to understand the description of memory and inheritance of various processes [4,5]. The book [6] plays an important role in the area of applied fractional calculus. In recent years, researchers in the field of physics, chemistry, Neural Networks, economic and mathematical modeling, biological problems and engineering have been very much attracted to fractional calculus [7], because FC interprets the whole function geometrically and globalizes its entire function.

Mosquito-borne diseases are primarily spread by female mosquitoes while taking a blood meal from living organisms such as humans, animals and birds. A parasite, virus, or bacteria-infected female mosquito can transmit those foreign agents to humans [8]. For instance, the Dengue virus, Zika virus, Yellow fever virus and Chikungunya are transmitted from infected human to uninfected human via primary vector Aedes Aegypti mosquitoes. Currently, the secondary vector for the above-mentioned diseases is Aedes Albopictus [9–11]. In recent years, the death rate due to mosquito-borne diseases has increased dramatically [8]. Gubler et al. [12,13] and Ong et al. [14] explained that dengue and dengue hemorrhagic fever are a more common issue for public health. According to the World Health Organization (WHO) [15], per annum, mosquito-borne diseases cause more than 40,000 deaths and 96 million asymptomatic cases in 129 countries.

Currently, there are several methods to control Aedes Aegypti mosquitoes such as insecticide spraying, sterile insect technique, incompatible insect technique, combined sterile insect technique, and genetic modifications. In [16,17], the authors proposed that the Sterile insect technique is likely to be used in mosquito-borne disease control. The authors of [18], analyzed that the particular transgenic strain can simulate the female-specific flightless phenotype to increase the sterilization in male mosquitoes. In [19,20], the authors discussed that the safe and effective replacement of vector population by genetically modified mosquitoes will play a significant role in mosquito-borne disease control. Furthermore, some other types of mosquito control strategies, such as making changes in feeding behaviors, intervention strategies, using bed nets and mosquito repellents, are also tested [21,22].

A novel Aedes Aegypti suppression technique using the life-shortening bacterium Wolbachia plays an important role [23–25]. It is an endosymbiotic bacterium that is reported in nearly 60 percent of insect species by Wolbach (1924) [26]. The World Mosquito Program (WMP) [27] from Australia currently release Wolbachia infected mosquitoes over 10 countries, such as countries in Latin America, India, Sri Lanka, Vietnam, Indonesia and cities in Oceania. In that research, they found that Wolbachia is a self-sustaining bacterium and in the presence of Wolbachia infected mosquitoes there is zero possibility of having Dengue. The Wolbachia releasing strategy is more powerful than that of the above-mentioned control strategies in the sense that it is self-sustaining, affordable, only needs a small amount of release, the area covered is larger than the released area, and the most important thing is it is not harmful to human health. The authors of [28–31] discussed that Wolbachia can restrict the virus particles of various diseases. We know that the virus is transmitted from infected humans to uninfected humans via female mosquitoes. Meanwhile, if a virus-infected mosquito carries Wolbachia strain, then the virus cannot be transmitted to an uninfected human. Because this Wolbachia strain blocks the virus particles inside the salivary gland of mosquitoes (Ref. Figure 1).

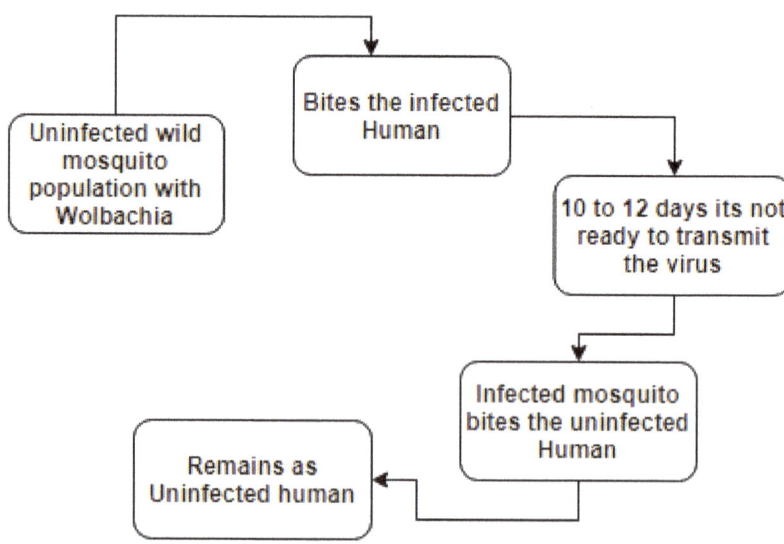

Figure 1. Mechanism of Wolbachia among mosquitoes and human.

The Wolbachia infection is introduced into wild mosquitoes population through two major processes such as microinjection and Introgression [32].

Micro injection: In this process, Wolbachia strains are microinjected into aquatic stages such as eggs, larvae and pupae.
Introgression: In this process, the Wolbachia strains are carried out to next generation through mating. If Wolbachia infected female mated with Wolbachia infected or uninfected male, then the produced offsprings have the Wolbachia strain (Called CI rescue). Suppose the Wolbachia uninfected female mated with a Wolbachia infected male then there is no viable progeny. Finally, if a non-Wolbachia female mated with a non-Wolbachia male then there is no Wolbachia infection in the offspring.

To understand the introgression process, one can refer to Figure 2.

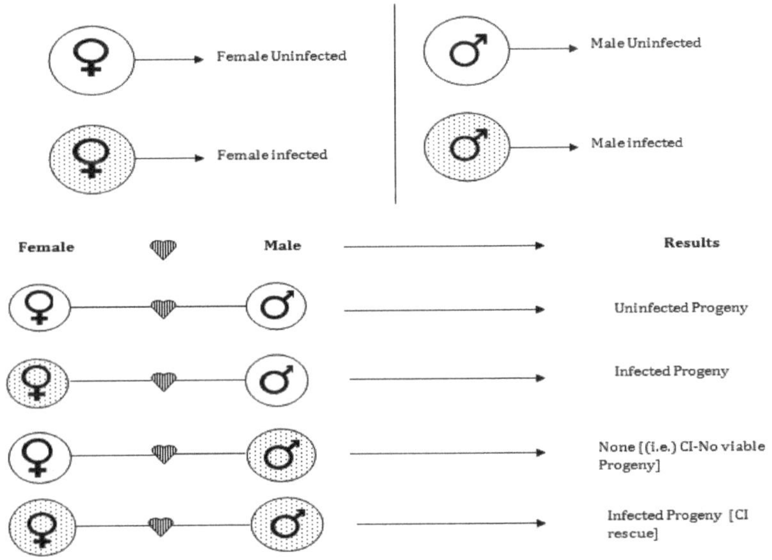

Figure 2. Block diagram representing the mechanism of Wolbachia infection in mosquitoes.

Furthermore, some existing mathematical models consider Wolbachia as a control agent for mosquito-borne diseases. In [33], the author proposed a deterministic model to control mosquito-borne diseases up to 90% via Wolbachia spread, also the author considered both human and mosquito populations to create a mathematical model. In [30,34], the authors proposed a mathematical model depicting the life stages of mosquitoes with Wolbachia and proved that Wolbachia has an excellent quality to control dengue virus spread. In [35], the authors analyzed the integer ordered mathematical model consisting of only four stages (aquatic stage with and without Wolbachia and adult female mosquitoes with and without Wolbachia), which considered the imperfect maternal transmission and Wolbachia invasion. In [36], the two sex mathematical model is discussed to analyze the persistence of Wolbachia. In [37], the age and bite structured mathematical model is proposed and performed the mathematical analysis. In [38], the authors discussed the linear feedback control strategy of a mathematical model containing only three stages such as aquatic, female Wolbachia infected and uninfected mosquitoes. In this, the author analyzed the Wolbachia infected mosquitoes release into the seasonal environment. In [39], the authors presented a mathematical model to depict the mechanism of the virus inside both humans and mosquitoes. In this work, the author utilized two various types of controls like vaccination for humans and Wolbachia infected mosquitoes' release for mosquitoes. The pontriyagin maximum principle was utilized to analyze the optimal control of the proposed mathematical model. In [40], the authors discussed the Wolbachia infection among Aedes Aegypti mosquitoes via delay differential equations. In that work, the author proposed the delay dependent stability criteria of the proposed model by utilizing the results from spectrum analysis. In [41], the authors proposed an age structured fractional order mathematical model to control the Aedes Aegypti mosquitoes via Wolbachia bacterium using the Linear Matrix Inequality (LMI) approach.

As per the practical results of [27], Wolbachia should be released into every stage to get the optimal control in a short period. Also, by utilizing fractional calculus we can get the memory property and inheritance of this process. In nature, Wolbachia infected mosquitoes may lose the Wolbachia infection. Because of this, invasion in Wolbachia is unavoidable. Motivation by the above discussions, our contributions are listed below:

- A novel mathematical model, which considers the total of ten stages in Aedes Aegypti mosquitoes (combining both Wolbachia infected and Wolbachia uninfected) is proposed and the possible optimal stages to release the Wolbachia are discussed, and the most important concept of Wolbachia invasion and Wolbachia gain are adopted.
- The Wolbachia free equilibrium, Wolbachia present Equilibrium, Zero mosquitoes, and both Wolbachia and Non-Wolbachia mosquitoes co-existence equilibrium are derived. And utilizing fixed point theory results, the Existence and Uniqueness results of the Wolbachia invasive model are proposed. To attain optimal control, we utilized an impulsive control strategy.
- We perform global Mittag-Leffler stability analysis of the proposed model via Linear Matrix Inequality (LMI) theory and Lyapunov theory.
- In the end, by utilizing the data from the published literature, we have presented the numerical simulation of the proposed model using MATLAB software.

The rest of the paper is arranged as follows—in Section 2, we provide some basic Definitions, Lemmas and Theorems. In Section 3, the fractional order complete mathematical model describes the interaction between Wolbachia infected and Non-Wolbachia mosquitoes is presented. In Section 4, the possible equilibrium points are presented. In Section 5, the Wolbachia invasive and gain model with impulsive control is presented. In Section 6, the existence and uniqueness results are analyzed and the global Mittag-Leffler stability results are derived in Section 7. In Section 8, the numerical simulation results are presented. In Section 9, the work is concluded.

Notations. $\mathbb{N}$ denotes the space of all natural numbers, $\mathbb{R}$ denotes the space of all real numbers, $\mathbb{C}$ denotes the space of all complex numbers, $\mathbb{R}^n$ denotes the space of n-dimensional Euclidean space, $\mathbb{Z}^+$ denotes the space of all positive integers. Moreover, $Re(\cdot)$ denotes the real part of a complex number and $[.]$ denotes the integer part of a number. $*$ denotes the corresponding symmetric terms in a symmetric matrix. Also, ${}_k^cD_t^\alpha(\cdot)$ and ${}_k^cI_t^\alpha(\cdot)$ denotes the derivative and anti derivative of order α with respect to t respectively, c denotes that its in Caputo sense, k denotes the initial condition and $\Gamma(\cdot)$ denotes the Gamma function.

2. Preliminaries

In this section, we provide some basic Definitions, Lemmas and Theorems, which are used to attain our results.

Definition 1. *Ref. [4] The most important basic function in fractional calculus is the gamma function. It is defined as follows:*

$$\Gamma(z) = \int_0^\infty e^{-s} s^{z-1}\, \mathrm{d}\, s,$$

with $Re(z) > 0$.

Definition 2. *Ref. [1] The Caputo fractional derivative of a continuous function $f(t)$ over $[k,T]$ of order $\alpha \in \mathbb{C}$ (with $Re(\alpha) > 0$, $\alpha \notin \mathbb{N}$) is*

$${}_k^cD_t^\alpha f(t) = \frac{1}{\Gamma(n-\alpha)}\left[\int_k^t (t-\eta)^{n-\alpha-1}\frac{d^n}{d\eta^n}f(\eta)d\eta\right], \tag{1}$$

where, $n = [Re(\alpha)] + 1$.
If $0 < Re(\alpha) < 1$, *then the expression* (1), *can be rewritten as*

$$ {}_{k}^{c}D_{t}^{\alpha}f(t) = \frac{1}{\Gamma(1-\alpha)}\left[\int_{k}^{t}\frac{f'(\eta)d\eta}{(t-\eta)^{\alpha}}\right]. \quad (2)$$

Since, $n = 1$ for all $0 < Re(\alpha) < 1$.

Definition 3. *Ref. [42] The Caputo sense fractional integral of a continuous function f on* $L^1([0,T],\mathbb{R})$ *over* $\alpha \in (0,1]$ *with respect to t is defined as*

$$ {}_{0}^{c}I_{t}^{\alpha}f(t) = \frac{1}{\Gamma(\alpha)}\int_{0}^{t}(t-\eta)^{\alpha-1}f(\eta)d\eta. \quad (3)$$

The two parameter Mittag-Leffler function is defined as follows:[4]

$$E_{a,b}(z) = \sum_{l=0}^{\infty}\frac{z^l}{\Gamma(al+b)},$$

where, $z \in \mathbb{C}$, $a > 0$, and $b > 0$. If $b = 1$ the $E_a(z) = \sum_{l=0}^{\infty}\frac{z^l}{\Gamma(al+1)}$. If both $a = 1$ and $b = 1$, the $E_{1,1}(z) = e^z$.

Lemma 1 (Schur Complement [43])**.** *Let us denote three* $n \times n$ *matrices as* Ψ_1, Ψ_2, Ψ_3, *where* $\Psi_1 = \Psi_1^\top$ *and* $\Psi_2 = \Psi_2^\top > 0$. *Then* $\Psi_1 + \Psi_3^\top\Psi_2^{-1}\Psi_3 < 0$ *if and only if* $\begin{bmatrix}\Psi_1 & \Psi_3^\top \\ \Psi_3 & -\Psi_2\end{bmatrix} < 0$ *(or)* $\begin{bmatrix}-\Psi_2 & \Psi_3 \\ \Psi_3^\top & \Psi_1\end{bmatrix} < 0.$

Lemma 2. *Ref. [44] For any scalar* $\epsilon > 0$, $A, N \in \mathbb{R}^n$ *and matrix* P_1, *then*

$$A^\top P_1 N \leq \frac{1}{2\epsilon}A^\top P_1 P_1^\top A + \frac{\epsilon}{2}N^\top N.$$

Let us consider the fractional order dynamical system with impulse of type,

$$\begin{aligned} {}_{k}^{c}D_{t}^{\alpha}x(t) &= -A_1x(t) + A_2f(x(t)), t \neq t_\theta, \theta = 1,2,\cdots,m, \\ \Delta x(t_\theta) &= x(t_\theta^+) - x(t_\theta^-) = \delta_\theta(x(t_\theta)), t = t_\theta, \theta = 1,2,\cdots,m, \end{aligned} \quad (4)$$

with initial condition $x(t_0) = x_0 \in \mathbb{Z}^+$, where the n states is defined by $x(t) = [x_1(t), x_2(t), x_3(t), \cdots, x_n(t)]^\top \in \mathbb{R}^n$ and $f(x(t)) = [f(x_1(t)), f(x_2(t)), f(x_3(t)), \cdots, f(x_n(t))]^\top$ be a function, A_1 and A_2 are constant coefficient matrices with the impulsive operator $\delta_\theta : \mathbb{R}^n \to \mathbb{R}^n$.

Definition 4. *Ref. [44] The system* (4), *is said to be globally Mittag-Leffler stable at its equilibrium points, if the following hold:*

$$||x(t) - x^*|| \leq [h(x_0 - x^*)E_\alpha(-\kappa t^a)]^b,$$

where x^* *is an equilibrium point,* $0 < \alpha < 1$, $\kappa \geq 0$ *and* $a, b > 0$. *Moreover,* $h(0) = 0$, $h(x) \geq 0$ *and* $h(x)$ *is locally Lipschitz with Lipchitz constant* h_0.

Lemma 3. *Ref. [45] Let us consider the fractional order system with impulsive control of type* (4). *Suppose* $f(0) = 0$, $t > 0$ *and* $\delta_\theta(0) = 0$, $\theta = 1,2,3,\cdots,m$. *If there exists a positive definite function V such that the following hold:*

(1) *There exists positive constants* α_1 *and* α_2

$$\alpha_1||x(t)|| \quad \leq \quad V(t) \leq \alpha_2||x(t)||, x(t) \in \mathbb{R}^n.$$

(2) ${}_0^c D_t^\alpha V(t) \leq -\epsilon_1 V(t), t \neq t_\theta, \theta = 1, 2, 3, \cdots, m$ *for any scalar* ϵ_1.
(3) $V(t_\theta^+) \leq V(t_\theta), t = t_\theta, \theta = 1, 2, 3, \cdots, m.$
then the equilibrium point of the system (4) *is globally Mittag-Leffler stable.*

Definition 5. *Ref. [46] A map* $\nu : H \to H$, *H compact Banach space, is said to be a contraction mapping if there exists* $h \in (0,1)$ *such that*

$$||\nu(m_1) - \nu(m_2)|| \quad \leq \quad h||m_1 - m_2||$$

for every $m_1, m_2 \in H$.

Theorem 1 (Contraction Mapping Theorem). *Ref. [46] Suppose H is a complete metric space and* $\nu : H \to H$ *is a contraction mapping. Then,* ν *has a unique fixed point.*

3. Model Formulation

In this section, a novel mathematical model is proposed to expose the transmission dynamics of the gram negative bacteria Wolbachia among Aedes Aegypti mosquitoes. While constructing the model we have considered the total of 10 stages such as non-Wolbachia eggs(W_e), non- Wolbachia larvae (W_l), non-Wolbachia pupae (W_p), non-Wolbachia adult female (W_f), non- Wolbachia adult male (W_a), Wolbachia infected eggs (I_e), Wolbachia infected larvae (I_l), Wolbachia infected pupae (I_p), Wolbachia infected adult female (I_f), Wolbachia infected adult male (I_a). The total population at time t is denoted as $T = W_e(t) + W_l(t) + W_p(t) + W_f(t) + W_a(t) + I_e(t) + I_l(t) + I_p(t) + I_f(t) + I_a(t)$. The eggs with zero Wolbachia infection are produced at the rate Λ_{w_e} by the mating process between non-Wolbachia female (W_f) and non-Wolbachia male (W_a). There is no other possibilities of having a non-Wolbachia eggs. Therefore, the reproduction rate of non-Wolbachia mosquitoes can be calculated by the term $\frac{\Lambda_{we} W_f W_a}{T}$. Along with this, the terms λ_{w_e} (natural mortality rate of non-Wolbachia eggs) and γ_{w_e} (maturation rate of non-Wolbachia eggs) denotes the limitations in the growth of wild mosquito eggs. At the same time, after release of Wolbachia infected mosquitoes (in both aquatic and ariel stage) in a common environment, the production of Wolbachia infected mosquito eggs $I_e(t)$, depends on mating between Wolbachia infected female $I_f(t)$ and non-Wolbachia male $W_a(t)$ and from mating between Wolbachia infected female $I_f(t)$ and Wolbachia infected male $I_a(t)$. Through this, the birth rate of Wolbachia infected mosquito eggs population $I_e(t)$ with the reproduction rate Λ_{i_e} is

$$\frac{\Lambda_{i_e}(I_f W_a + I_f I_a)}{T} \quad = \quad \frac{\Lambda_{i_e} I_f (W_a + I_a)}{T}.$$

Similarly, the increase in the growth of Wolbachia infected eggs is limited by the natural mortality rate λ_{i_e} and the maturation rate γ_{i_e} (That is, the rate in which the corresponding compartment moved into the next stage).

Furthermore, the quantity $(1-\alpha)\gamma_{i_e} I_e$ is added to the wild mosquito larvae population. Because the term α and $(1-\alpha)$ denotes the probability of getting larvae with and without Wolbachia respectively. Similarly, β and $(1-\beta)$ denotes the probability of getting pupae with and without Wolbachia respectively, ϵ and $(1-\epsilon)$ denotes the probability rate of having Wolbachia infection in adult mosquitoes by introgression. That is, ϵ be the probability of getting Wolbachia infected adults (with ρ_{i_w} = probability of getting male and $(1-\rho_{i_w})$ = probability of getting female). Because of these reasons, the terms $(1-\alpha)\gamma_{i_e} I_e$, $(1-\beta)\gamma_{i_l} I_l$, $(1-\epsilon)\gamma_{i_p}\rho_{i_w} I_p$ and $(1-\epsilon)\gamma_{i_p}(1-\rho_{i_w}) I_p$ are added to the corresponding stages and similarly, the terms $\alpha\gamma_{i_e} I_e$, $\beta\gamma_{i_l} I_l$ and $\epsilon\gamma_{i_p} I_{i_p}$ are removed from the corresponding stages. The parameter description of the system of Equation (5) is presented in Table 1.

Table 1. Description of parameters involved in system of Equation (5).

Parameter	Description
$\Lambda_{w_e}, \Lambda_{i_e}$	Reproduction rate of non-Wolbachia mosquitoes and Wolbachia infected mosquitoes respectively
λ_{w_e}	The natural death rate of eggs without Wolbachia infection
λ_{w_l}	The natural death of larvae without Wolbachia infection
λ_{w_p}	The natural death of pupae without Wolbachia infection
λ_{w_f}	The natural death of adult female mosquitoes without Wolbachia infection
λ_{w_a}	The natural death of adult male mosquitoes without Wolbachia infection
λ_{i_e}	The natural death of eggs with Wolbachia infection
λ_{i_l}	The natural death of larvae with Wolbachia infection
λ_{i_p}	The natural death of pupae with Wolbachia infection
λ_{i_f}	The natural death of adult female mosquitoes with Wolbachia infection
λ_{i_a}	The natural death of infected adult male mosquitoes with Wolbachia infection
γ_{w_e}	The rate at which the fraction of non-Wolbachia eggs matured into non-Wolbachia larvae
γ_{w_l}	The rate at which the fraction of non-Wolbachia larvae matured into non-Wolbachia pupae
γ_{w_p}	The rate at which the fraction of non-Wolbachia pupae matured into non-Wolbachia
	immature female or male
γ_{i_e}	The rate at which the fraction of the Wolbachia infected mosquito eggs
	matured into Wolbachia infected or uninfected larvae
γ_{i_l}	The rate at which the fraction of the Wolbachia infected mosquito larvae
	matured into Wolbachia infected or uninfected pupae
γ_{i_p}	The rate at which the fraction of the Wolbachia infected mosquito pupae
	matured into Wolbachia infected or uninfected adults
ρ	The probability of having male or female mosquitoes

From the above facts, the novel mathematical model that describes the transmission dynamics of Wolbachia among Aedes Aegypti mosquitoes is proposed as follows:

$$\begin{cases} {}_0^cD_t^\alpha W_e &= \dfrac{\Lambda_{w_e} W_f W_a}{T} - \lambda_{w_e} W_e - \gamma_{w_e} W_e \\ {}_0^cD_t^\alpha W_l &= \gamma_{w_e} W_e - \lambda_{w_l} W_l - \gamma_{w_l} W_l + (1-\alpha)\gamma_{i_e} I_e \\ {}_0^cD_t^\alpha W_p &= \gamma_{w_l} W_l - \lambda_{w_p} W_p - \gamma_{w_p} W_p + (1-\beta)\gamma_{i_l} I_l \\ {}_0^cD_t^\alpha W_f &= \rho\gamma_{w_p} W_p - \lambda_{w_f} W_f + (1-\epsilon)\gamma_{i_p}\rho_{i_w} I_p \\ {}_0^cD_t^\alpha W_a &= (1-\rho)\gamma_{w_p} W_p - \lambda_{w_a} W_a + (1-\epsilon)\gamma_{i_p}(1-\rho_{i_w}) I_p \\ {}_0^cD_t^\alpha I_e &= \dfrac{\Lambda_{i_e} I_f (W_a + I_a)}{T} - \lambda_{i_e} I_e - \alpha\gamma_{i_e} I_e \\ {}_0^cD_t^\alpha I_l &= \alpha\gamma_{i_e} I_e - \lambda_{i_l} I_l - \beta\gamma_{i_l} I_l \\ {}_0^cD_t^\alpha I_p &= \beta\gamma_{i_l} I_l - \lambda_{i_p} I_p - \epsilon\gamma_{i_p} I_p \\ {}_0^cD_t^\alpha I_f &= \rho_i\epsilon\gamma_{i_p} I_p - \lambda_{i_f} I_f \\ {}_0^cD_t^\alpha I_a &= (1-\rho_i)\epsilon\gamma_{i_p} I_p - \lambda_{i_a} I_a. \end{cases} \tag{5}$$

The dynamics of the population can be easily understand by the schematic diagram Figure 3 and the parameters are described in Table 1.

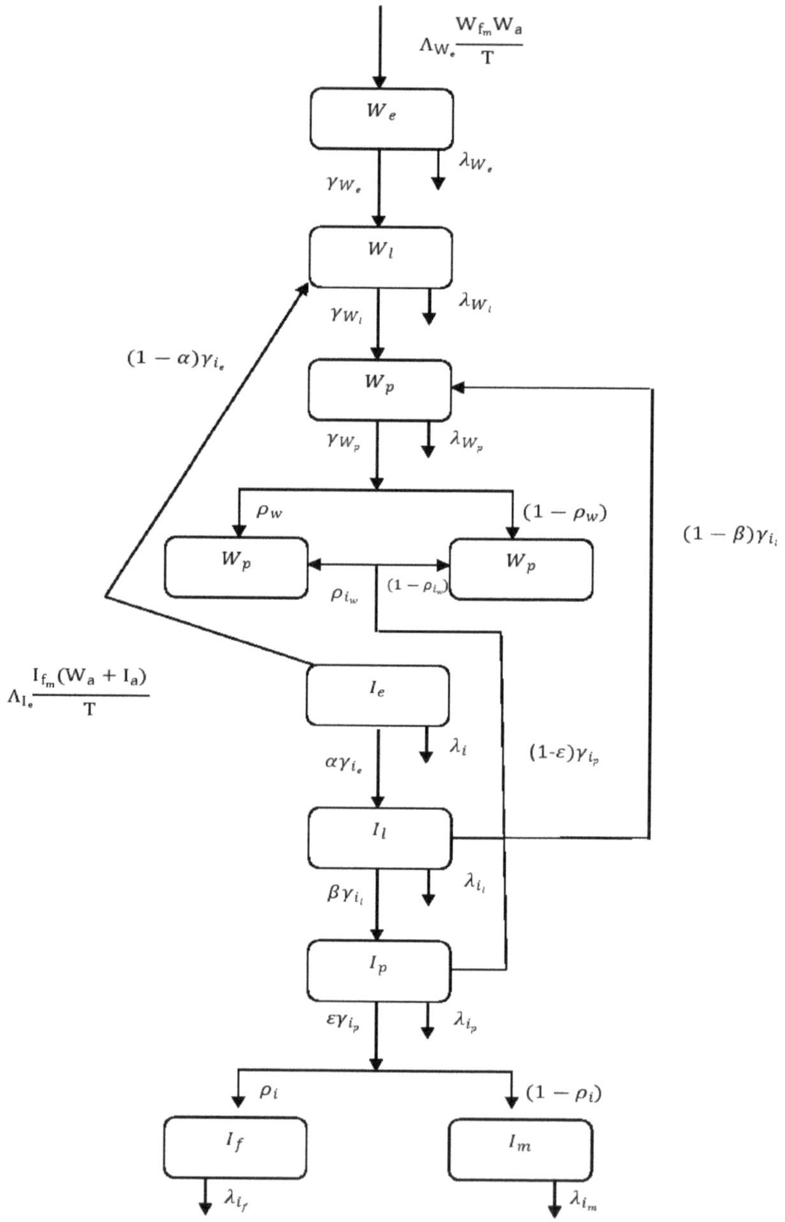

Figure 3. Schematic representation Wolbachia spread dynamics among Aedes Aegypti mosquitoes.

4. Equilibrium Points

In this section, we can find the four cases of possible equilibrium points such as wild mosquitoes only, Wolbachia mosquitoes only, co-existence of both population and zero mosquitoes.

4.1. Zero Mosquitoes

Suppose there is no mosquitoes, then the equilibrium point can be written as $P_1 = (0,0,0,0,0,0,0,0,0,0)$. This is trivial but does not exists in nature.

4.2. Wolbachia Infected Mosquitoes Free Equilibrium

Suppose, there is no Wolbachia infected mosquitoes population then the possible equilibrium can be written as

$$P_2 = (W_{e_1}^*, W_{l_1}^*, W_{p_1}^*, W_{f_1}^*, W_{a_1}^*, 0,0,0,0,0),$$

where,

$$\begin{aligned}
W_{e_1}^* &= \frac{T\lambda_{w_f}\lambda_{w_a}(\lambda_{w_e}+\gamma_{w_e})(\lambda_{w_l}+\gamma_{w_l})^2(\lambda_{w_p}+\gamma_{w_p})^2}{\rho(1-\rho)\Lambda_{w_e}\gamma_{w_p}^2\gamma_{w_e}^2\gamma_{w_l}^2} \\
W_{l_1}^* &= \frac{\gamma_{w_e}}{\lambda_{w_l}+\gamma_{w_l}}W_{e_1}^* \\
W_{p_1}^* &= \frac{\gamma_{w_l}\gamma_{w_e}}{(\lambda_{w_l}+\gamma_{w_l})(\lambda_{w_p}+\gamma_{w_p})}W_{e_1}^* \\
W_{f_1}^* &= \frac{\rho\gamma_{w_p}\gamma_{w_e}\gamma_{w_l}}{\lambda_{w_f}(\lambda_{w_f}+\gamma_{w_f})(\lambda_{w_p}+\gamma_{w_p})}W_{e_1}^* \\
W_{a_1}^* &= \frac{(1-\rho)\gamma_{w_p}\gamma_{w_e}}{\lambda_{w_e}(\lambda_{w_l}+\gamma_{w_l})(\lambda_{w_p}+\gamma_{w_p})}W_{e_1}^*
\end{aligned}$$

4.3. Wild Mosquitoes Free Equilibrium

After the successful replacement of Wolbachia uninfected mosquitoes by Wolbachia infected mosquitoes the equilibrium point can be represented by

$$P_3 = (0,0,0,0,0, I_{e_2}^*, I_{l_2}^*, I_{p_2}^*, I_{f_2}^*, I_{a_2}^*),$$

where,

$$\begin{aligned}
I_{e_2}^* &= \frac{(\lambda_{i_l}+\beta\gamma_{i_l})(\lambda_{i_p}+\epsilon\gamma_{i_p})}{\alpha\beta\gamma_{i_e}\gamma_{i_l}}I_{p_2}^* \\
I_{l_2}^* &= \frac{(\lambda_{i_p}+\epsilon\gamma_{i_p})}{\beta\gamma_{i_l}}I_{p_2}^* \\
I_{p_2}^* &= \frac{T\lambda_{i_f}\lambda_{i_a}(\lambda_{i_e}+\alpha\gamma_{i_e})(\lambda_{i_l}+\beta\gamma_{i_l})(\lambda_{i_p}+\epsilon\gamma_{i_p})}{\Lambda_{i_e}\alpha\beta\rho_i(1-\rho_i)\epsilon^2\gamma_{i_p}^2\gamma_{i_e}\gamma_{i_l}} \\
I_{f_2}^* &= \frac{\rho_i\epsilon\gamma_{i_p}}{\lambda_{i_f}}I_{p_2}^* \\
I_{a_2}^* &= \frac{(1-\rho_i)\epsilon\gamma_{i_p}}{\lambda_{i_a}}I_{p_2}^*.
\end{aligned}$$

4.4. Both Wolbachia Infected Mosquitoes and Non-Wolbachia Mosquitoes Co-Existence Equilibrium

If both Wolbachia infected and Wolbachia uninfected mosquitoes present in common environment, then the equilibrium point is

$$S_n = \left\{W_{e_n}^*, W_{l_n}^*, W_{p_n}^*, W_{f_n}^*, W_{a_n}^*, I_{e_n}^*, I_{l_n}^*, I_{p_n}^*, I_{f_n}^*, I_{a_n}^*\right\}, n = 3,4.$$

$$W^*_{e_n} = \left(\frac{\lambda_{w_l}+\gamma_{w_l}}{\gamma_{w_e}}\right)\left(\frac{\lambda_{w_p}+\gamma_{w_p}}{\gamma_{w_l}}\right)\left(\frac{TB_1B_2B_3\lambda_{i_f}\lambda_{w_a}}{\Lambda_{i_e}(1-\rho)\rho_i\gamma_{w_p}}\right) - \frac{I^*_{a_n}}{\gamma_{w_e}}$$
$$\left[B_4(\lambda_{w_l}+\gamma_{w_l})\left(\frac{\lambda_{w_p}+\gamma_{w_p}}{\gamma_{w_l}}\right) + (\lambda_{w_l}+\gamma_{w_l})\left(\frac{(1-\beta)\gamma_{i_l}}{\gamma_{w_l}}\right)\left(\frac{\lambda_{i_a}B_1}{\beta\gamma_{i_l}(1-\rho_i)}\right) + \frac{(1-\alpha)\gamma_{i_e}\lambda_{i_a}B_1B_2}{\alpha\gamma_{i_e}(1-\rho_i)}\right]$$

$$W^*_{l_n} = \left(\frac{\lambda_{w_p}+\gamma_{w_p}}{\gamma_{w_l}}\right)\left[\frac{TB_1B_2B_3\lambda_{i_f}\lambda_{w_a}}{\Lambda_{i_e}(1-\rho)\rho_i\gamma_{w_p}} - B_4I^*_{a_n}\right] - \left(\frac{(1-\beta)\gamma_{i_l}}{\gamma_{w_l}}\right)\left[\frac{\lambda_{i_a}B_1}{\beta\gamma_{i_l}(1-\rho_i)}\right]$$

$$W^*_{p_n} = \left[\frac{TB_1B_2B_3\lambda_{i_f}\lambda_{w_a}}{\Lambda_{i_e}(1-\rho)\rho_i\gamma_{w_p}} - B_4I^*_{a_n}\right]$$

$$W^*_{f_n} = \frac{\rho\gamma_{w_p}}{\lambda_{w_f}}\left[\frac{TB_1B_2B_3\lambda_{i_f}\lambda_{w_a}}{\Lambda_{i_e}(1-\rho)\rho_i\gamma_{w_p}} - B_4I^*_{a_n}\right] + \frac{(1-\epsilon)\rho_{i_w}\lambda_{i_a}}{\epsilon\lambda_{w_f}(1-\rho_i)}I^*_{a_n}$$

$$W^*_{a_n} = \frac{TB_1B_2B_3\lambda_{i_f}}{\rho_i\Lambda_{i_e}} - I^*_{a_n}$$

$$I^*_{e_n} = \frac{B_1B_2\lambda_{i_a}I^*_{a_n}}{\alpha\gamma_{i_e}(1-\rho_i)}$$

$$I^*_{p_n} = \frac{\lambda_{i_a}}{(1-\rho_i)\epsilon\gamma_{i_p}}I^*_{a_n}$$

$$I^*_{f_n} = \frac{\rho_i\lambda_{i_a}I^*_{a_n}}{(1-\rho_i)\lambda_{i_f}}$$

with $I^*_{a_3} > I^*_{a_4}$, both roots can be found from the quadratic equation

$$a_1 I_a^{*^2} + a_2 I_a^* + a_3 = 0,$$

where,

$$a_1 = \frac{\Lambda_{w_e}\rho B_4\gamma_{w_p}}{T\lambda_{w_f}};$$
$$a_2 = \left(\frac{\lambda_{w_e}+\gamma_{w_e}}{T\lambda_{w_f}}\right)\left(\frac{\lambda_{w_e}\lambda_{i_f}\rho B_1B_2B_3}{\rho_i\Lambda_{i_e}\lambda_{w_f}}\right)\left(\frac{\lambda_{w_a}}{(1-\rho)} + B_4\gamma_{w_p}\right)$$
$$\left(\frac{(\lambda_{w_l}+\gamma_{w_l})(\lambda_{w_p}+\gamma_{w_p})B_4}{\gamma_{w_l}} + \frac{(\lambda_{w_l}+\gamma_{w_l})(1-\beta)\lambda_{i_a}B_1}{\gamma_{w_l}\beta(1-\rho_i)} + \frac{(1-\alpha)\lambda_{i_a}B_1B_2}{\alpha(1-\rho_i)}\right);$$
$$a_3 = \frac{\Lambda_{w_e}\rho TB_1^2B_2^2B_3^2\lambda_{w_a}}{\Lambda_{i_e}^2(1-\rho)\rho_i^2\lambda_{w_f}}.$$

Here,

$$
\begin{aligned}
B_1 &= 1+\frac{\lambda_{i_p}}{\epsilon\gamma_{i_p}}; \\
B_2 &= 1+\frac{\lambda_{i_l}}{\beta\gamma_{i_l}}; \\
B_3 &= 1+\frac{\lambda_{i_e}}{\alpha\gamma_{i_e}}; \\
B_4 &= 1+\frac{(1-\epsilon)(1-\rho_{i_w})\lambda_{i_a}}{(1-\rho)(1-\rho_i)\epsilon\gamma_{w_p}}.
\end{aligned}
$$

For more details about the calculations of Section 4, kindly refer the Appendix A section.

5. Wolbachia Invasion Model

We considered the possibility of Wolbachia loss in adult mosquitoes and possibility of Wolbachia gain in aquatic stage mosquitoes. Then Equation (5), can be rewritten as

$$
\begin{cases}
{}_0^cD_t^\alpha W_e(t) &= \dfrac{\Lambda_{w_e} W_f W_a}{T} - \lambda_{w_e} W_e - \gamma_{w_e} W_e - \eta_1 I_e \\
{}_0^cD_t^\alpha W_l(t) &= \gamma_{w_e} W_e - \lambda_{w_l} W_l - \gamma_{w_l} W_l + (1-\alpha)\gamma_{i_e} I_e - \eta_2 I_l \\
{}_0^cD_t^\alpha W_p(t) &= \gamma_{w_l} W_l - \lambda_{w_p} W_p - \gamma_{w_p} W_p + (1-\beta)\gamma_{i_l} I_l - \eta_3 I_p \\
{}_0^cD_t^\alpha W_f(t) &= \rho\gamma_{w_p} W_p - \lambda_{w_f} W_f + (1-\epsilon)\gamma_{i_p}\rho_{i_w} I_p + \eta_4 W_f \\
{}_0^cD_t^\alpha W_a(t) &= (1-\rho)\gamma_{w_p} W_p - \lambda_{w_a} W_a + (1-\epsilon)\gamma_{i_p}(1-\rho_{i_w}) I_p + \eta_5 W_a \\
{}_0^cD_t^\alpha I_e(t) &= \dfrac{\Lambda_{i_e} I_f (W_a + I_a)}{T} - \lambda_{i_e} I_e - \alpha\gamma_{i_e} I_e + \eta_1 I_e \\
{}_0^cD_t^\alpha I_l(t) &= \alpha\gamma_{i_e} I_e - \lambda_{i_l} I_l - \beta\gamma_{i_l} I_l + \eta_2 I_l \\
{}_0^cD_t^\alpha I_p(t) &= \beta\gamma_{i_l} I_l - \lambda_{i_p} I_p - \epsilon\gamma_{i_p} I_p + \eta_3 I_p \\
{}_0^cD_t^\alpha I_f(t) &= \rho_i\epsilon\gamma_{i_p} I_p - \lambda_{i_f} I_f - \eta_4 W_f \\
{}_0^cD_t^\alpha I_a(t) &= (1-\rho_i)\epsilon\gamma_{i_p} I_p - \lambda_{i_a} I_a - \eta_5 W_a,
\end{cases}
\tag{6}
$$

where η_1, η_2 and η_3 all are the rates at which the non-Wolbachia aquatic population gain Wolbachia infected mosquitoes infection and η_4 & η_5 are the rates at which the Wolbachia infected mosquitoes losses their Wolbachia infection.

Impulsive control plays an predominant role in dynamical systems such as Neural Networks [47,48], non–linear delay dynamic systems [49–51] and so forth. To optimize the Wolbachia release, we can release the Wolbachia infected eggs, larvae and pupae in the form of 'Zancu kit' and Wolbachia infected adult female and male mosquitoes (introgression) impulsively. The situation should be monitored weekly once by Biogents trap (BG trap or BG sentinel trap). While monitoring, if there is less number of Wolbachia infected mosquitoes then in that situation we should release Wolbachia infected mosquitoes impulsively.

The mathematical model which describes the transmission dynamics of Wolbachia among Aedes Aegypti mosquitoes along with Wolbachia invasion and impulsive control is defined as follows:

When $t \neq t_\theta$ for $\theta = 1, 2, ...m$,

$$\begin{cases} {}_0^cD_t^\alpha W_e(t) &= \dfrac{\Lambda_{w_e} W_f W_a}{T} - \lambda_{w_e} W_e - \gamma_{w_e} W_e - \eta_1 I_e \\ {}_0^cD_t^\alpha W_l(t) &= \gamma_{w_e} W_e - \lambda_{w_l} W_l - \gamma_{w_l} W_l + (1-\alpha)\gamma_{i_e} I_e - \eta_2 I_l \\ {}_0^cD_t^\alpha W_p(t) &= \gamma_{w_l} W_l - \lambda_{w_p} W_p - \gamma_{w_p} W_p + (1-\beta)\gamma_{i_l} I_l - \eta_3 I_p \\ {}_0^cD_t^\alpha W_f(t) &= \rho\gamma_{w_p} W_p - \lambda_{w_f} W_f + (1-\epsilon)\gamma_{i_p}\rho_{i_w} I_p + \eta_4 W_f \\ {}_0^cD_t^\alpha W_a(t) &= (1-\rho)\gamma_{w_p} W_p - \lambda_{w_a} W_a + (1-\epsilon)\gamma_{i_p}(1-\rho_{i_w}) I_p + \eta_5 W_a \\ {}_0^cD_t^\alpha I_e(t) &= \dfrac{\Lambda_{i_e} I_f (W_a + I_a)}{T} - \lambda_{i_e} I_e - \alpha\gamma_{i_e} I_e + \eta_1 I_e \\ {}_0^cD_t^\alpha I_l(t) &= \alpha\gamma_{i_e} I_e - \lambda_{i_l} I_l - \beta\gamma_{i_l} I_l + \eta_2 I_l \\ {}_0^cD_t^\alpha I_p(t) &= \beta\gamma_{i_l} I_l - \lambda_{i_p} I_p - \epsilon\gamma_{i_p} I_p + \eta_3 I_p \\ {}_0^cD_t^\alpha I_f(t) &= \rho_i\epsilon\gamma_{i_p} I_p - \lambda_{i_f} I_f - \eta_4 W_f \\ {}_0^cD_t^\alpha I_a(t) &= (1-\rho_i)\epsilon\gamma_{i_p} I_p - \lambda_{i_a} I_a - \eta_5 W_a \end{cases} \tag{7}$$

When $t = t_\theta$ for $\theta = 1, 2, ...m$,

$$\begin{cases} \Delta W_e(t) = 0 \\ \Delta W_l(t) = 0 \\ \Delta W_p(t) = 0 \\ \Delta W_f(t) = 0 \\ \Delta W_a(t) = 0 \\ \Delta I_e(t) = \delta_1 I_e(t_\theta) \\ \Delta I_l(t) = \delta_2 I_l(t_\theta) \\ \Delta I_p(t) = \delta_3 I_p(t_\theta) \\ \Delta I_f(t) = \delta_4 I_f(t_\theta) \\ \Delta I_a(t) = \delta_5 I_a(t_\theta), \end{cases}$$

with initial conditions,

$$\begin{aligned} W_e(t_0) &= W_{e_0}; W_l(t_0) = W_{l_0}; W_{t_0}(0) = W_{p_0}; W_{t_0}(0) = W_{f_0}; W_{t_0}(0) = W_{a_0}; \\ I_e(t_0) &= I_{e_0}; I_l(t_0) = I_{l_0}; I_p(t_0) = I_{p_0}; I_f(t_0) = I_{f_0}; I_a(t_0) = I_{a_0}; \end{aligned}$$

where W_{e_0}, W_{l_0}, W_{p_0}, W_{f_0}, W_{a_0}, I_{e_0}, I_{l_0}, I_{p_0}, I_{f_0} and I_{a_0} all are positive integers. Moreover,

$$\begin{aligned} \Delta W_e(t_\theta) &= W_e(t_\theta^+) - W_e(t_\theta^-) \\ \Delta W_l(t_\theta) &= W_l(t_\theta^+) - W_l(t_\theta^-) \\ \Delta W_p(t_\theta) &= W_p(t_\theta^+) - W_p(t_\theta^-) \\ \Delta W_f(t_\theta) &= W_f(t_\theta^+) - W_f(t_\theta^-) \\ \Delta W_a(t_\theta) &= W_a(t_\theta^+) - W_a(t_\theta^-) \\ \Delta I_e(t_\theta) &= I_e(t_\theta^+) - I_e(t_\theta^-) \\ \Delta I_l(t_\theta) &= I_l(t_\theta^+) - I_l(t_\theta^-) \\ \Delta I_p(t_\theta) &= I_p(t_\theta^+) - I_p(t_\theta^-) \\ \Delta I_f(t_\theta) &= I_f(t_\theta^+) - I_f(t_\theta^-) \\ \Delta I_a(t_\theta) &= I_a(t_\theta^+) - I_a(t_\theta^-), \end{aligned}$$

with $t_1 < t_2 < t_3 \cdots < t_m$. Let us assume that,

$$M(t) = \begin{bmatrix} W_e(t) & W_l(t) & W_p(t) & W_f(t) & W_a(t) & I_e(t) & I_l(t) & I_p(t) & I_f(t) & I_a(t) \end{bmatrix}^\top;$$

$$M^* = \begin{bmatrix} W_e^* & W_l^* & W_p^* & W_f^* & W_a^* & I_e^* & I_l^* & I_p^* & I_f^* & I_a^* \end{bmatrix}^\top;$$

$$\begin{aligned}\Delta M(t_\theta) = & \big[\Delta W_e(t) \quad \Delta W_l(t_\theta) \quad \Delta W_p(t_\theta) \quad \Delta W_f(t_\theta) \quad \Delta W_a(t_\theta) \\ & \Delta I_e(t_\theta) \quad \Delta I_l(t_\theta) \quad \Delta I_p(t_\theta) \quad \Delta I_f(t_\theta) \quad \Delta I_a(t_\theta)\big]^\top;\end{aligned}$$

Therefore, (7) can be rewritten as,

$$\begin{cases} {}_0^cD_t^\alpha M(t) & = -W_1 M(t) + g(M(t)), t \neq t_\theta, \theta = 1,2,3,\cdots,m \\ \Delta M(t_\theta) & = M(t_\theta^+) - M(t_\theta^-) = \delta_\theta M(t_\theta), t = t_\theta, \theta = 1,2,3,\cdots,m \\ M(t_0) & = M_0 \in \mathbb{Z}^+, \end{cases} \tag{8}$$

where,

$$W_1 = -\begin{bmatrix} \lambda_{w_e} + \gamma_{w_e} & 0 & 0 & 0 & 0 & \eta_1 & 0 \\ -\gamma_{w_e} & \lambda_{w_l} + \gamma_{w_l} & 0 & 0 & 0 & -(1-\alpha)\gamma_{i_e} & \eta_2 \\ 0 & -\gamma_{w_l} & \lambda_{w_p} + \gamma_{w_p} & 0 & 0 & 0 & -(1-\beta)\gamma_{i_l} \\ 0 & 0 & -\rho\gamma_{w_p} & \lambda_{w_f} - \eta_4 & 0 & 0 & 0 \\ 0 & 0 & -(1-\rho)\gamma_{w_p} & 0 & \lambda_{w_a} - \eta_5 & 0 & 0 \\ 0 & 0 & 0 & 0 & 0 & \lambda_{i_e} + \alpha\gamma_{i_e} - \eta_1 & 0 \\ 0 & 0 & 0 & 0 & 0 & -\alpha\gamma_{i_e} & \lambda_{i_l} + \beta\gamma_{i_l} - \eta_2 \\ 0 & 0 & 0 & 0 & 0 & 0 & -\beta\gamma_{i_l} \\ 0 & 0 & 0 & \eta_4 & 0 & 0 & 0 \\ 0 & 0 & 0 & 0 & \eta_5 & 0 & 0 \end{bmatrix}$$

$$\begin{bmatrix} 0 & 0 & 0 \\ 0 & 0 & 0 \\ \eta_3 & 0 & 0 - (1-\epsilon)\gamma_{i_p}\rho_{i_w} \quad 0 \quad 0 \\ -(1-\epsilon)\gamma_{i_p}(1-\rho_{i_w}) & 0 & 0 \\ 0 & 0 & 0 \\ 0 & 0 & 0 \\ \lambda_{i_p} + \epsilon\gamma_{i_p} - \eta_3 & 0 & 0 \\ -\rho_i\epsilon\gamma_{i_p} & \lambda_{i_f} & 0 \\ -(1-\rho_i\epsilon\gamma_{i_p}) & 0 & \lambda_{i_a} \end{bmatrix};$$

$$g(M(t)) = \begin{bmatrix} \frac{\Lambda_{w_e} m_4 m_5}{T} \\ 0 \\ 0 \\ 0 \\ 0 \\ \frac{\Lambda_{i_e} m_9 (m_5 + m_{10})}{T} \\ 0 \\ 0 \\ 0 \\ 0 \end{bmatrix}; \quad \delta_\theta = \begin{bmatrix} 0 & 0 & 0 & 0 & 0 & 0 & 0 & 0 & 0 & 0 \\ 0 & 0 & 0 & 0 & 0 & 0 & 0 & 0 & 0 & 0 \\ 0 & 0 & 0 & 0 & 0 & 0 & 0 & 0 & 0 & 0 \\ 0 & 0 & 0 & 0 & 0 & 0 & 0 & 0 & 0 & 0 \\ 0 & 0 & 0 & 0 & 0 & 0 & 0 & 0 & 0 & 0 \\ 0 & 0 & 0 & 0 & 0 & \delta_1 & 0 & 0 & 0 & 0 \\ 0 & 0 & 0 & 0 & 0 & 0 & \delta_2 & 0 & 0 & 0 \\ 0 & 0 & 0 & 0 & 0 & 0 & 0 & \delta_3 & 0 & 0 \\ 0 & 0 & 0 & 0 & 0 & 0 & 0 & 0 & \delta_4 & 0 \\ 0 & 0 & 0 & 0 & 0 & 0 & 0 & 0 & 0 & \delta_5 \end{bmatrix};$$

6. Existence and Uniqueness of Solution

By utilizing the results from fixed point theory, the existence and uniqueness results for the system of Equation (7) were derived in this section.

Let $C_{n,m} = H$ be the Banach space of all bounded continuous function defined on $[n, m] \in \mathbb{R}$.

For the sake of simplicity, let

$$\begin{cases} {}_0^cD_t^\alpha W_e(t) &= K_1(t, m_1(t), m_2(t), \cdots, m_{10}(t)) \\ {}_0^cD_t^\alpha W_l(t) &= K_2(t, m_1(t), m_2(t), \cdots, m_{10}(t)) \\ {}_0^cD_t^\alpha W_p(t) &= K_3(t, m_1(t), m_2(t), \cdots, m_{10}(t)) \\ {}_0^cD_t^\alpha W_f(t) &= K_4(t, m_1(t), m_2(t), \cdots, m_{10}(t)) \\ {}_0^cD_t^\alpha W_a(t) &= K_5(t, m_1(t), m_2(t), \cdots, m_{10}(t)) \\ {}_0^cD_t^\alpha I_e(t) &= K_6(t, m_1(t), m_2(t), \cdots, m_{10}(t)) \\ {}_0^cD_t^\alpha I_l(t) &= K_7(t, m_1(t), m_2(t), \cdots, m_{10}(t)) \\ {}_0^cD_t^\alpha I_p(t) &= K_8(t, m_1(t), m_2(t), \cdots, m_{10}(t)) \\ {}_0^cD_t^\alpha I_f(t) &= K_9(t, m_1(t), m_2(t), \cdots, m_{10}(t))) \\ {}_0^cD_t^\alpha I_a(t) &= K_{10}(t, m_1(t), m_2(t), \cdots, m_{10}(t)). \end{cases} \tag{9}$$

where, $m_1(t) = W_e(t)$, $m_2(t) = W_l(t)$, $m_3(t) = W_p(t)$, $m_4(t) = W_f(t)$, $m_5(t) = W_a(t)$, $m_6(t) = I_e(t)$, $m_7(t) = I_l(t)$, $m_8(t) = I_p(t)$, $m_9(t) = I_f(t)$ and $m_{10}(t) = I_a(t)$. Moreover, let us assume that,

$$\begin{cases} K_1(t, m_1(t), m_2(t), \cdots, m_{10}(t)) &= \dfrac{\Lambda_{w_e} W_f W_a}{T} - \lambda_{w_e} W_e - \gamma_{w_e} W_e - \eta_1 I_e \\ K_2(t, m_1(t), m_2(t), \cdots, m_{10}(t)) &= \gamma_{w_e} W_e - \lambda_{w_l} W_l - \gamma_{w_l} W_l + (1-\alpha)\gamma_{i_e} I_e - \eta_2 I_l \\ K_3(t, m_1(t), m_2(t), \cdots, m_{10}(t)) &= \gamma_{w_l} W_l - \lambda_{w_p} W_p - \gamma_{w_p} W_p + (1-\beta)\gamma_{i_l} I_l - \eta_3 I_p \\ K_4(t, m_1(t), m_2(t), \cdots, m_{10}(t)) &= \rho\gamma_{w_p} W_p - \lambda_{w_f} W_f + (1-\epsilon)\gamma_{i_p}\rho_{i_w} I_p + \eta_4 W_f \\ K_5(t, m_1(t), m_2(t), \cdots, m_{10}(t)) &= (1-\rho)\gamma_{w_p} W_p - \lambda_{w_a} W_a + (1-\epsilon)\gamma_{i_p}(1-\rho_{i_w}) I_p + \eta_5 W_a \\ K_6(t, m_1(t), m_2(t), \cdots, m_{10}(t)) &= \dfrac{\Lambda_{i_e} I_f (W_a + I_a)}{T} - \lambda_{i_e} I_e - \alpha\gamma_{i_e} I_e + \eta_1 I_e \\ K_7(t, m_1(t), m_2(t), \cdots, m_{10}(t)) &= \alpha\gamma_{i_e} I_e - \lambda_{i_l} I_l - \beta\gamma_{i_l} I_l + \eta_2 I_l \\ K_8(t, m_1(t), m_2(t), \cdots, m_{10}(t)) &= \beta\gamma_{i_l} I_l - \lambda_{i_p} I_p - \epsilon\gamma_{i_p} I_p + \eta_3 I_p \\ K_9(t, m_1(t), m_2(t), \cdots, m_{10}(t)) &= \rho_i\epsilon\gamma_{i_p} I_p - \lambda_{i_f} I_f - \eta_4 W_f \\ K_{10}(t, m_1(t), m_2(t), \cdots, m_{10}(t)) &= (1-\rho_i)\epsilon\gamma_{i_p} I_p - \lambda_{i_a} I_a - \eta_5 W_a. \end{cases} \tag{10}$$

By the Definition 3 of, fractional order anti derivative in Caputo sense, we have

$$\begin{cases} W_e(t) - W_e(0) &= \dfrac{1}{\Gamma(\alpha)} \displaystyle\int_0^t (t-\eta)^{\alpha-1} K_1(\eta, m_1(\eta), m_2(\eta), \cdots, m_{10}(\eta)) d\eta \\ W_l(t) - W_l(0) &= \dfrac{1}{\Gamma(\alpha)} \displaystyle\int_0^t (t-\eta)^{\alpha-1} K_2(\eta, m_1(\eta), m_2(\eta), \cdots, m_{10}(\eta)) d\eta \\ W_p(t) - W_p(0) &= \dfrac{1}{\Gamma(\alpha)} \displaystyle\int_0^t (t-\eta)^{\alpha-1} K_3(\eta, m_1(\eta), m_2(\eta), \cdots, m_{10}(\eta)) d\eta \\ W_f(t) - W_f(0) &= \dfrac{1}{\Gamma(\alpha)} \displaystyle\int_0^t (t-\eta)^{\alpha-1} K_4(\eta, m_1(\eta), m_2(\eta), \cdots, m_{10}(\eta)) d\eta \\ W_a(t) - W_a(0) &= \dfrac{1}{\Gamma(\alpha)} \displaystyle\int_0^t (t-\eta)^{\alpha-1} K_5(\eta, m_1(\eta), m_2(\eta), \cdots, m_{10}(\eta)) d\eta \end{cases}$$

$$\begin{cases} I_e(t) - I_e(0) &= \frac{1}{\Gamma(\alpha)} \int_0^t (t-\eta)^{\alpha-1} K_6(\eta, m_1(\eta), m_2(\eta), \cdots, m_{10}(\eta)) d\eta \\ I_l(t) - I_l(0) &= \frac{1}{\Gamma(\alpha)} \int_0^t (t-\eta)^{\alpha-1} K_7(\eta, m_1(\eta), m_2(\eta), \cdots, m_{10}(\eta)) d\eta \\ I_p(t) - I_p(0) &= \frac{1}{\Gamma(\alpha)} \int_0^t (t-\eta)^{\alpha-1} K_8(\eta, m_1(\eta), m_2(\eta), \cdots, m_{10}(\eta)) d\eta \\ I_f(t) - I_f(0) &= \frac{1}{\Gamma(\alpha)} \int_0^t (t-\eta)^{\alpha-1} K_9(\eta, m_1(\eta), m_2(\eta), \cdots, m_{10}(\eta)) d\eta \\ I_a(t) - I_a(0) &= \frac{1}{\Gamma(\alpha)} \int_0^t (t-\eta)^{\alpha-1} K_{10}(\eta, m_1(\eta), m_2(\eta), \cdots, m_{10}(\eta)) d\eta. \end{cases}$$

This implies that,

$$\begin{cases} W_e(t) &= W_e(0) + \frac{1}{\Gamma(\alpha)} \int_0^t (t-\eta)^{\alpha-1} K_1(\eta, m_1(\eta), m_2(\eta), \cdots, m_{10}(\eta)) d\eta \\ W_l(t) &= W_l(0) + \frac{1}{\Gamma(\alpha)} \int_0^t (t-\eta)^{\alpha-1} K_2(\eta, m_1(\eta), m_2(\eta), \cdots, m_{10}(\eta)) d\eta \\ W_p(t) &= W_p(0) + \frac{1}{\Gamma(\alpha)} \int_0^t (t-\eta)^{\alpha-1} K_3(\eta, m_1(\eta), m_2(\eta), \cdots, m_{10}(\eta)) d\eta \\ W_f(t) &= W_f(0) + \frac{1}{\Gamma(\alpha)} \int_0^t (t-\eta)^{\alpha-1} K_4(\eta, m_1(\eta), m_2(\eta), \cdots, m_{10}(\eta)) d\eta \\ W_a(t) &= W_a(0) + \frac{1}{\Gamma(\alpha)} \int_0^t (t-\eta)^{\alpha-1} K_5(\eta, m_1(\eta), m_2(\eta), \cdots, m_{10}(\eta)) d\eta \\ I_e(t) &= I_e(0) + \frac{1}{\Gamma(\alpha)} \int_0^t (t-\eta)^{\alpha-1} K_6(\eta, m_1(\eta), m_2(\eta), \cdots, m_{10}(\eta)) d\eta \\ I_l(t) &= I_l(0) + \frac{1}{\Gamma(\alpha)} \int_0^t (t-\eta)^{\alpha-1} K_7(\eta, m_1(\eta), m_2(\eta), \cdots, m_{10}(\eta)) d\eta \\ I_p(t) &= I_p(0) + \frac{1}{\Gamma(\alpha)} \int_0^t (t-\eta)^{\alpha-1} K_8(\eta, m_1(\eta), m_2(\eta), \cdots, m_{10}(\eta)) d\eta \\ I_f(t) &= I_f(0) + \frac{1}{\Gamma(\alpha)} \int_0^t (t-\eta)^{\alpha-1} K_9(\eta, m_1(\eta), m_2(\eta), \cdots, m_{10}(\eta)) d\eta \\ I_a(t) &= I_a(0) + \frac{1}{\Gamma(\alpha)} \int_0^t (t-\eta)^{\alpha-1} K_{10}(\eta, m_1(\eta), m_2(\eta), \cdots, m_{10}(\eta)) d\eta. \end{cases} \tag{11}$$

Now, we define Equation (11) as

$$M(t) = M(0) + \frac{1}{\Gamma(\alpha)} \int_0^t (t-\eta)^{\alpha-1} K(\eta, m_1(\eta), m_2(\eta), \cdots, m_{10}(\eta)) d\eta,$$

where

$$M(t) = \begin{pmatrix} W_e(t) \\ W_l(t) \\ W_p(t) \\ W_f(t) \\ W_a(t) \\ I_e(t) \\ I_l(t) \\ I_p(t) \\ I_f(t) \\ I_a(t) \end{pmatrix} \textit{ and } K(\eta, m_1(\eta), m_2(\eta), \cdots, m_{10}(\eta)) = \begin{pmatrix} K_1(\eta, m_1(\eta), m_2(\eta), \cdots, m_{10}(\eta)) \\ K_2(\eta, m_1(\eta), m_2(\eta), \cdots, m_{10}(\eta)) \\ K_3(\eta, m_1(\eta), m_2(\eta), \cdots, m_{10}(\eta)) \\ K_4(\eta, m_1(\eta), m_2(\eta), \cdots, m_{10}(\eta)) \\ K_5(\eta, m_1(\eta), m_2(\eta), \cdots, m_{10}(\eta)) \\ K_6(\eta, m_1(\eta), m_2(\eta), \cdots, m_{10}(\eta)) \\ K_7(\eta, m_1(\eta), m_2(\eta), \cdots, m_{10}(\eta)) \\ K_8(\eta, m_1(\eta), m_2(\eta), \cdots, m_{10}(\eta)) \\ K_9(\eta, m_1(\eta), m_2(\eta), \cdots, m_{10}(\eta)) \\ K_{10}(\eta, m_1(\eta), m_2(\eta), \cdots, m_{10}(\eta)) \end{pmatrix}.$$

Let us define $C_{n,m}$ as $C_{n,m} = \mathfrak{F}_n(t_0) \times H_m(s)$ where,

$$s = \min\{W_{e_0}, W_{l_0}, W_{p_0}, W_{f_0}, W_{a_0}, I_{e_0}, I_{l_0}, I_{p_0}, I_{f_0}, I_{a_0}\}$$

and

$$\begin{aligned} \mathfrak{F}_n(t_0) &= [t_0 - n, t_0 + n] \\ H_m(s) &= [s - m, s + m]. \end{aligned}$$

Along with this, we assumed that

$$\begin{aligned} R \;=\; & \max_{C_{n,m}}\{\sup_{C_{n,m}} ||\mathfrak{F}_1||, \sup_{C_{n,m}} ||\mathfrak{F}_2||, \sup_{C_{n,m}} ||\mathfrak{F}_3||, \sup_{C_{n,m}} ||\mathfrak{F}_4||, \sup_{C_{n,m}} ||\mathfrak{F}_5||, \sup_{C_{n,m}} ||\mathfrak{F}_6||, \sup_{C_{n,m}} ||\mathfrak{F}_7||, \\ & \sup_{C_{n,m}} ||\mathfrak{F}_8||, \sup_{C_{n,m}} ||\mathfrak{F}_9||, \sup_{C_{n,m}} ||\mathfrak{F}_{10}||\}. \end{aligned}$$

Let us define the norm at infinity as follows:

$$||\Psi||_\infty = \sup_{t \in \mathfrak{F}_n} |\Psi(t)|.$$

Here, the operator ν: $C_{n,m} \to C_{n,m}$ is defined by

$$\nu(M(t)) = M(0) + \frac{1}{\Gamma(\alpha)} \int_0^t (t-\eta)^{\alpha-1} K(\eta, m_1(\eta), m_2(\eta), \cdots, m_{10}(\eta)) d\eta. \tag{12}$$

To prove ν is well defined operator, we should prove that

$$||\nu M(t) - M(0)||_\infty < \begin{pmatrix} m \\ m \\ m \\ m \\ m \\ m \\ m \\ m \\ m \\ m \end{pmatrix}$$

Now, let

$$\begin{aligned} ||\nu_1 W_e(t) - W_e(t)||_\infty &= ||\frac{1}{\Gamma(\alpha)} \int_0^t (t-\eta)^{\alpha-1} K_1(\eta, y_1(\eta), y_2(\eta), \cdots, y_{10}(\eta)) d\eta||_\infty \\ &\le \frac{1}{\Gamma(\alpha)} \int_0^t (t-\eta)^{\alpha-1} ||K_1(\eta, y_1(\eta), y_2(\eta), \cdots, y_{10}(\eta))||_\infty d\eta \\ &\le \frac{R}{\Gamma(\alpha)} \int_0^t (t-\eta)^{\alpha-1} d\eta \\ &\le \frac{Rn^\alpha}{\Gamma(\alpha+1)}, \end{aligned}$$

where,

$$n < \left(\frac{m\Gamma(\alpha+1)}{R}\right)^{1/n}.$$

As well as, we can prove that the other equations of (6) can satisfies this inequality.

That is, the operator ν is well-defined if

$$n < \left(\frac{m\Gamma(\alpha+1)}{R}\right)^{1/n}.$$

Now, we should prove that the operator ν satisfies the Lipschitz condition. That is,

$$||\nu_{M_1} - \nu_{M_2}||_\infty < h||M_1 - M_2||_\infty$$

To prove this, let

$$\begin{aligned}
||\nu_1 W_{e_1} - \nu_1 W_{e_2}|| &= ||\frac{1}{\Gamma(\alpha)}\int_0^t (t-\eta)^{\alpha-1}K_1(W_{e_1}, m_2(\eta), \cdots, m_{10}(\eta)\eta)d\eta \\
&\quad -\frac{1}{\Gamma(\alpha)}\int_0^t (t-\eta)^{\alpha-1}K_1(W_{e_2}, m_2(\eta), \cdots, m_{10}(\eta), \eta)d\eta||_\infty \\
&= \frac{1}{\Gamma(\alpha)}||\int_0^t K_1(\eta, W_{e_1}, m_2(\eta), \cdots, m_{10}(\eta)) \\
&\quad -K_1(\eta, W_{e_2}, m_2(\eta), \cdots, m_{10}(\eta))(t-\eta)^{\alpha-1}d\eta|| \\
&\leq \frac{1}{\Gamma(\alpha)}\int_0^t ||K_1(\eta, W_{e_1}, m_2(\eta), \cdots, m_{10}(\eta)) \\
&\quad -K_1(\eta, W_{e_2}, y_2, \cdots, y_{10})||(t-\eta)^{\alpha-1}d\eta \\
&\leq \frac{1}{\Gamma(\alpha)}\int_0^t ||\frac{\omega}{N}(\mathfrak{F}(t)(W_{e_1} - W_{e_2}))||(t-\eta)^{\alpha-1}d\eta \\
&\leq \frac{n^\alpha|\omega|||\mathfrak{F}(t)||_\infty}{N\Gamma(\alpha+1)}||W_{e_1} - W_{e_2}||_\infty \\
&\leq h_1||W_{e_1} - W_{e_2}||_\infty,
\end{aligned}$$

with $h_1 = \frac{n^\alpha|\omega|||\mathfrak{F}(t)||_\infty}{N\Gamma(\alpha+1)}$.

Similarly, we can prove that

$$\begin{aligned}
||\nu W_{l_1} - \nu W_{l_2}|| &\leq h_2||W_{l_1} - W_{l_2}|| \\
||\nu W_{p_1} - \nu W_{p_2}|| &\leq h_3||W_{p_1} - W_{p_2}|| \\
||\nu W_{f_1} - \nu W_{f_2}|| &\leq h_4||W_{f_1} - W_{f_2}|| \\
||\nu W_{a_1} - \nu W_{a_2}|| &\leq h_5||W_{a_1} - W_{a_2}|| \\
||\nu I_{e_1} - \nu I_{e_2}|| &\leq h_6||I_{e_1} - I_{e_2}|| \\
||\nu I_{l_1} - \nu I_{l_2}|| &\leq h_7||I_{l_1} - I_{l_2}|| \\
||\nu I_{p_1} - \nu I_{p_2}|| &\leq h_8||I_{p_1} - I_{p_2}|| \\
||\nu I_{f_1} - \nu I_{f_2}|| &\leq h_9||I_{f_1} - I_{f_2}|| \\
||\nu I_{a_1} - \nu I_{a_2}|| &\leq h_{10}||I_{a_1} - I_{a_2}||.
\end{aligned}$$

By the definition of Contraction mapping Definition 5, the map ν is a contraction map if $0 < h_i < 1$ for all $i = 1, 2, 3, \cdots, 10$. Therefore, ν is a contraction mapping on a compact Banach space H. Then by Contraction mapping Theorem 1, ν has a solution and it is unique.

This implies that, the system of Equation (7) has a solution and its unique.

7. Stability Analysis

In the present section, the global Mittag-Leffler stability results were derived via LMI (Linear Matrix Inequality) approach and Lyapunov method.

Assumption (A1): Assume that the function $g(M(t))$ satisfies the following: For any $e_1, e_2 \in \mathbb{R}^n$ there exists $S_1 \in \mathbb{R}^{n\times n}$, such that $||g(e_1) - g(e_2)|| \leq ||S_1(e_1 - e_2)||$.

Theorem 2. *Assume that the system* (8) *satisfies the assumption (A1) and the impulsive operator satisfies that*

$$\delta_\theta(M(t_\theta)) = -\bar{\delta}(M(t_\theta) - M^*), \theta = 1, 2, \cdots, m,$$

where M^* is an equilibrium point of system (8).
The system (8) *is said to be globally Mittag-Leffler stable if there exists a positive definite matrix Q and positive scalars ξ and γ_1 such that the following inequalities hold:*

$$Q^{\frac{-1}{2}}[\gamma_1 + \bar{\delta}]^\top Q[\gamma_1 + \bar{\delta}]Q^{\frac{-1}{2}} \leq \gamma_1 \tag{13}$$

and

$$\tilde{\Omega} = \begin{bmatrix} -2QW_1 & Q & \xi S_1 \\ * & -\xi & 0 \\ * & * & -\xi. \end{bmatrix} < 0. \tag{14}$$

Proof. Let us consider the system (8) with the initial condition $M(t_0) = M_0 \in \mathbb{Z}^+$ and an equilibrium point M^*. By using the transformation, $\mathcal{N}(t) = M(t) - M^*$, then the system (8) is transformed into

$$\begin{aligned} {}_0^C D_t^\alpha \mathcal{N}(t) &= -W_1\mathcal{N}(t) + \bar{g}(\mathcal{N}(t)), t \neq t_\theta, \theta = 1, 2, 3, ...m \\ \Delta\mathcal{N}(t_\theta) &= \mathcal{N}(t_\theta^+) - \mathcal{N}(t_\theta^-) = -\bar{\delta}\mathcal{N}(t_\theta), t = t_\theta, \theta = 1, 2, 3, ...m \\ \mathcal{N}(t_0) &= \mathcal{N}_0 \in \mathbb{Z}^+. \end{aligned} \tag{15}$$

where, $\mathcal{N}(t) = (\mathcal{N}_1, \mathcal{N}_2, \mathcal{N}_3, ..., \mathcal{N}_{10})^\top$ and $\bar{g}(\mathcal{N}(t)) = (\bar{g}(\mathcal{N}_1), \bar{g}(\mathcal{N}_2), .., \bar{g}(\mathcal{N}_{10}))^\top$ and $\mathcal{N}_0 = M_0 - M^*$. Let us consider a Lyapunov function as:

$$V(t) = \mathcal{N}^\top(t)Q\mathcal{N}(t), \tag{16}$$

where Q is a positive definite matrix. Now, the time derivative of $V(t)$ along with the trajectories of the system (16) is

$$\begin{aligned} {}_0^C D_t^\alpha V(t) &\leq 2\mathcal{N}^\top(t)Q{}_0^C D_t^\alpha \mathcal{N}(t) \\ &= \mathcal{N}^\top(t)2Q[-W_1\mathcal{N}(t) + \bar{g}(\mathcal{N}(t))] \\ &= \mathcal{N}^\top(t)(-2QW_1)\mathcal{N}(t) + \mathcal{N}^\top(t)(2Q)\bar{g}(\mathcal{N}(t)) \end{aligned} \tag{17}$$

By Lemma 2,

$$\mathcal{N}^\top(t)(2Q)\bar{g}(\mathcal{N}(t)) \leq \frac{1}{\xi}\mathcal{N}^\top(t)(QQ^\top)\mathcal{N}(t) + \xi\bar{g}^\top(\mathcal{N}(t))\bar{g}(\mathcal{N}(t)). \tag{18}$$

By assumption (A1),

$$\begin{aligned} \bar{g}^\top(\mathcal{N}(t))\bar{g}(\mathcal{N}(t)) &= \langle \bar{g}(M(t)) - \bar{g}(M^*), \bar{g}(M(t)) - \bar{g}(M^*)\rangle \\ &= \langle (\bar{g}(\mathcal{N}(t) + M^*)) - \bar{g}(M^*), \bar{g}(\mathcal{N}(t) + M^*) - \bar{g}(M^*)\rangle \\ &\leq \mathcal{N}^\top(t)S_1^\top S_1\mathcal{N}(t). \end{aligned} \tag{19}$$

Combine (18) and (19) and substitute in (17) we have,

$$
\begin{aligned}
{}_0^C D_t^\alpha V(t) &\leq \mathcal{N}^\top(t)(-2QW_1)\mathcal{N}(t) + \mathcal{N}^\top(t)(\xi^{-1}QQ^\top)\mathcal{N}(t) + \mathcal{N}^\top(t)(\xi S_1^\top S_1)\mathcal{N}(t) \\
&= \mathcal{N}^\top(t)[-2QW_1 + \xi^{-1}QQ^\top + \xi S_1^\top S_1]\mathcal{N}(t). \qquad (20)
\end{aligned}
$$

Let, $\Omega = -2QW_1 + \xi^{-1}QQ^\top + \xi S_1^\top S_1$ and Ω can be rewritten as

$$
\Omega = \begin{bmatrix} -2QW_1 & Q & S_1 \\ * & -\xi & 0 \\ * & * & -\xi^{-1} \end{bmatrix}.
$$

Now, pre and post multiply Ω by diag$\{I, I, \xi\}$, we get

$$
\tilde{\Omega} = \begin{bmatrix} -2QW_1 & Q & \xi S_1 \\ * & -\xi & 0 \\ * & * & -\xi \end{bmatrix}. \qquad (21)
$$

By Schur compliment Lemma 1, $\tilde{\Omega} < 0$.
Furthermore, the Equation (20), can be modified as

$$
\begin{aligned}
{}_0^C D_t^\alpha V(t) &\leq \mathcal{N}^\top(t)\tilde{\Omega}\mathcal{N}(t) \\
&= -\mathcal{N}^\top(t)Q^{\frac{1}{2}}[-Q^{-\frac{1}{2}}\tilde{\Omega}Q^{-\frac{1}{2}}]Q^{\frac{1}{2}}\mathcal{N}(t))
\end{aligned}
$$

let, $\epsilon_1 = \lambda_{min}(-Q^{\frac{-1}{2}}\tilde{\Omega}Q^{\frac{-1}{2}})$ and we know that $V(t) = \mathcal{N}^\top(t)Q\mathcal{N}(t)$. This implies that,

$$
{}_0^C D_t^\alpha V(t) \leq -\epsilon_1 V(t). \qquad (22)
$$

For, $t_\theta = t, \theta = 1, 2, 3, \cdots m$

$$
\begin{aligned}
V(t_\theta^+) &= \mathcal{N}^\top(t_\theta^+)Q\mathcal{N}(t_\theta^+) \\
&= [\mathcal{N}(t_\theta^-) + \bar{\delta}\mathcal{N}(t_\theta^-)]^\top Q[\mathcal{N}(t_\theta^-) + \bar{\delta}\mathcal{N}(t_\theta^-)] \\
&= \mathcal{N}^\top(t_\theta^-)[\gamma_1 + \bar{\delta}]^\top Q[\gamma_1 + \bar{\delta}]\mathcal{N}(t_\theta^-) \\
&= \mathcal{N}^\top(t_\theta^-)Q^{\frac{1}{2}}[Q^{\frac{-1}{2}}\gamma_1 + \bar{\delta}]^\top Q[\gamma_1 + \bar{\delta}Q^{\frac{-1}{2}}]Q^{\frac{1}{2}}\mathcal{N}(t_\theta^-) \\
&\leq \mathcal{N}^\top(t_\theta^-)Q\mathcal{N}(t_\theta^-) = V(\mathcal{N}(t_\theta^-)) \\
V(t_\theta^+) &\leq V(t_\theta^-) \qquad (23)
\end{aligned}
$$

Therefore, we can easily prove that,

$$
\lambda_{min}(Q)||M(t)||^2 \leq V(t) \leq \lambda_{max}(Q)||M(t)||^2. \qquad (24)
$$

Conditions (22)–(24) satisfies the conditions of Lemma 3. Therefore by Lemma 3, our system (8) is globally Mittag-Leffler stable at its equilibrium point. □

8. Numerical Simulation

In this section, we provide an example to show the benefits of the proposed models (5)–(7). In this, we have analyzed three cases by published data mentioned in Table 2.

Table 2. Data from published literature.

Parameters	Description	Data
Λ_{w_e}	Reproduction rate of Wolbachia uninfected mosquitoes	1.25/day [52]
$\lambda_{w_e}, \lambda_{w_l}, \lambda_{w_p}$	The death rate of aquatic Wolbachia uninfected mosquitoes	$\frac{1}{7.78}$/day [53]
$\gamma_{w_e}, \gamma_{w_l}, \gamma_{w_p}$	The Maturation rate of Wolbachia uninfected mosquitoes	$\frac{1}{6.67}$/day [54]
$\lambda_{w_f}, \lambda_{w_a}$	The death rate of adult Wolbachia uninfected mosquitoes	$\frac{1}{14}$/day [53]
$\lambda_{i_e}, \lambda_{i_l}, \lambda_{i_p}$	The death rate of aquatic Wolbachia infected mosquitoes	$\frac{1}{7.78}$ /day [53]
$\lambda_{i_f}, \lambda_{i_a}$	The death rate of adult Wolbachia infected mosquitoes	$\frac{1}{7}$/day [24]
Λ_{i_e}	Reproduction rate of Wolbachia infected mosquitoes	$0.95 * \Lambda_{w_e}/day$ [52]
$\gamma_{i_e}, \gamma_{i_l}, \gamma_{i_p}$	The maturation rate of Wolbachia infected mosquitoes	$\frac{1}{6.67}$/day [24]

Case 1. In this case, we have analyzed the transmission dynamics of Wolbachia among Aedes Aegypti mosquitoes via substituting the values mentioned in Table 2.
For this consider the system (5), with initial conditions $W_{e_0} = 0.9$, $W_{l_0} = 0.9$, $W_{p_0} = 0.9$, $W_{f_0} = 0.3$, $W_{a_0} = 0.3$, $I_{e_0} = 0.9$, $I_{l_0} = 0.9$, $I_{p_0} = 0.9$, $I_{f_0} = 0.3$, $I_{a_0} = 0.3$, total population $T = 3000$, and the positive scalar used in Theorem 2 as $\xi = 0.8513$
The Figures 4–7 are depicts the dynamics of Equation (5) along with the parameters in Table 2 at various orders of α such as $\alpha = 0.28, 0.68, 0.98$ and 1. We can observe by simulation results that, there is a notable decrease in non-Wolbachia mosquitoes and increase in Wolbachia infected mosquitoes.

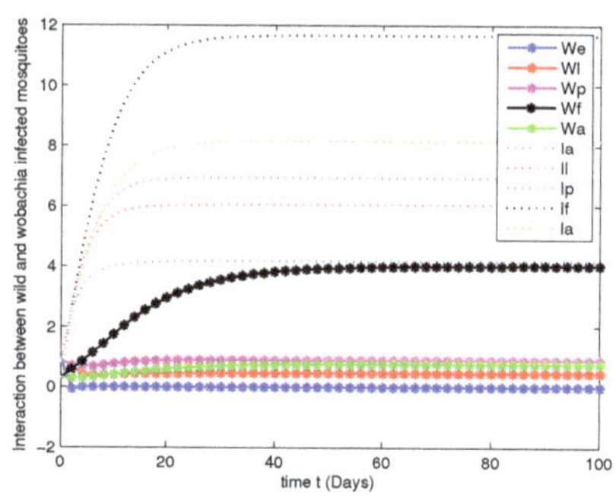

Figure 4. Population dynamics of both WU and WI mosquitoes at $\alpha = 0.28$.

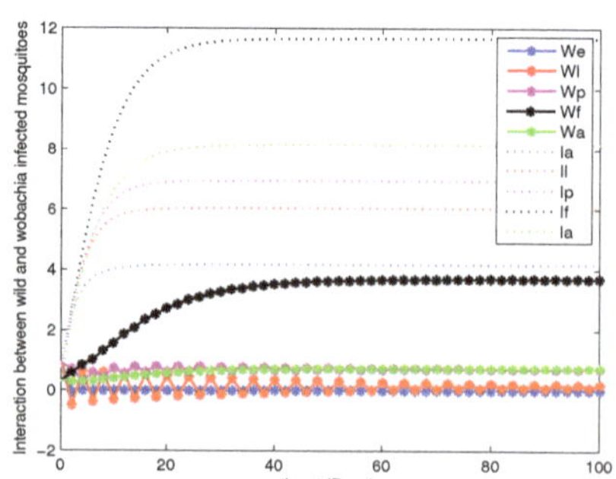

Figure 5. Population dynamics of both WU and WI mosquitoes at $\alpha = 0.68$.

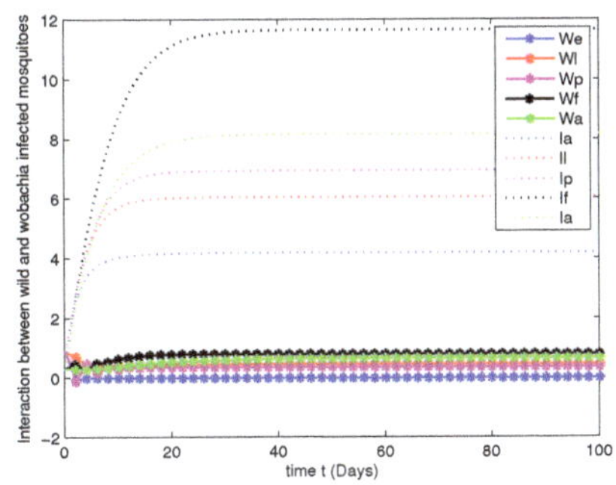

Figure 6. Population dynamics of both WU and WI mosquitoes at $\alpha = 0.98$.

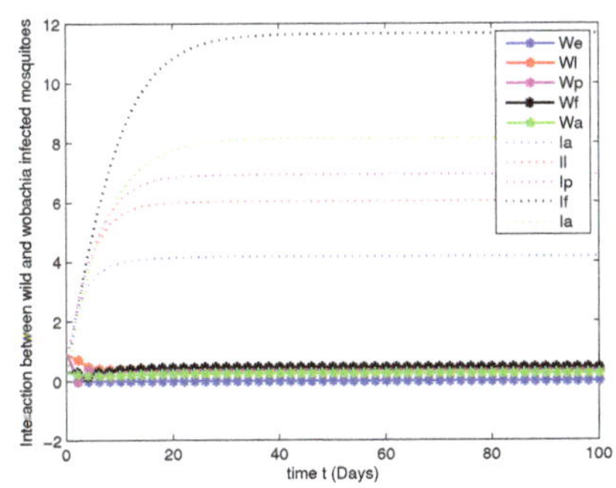

Figure 7. Population dynamics of both WU and WI mosquitoes at $\alpha = 1$.

Case 2. In this case, we have analyzed the merits and demerits of considering the Wolbachia invasion. For this consider the system of Equation (6) with parameters mentioned in Table 2. We have plotted (6) with initial conditions and total population as considered in Case 1. Along with this, the other parameters $\eta_1 = 0.03$, $\eta_2 = 0.03$, $\eta_3 = 0.03$, $\eta_4 = 0.5$ and $\eta_1 = 0.5$ are fitted.
Figures 8–11 are analyzed the dynamics of the system of Equation (6), with Wolbachia invasion and natural Wolbachia gain at various orders $\alpha = 0.28, 0.68, 0.98$ and 1. From this we can observe that , Wolbachia infected mosquitoes tends to annihilation before the eradication of non-Wolbachia mosquitoes. It will lead to the decay in natural CI rescue.

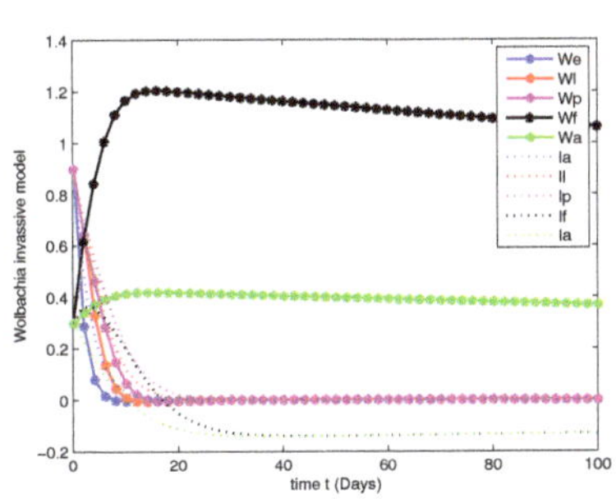

Figure 8. Population dynamics of Wolbachia invasive model at $\alpha = 0.28$.

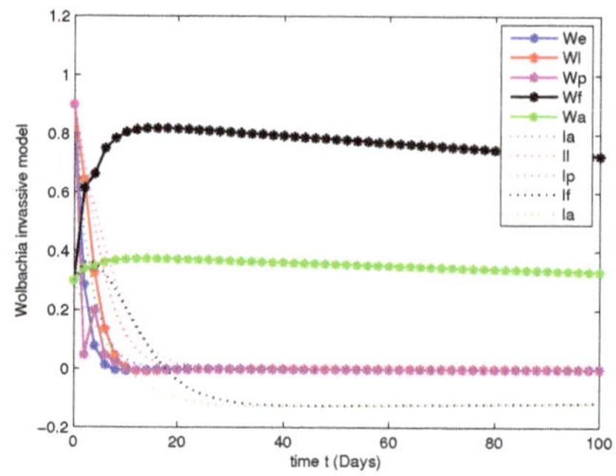

Figure 9. Population dynamics of Wolbachia invasive model at $\alpha = 0.68$.

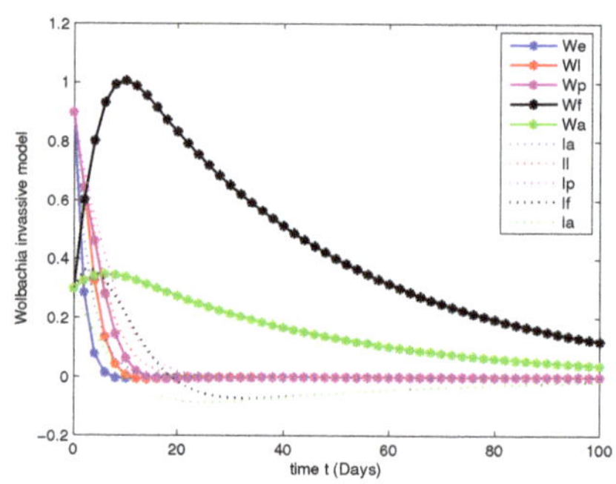

Figure 10. Population dynamics of Wolbachia invasive model at $\alpha = 0.98$.

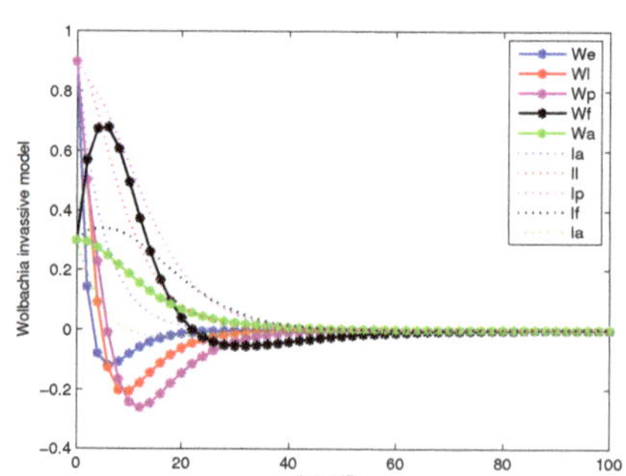

Figure 11. Population dynamics of Wolbachia invasive model at $\alpha = 0.28$.

Case 3. In this case, the decay due to the natural Wolbachia invasion is managed by releasing Wolbachia infected mosquitoes impulsively. For this case, along with the parameters mentioned in Table 2, we have fitted the values of impulsive control as $\delta_1 = 0.4$, $\delta_2 = 0.4$, $\delta_3 = 0.3$, $\delta_4 = 0.5$ and $\delta_5 = 0.5$, invasion rates are $\eta_1 = 0.03$, $\eta_2 = 0.03$, $\eta_3 = 0.03$, and gain rates are $\eta_4 = 0.5$ and $\eta_1 = 0.5$.
Figures 12–15 explicitly shows the dynamics of the systems of Equation (7) with impulsive control at orders $\alpha = 0.28, 0.68, 0.98$ and 1. From this we get that, at order $\alpha = 0.28$ the system leads to instability, when $\alpha = 0.68$ the system started to posses stable state and at $\alpha = 1$ the both population are annihilated at initial stage compared with Figures 7 and 11.

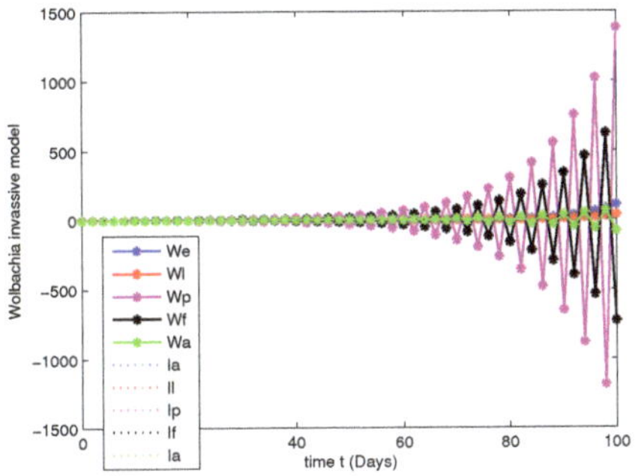

Figure 12. Population dynamics of Wolbachia invasive model after impulsive control at $\alpha = 0.28$.

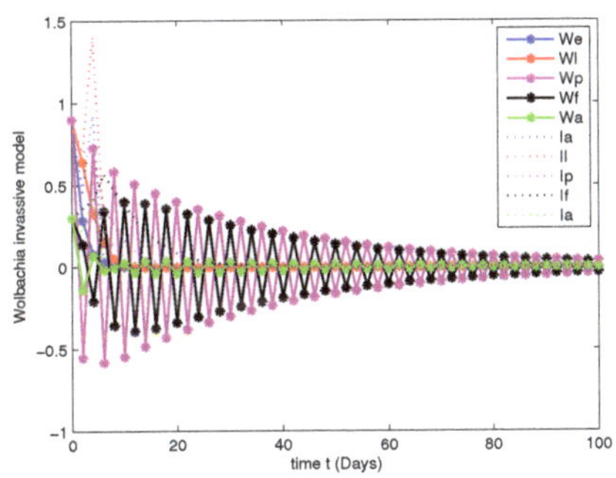

Figure 13. Population dynamics of Wolbachia invasive model after impulsive control at $\alpha = 0.68$.

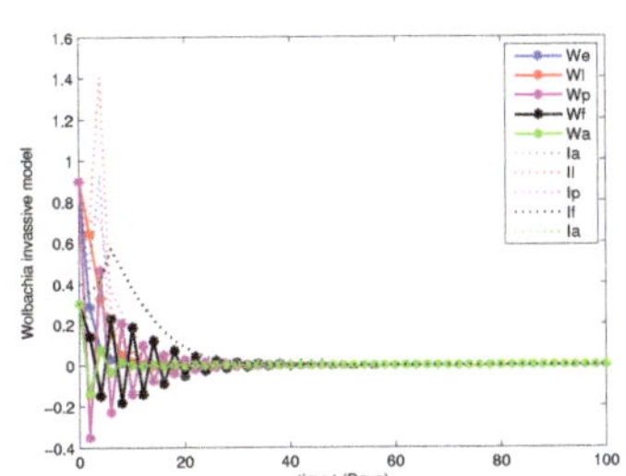

Figure 14. Population dynamics of Wolbachia invasive model after impulsive control at $\alpha = 0.98$.

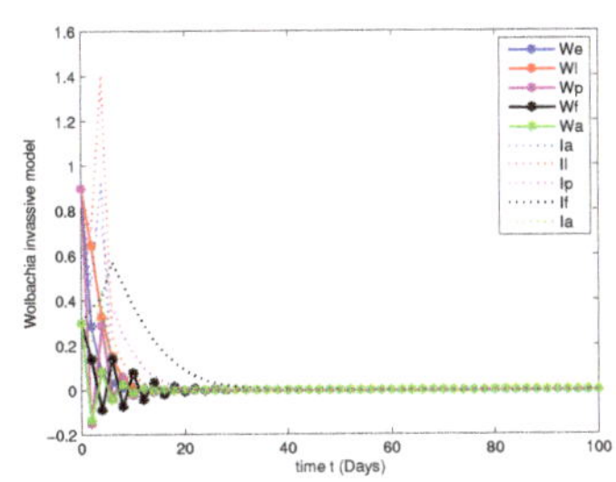

Figure 15. Population dynamics of Wolbachia invasive model after impulsive control at $\alpha = 1$.

By observing all the three cases, we can conclude that an impulsive control is an effective control strategy at Wolbachia invasion environment.

9. Conclusions

The effect of Wolbachia invasion and gain in vector population can lead to non-negligible in disease prevalence. Our impulsive control strategy shows that it is possible to control the transmission and invasion dynamics of Wolbachia bacterium. Our results shows that this method will increase the self-sustainability of Wolbachia bacterium among Aedes Aegypti mosquitoes. Another key result of the proposed fractional order model is, both mosquitoes population tends to annihilation after an impulsive controller synthesis. Further works on this model such as linearization, Lyapunov construction depicts that the created mathematical model is global Mittag-Leffler stable. In simulation performed here, depicts the effectiveness of the proposed model. In thus, we incorporated the real-world data from existing literature to compare the dynamical simulation of the 3 cases of model such as in the absence of Wolbachia invasion, the presence of Wolbachia invasion and the presence of Wolbachia invasion along with the impulsive control.

Author Contributions: Conceptualization, J.D. and R.R.; methodology, J.D. and R.R.; software, J.D. and R.R.; validation, J.D., R.R., J.A., M.N. and O.B.; formal analysis, J.D. and R.R.; investigation, J.D. and R.R.; resources, J.D. and R.R.; data curation, J.D. and R.R.; writing—original draft preparation, J.D. and R.R.; writing—review and editing, J.D. and R.R.; visualization, J.D., R.R., J.A., M.N. and O.B.; supervision, R.R., J.A., M.N. and O.B.; project administration, J.D. and R.R.; funding acquisition, J.A. All authors have read and agreed to the published version of the manuscript.

Funding: J. Alzabut would like to thank Prince Sultan University for supporting and funding this work through research group Nonlinear Analysis Methods in Applied Mathematics (NAMAM) group number RG-DES-2017-01.

Institutional Review Board Statement: Not applicable.

Informed Consent Statement: Not applicable.

Data Availability Statement: Not applicable.

Acknowledgments: This article has been written with the joint partial financial support of SERB-EEQ/2019/000365, the National Science Centre in Poland Grant DEC-2017/25/ B/ST7/02888, RUSA Phase 2.0 Grant No. F 24–51/2014-U, Policy (TN Multi-Gen), Dept.of Edn. Govt. of India, UGC-SAP (DRS-I) Grant No. F.510/8/DRS-I/ 2016(SAP-I), DST-PURSE 2nd Phase programme vide letter No. SR/ PURSE Phase 2/38 (G), DST (FIST - level I) 657876570 Grant No.SR/FIST/MS-I/ 2018/17.

Conflicts of Interest: The authors declare no conflict of interest.

Appendix A

Appendix A.1. Wolbachia Infected Mosquitoes Free Equilibrium

Suppose, there is no Wolbachia infected mosquitoes population then the possible equilibrium can be written as

$$P_2 = (W^*_{e_1}, W^*_{l_1}, W^*_{p_1}, W^*_{f_1}, W^*_{a_1}, 0, 0, 0, 0, 0)$$

where,

$$\begin{aligned}
W_{e_1}^* &= \frac{T\lambda_{w_f}\lambda_{w_a}(\lambda_{w_e}+\gamma_{w_e})(\lambda_{w_l}+\gamma_{w_l})^2(\lambda_{w_p}+\gamma_{w_p})^2}{\rho(1-\rho)\Lambda_{w_e}\gamma_{w_p}^2\gamma_{w_e}^2\gamma_{w_l}^2}\\
W_{l_1}^* &= \frac{\gamma_{w_e}}{\lambda_{w_l}+\gamma_{w_l}}W_{e_1}^*\\
W_{p_1}^* &= \frac{\gamma_{w_l}\gamma_{w_e}}{(\lambda_{w_l}+\gamma_{w_l})(\lambda_{w_p}+\gamma_{w_p})}W_{e_1}^*\\
W_{f_1}^* &= \frac{\rho\gamma_{w_p}\gamma_{w_e}\gamma_{w_l}}{\lambda_{w_f}(\lambda_{w_f}+\gamma_{w_f})(\lambda_{w_p}+\gamma_{w_p})}W_{e_1}^*\\
W_{a_1}^* &= \frac{(1-\rho)\gamma_{w_p}\gamma_{w_e}}{\lambda_{w_e}(\lambda_{w_l}+\gamma_{w_l})(\lambda_{w_p}+\gamma_{w_p})}W_{e_1}^*
\end{aligned}$$

These equilibrium points were derived by the following system of equations by putting $I_{e_1}^*=0$, $I_{l_1}^*=0$, $I_{p_1}^*=0$, $I_{f_1}^*=0$, $I_{a_1}^*=0$.

That is,

$$\begin{cases}
\dfrac{\Lambda_{w_e}W_{f_1}^*W_{a_1}^*}{T}-\lambda_{w_e}W_{e_1}^*-\gamma_{w_e}W_{e_1}^* &=0\\
\gamma_{w_e}W_{e_1}^*-\lambda_{w_l}W_{l_1}^*-\gamma_{w_l}W_{l_1}^*+(1-\alpha)\gamma_{i_e}I_{e_1}^* &=0\\
\gamma_{w_l}W_{l_1}^*-\lambda_{w_p}W_{p_1}^*-\gamma_{w_p}W_{p_1}^*+(1-\beta)\gamma_{i_l}I_{l_1}^* &=0\\
\rho\gamma_{w_p}W_{p_1}^*-\lambda_{w_f}W_{f_1}^*+(1-\epsilon)\gamma_{i_p}\rho_{i_w}I_{p_1}^* &=0\\
(1-\rho)\gamma_{w_p}W_{p_1}^*-\lambda_{w_a}W_{a_1}^*+(1-\epsilon)\gamma_{i_p}(1-\rho_{i_w})I_{p_1}^* &=0
\end{cases}$$

That is,

(i). By solving,

$$\gamma_{w_e}W_{e_1}^*-\lambda_{w_l}W_{l_1}^*-\gamma_{w_l}W_{l_1}^*+(1-\alpha)\gamma_{i_e}I_{e_1}^* \quad =0$$

We get the value of W_l^* as,

$$W_l^* \quad = \quad \frac{\gamma_{w_e}}{(\lambda_{w_l}+\gamma_{w_l})}W_e^*$$

(ii). By solving

$$\gamma_{w_l}W_{l_1}^*-\lambda_{w_p}W_{p_1}^*-\gamma_{w_p}W_{p_1}^*+(1-\beta)\gamma_{i_l}I_{l_1}^* \quad =0$$

We get the value of W_p^* as,

$$W_p^* \quad = \quad \frac{\gamma_{w_l}}{\lambda_{w_p}+\gamma_{w_p}}W_l^*$$

Substitute the value of W_l^* from (i),

$$W_p^* \quad = \quad \frac{\gamma_{w_l}\gamma_{w_e}}{(\lambda_{w_p}+\gamma_{w_p})(\lambda_{w_l}+\gamma_{w_l})}W_e^*$$

(iii). By solving

$$\rho\gamma_{w_p}W_{p_1}^*-\lambda_{w_f}W_{f_1}^*+(1-\epsilon)\gamma_{i_p}\rho_{i_w}I_{p_1}^* \quad =0$$

We get the value of W_f^* as,

$$W_f^* = \frac{\rho \gamma_{w_p}}{\lambda_{w_f}} W_p^*$$

Substitute the value of W_p^* from (ii),

$$W_f^* = \frac{\rho \gamma_{w_p} \gamma_{w_e} \gamma_{w_l}}{\lambda_{w_f}(\lambda_{w_p} + \gamma_{w_p})(\lambda_{w_l} + \gamma_{w_l})} W_e^*$$

(iv). By solving

$$(1-\rho)\gamma_{w_p} W_{p_1}^* - \lambda_{w_a} W_{a_1}^* + (1-\epsilon)\gamma_{i_p}(1-\rho_{i_w}) I_{p_1}^* = 0$$

We get the value of W_a^* as,

$$W_a^* = \frac{(1-\rho)\gamma_{w_p}}{\lambda_{w_a}} W_p^*$$

Substitute the value of W_p^* from (ii),

$$W_a^* = \frac{(1-\rho)\gamma_{w_p} \gamma_{w_l} \gamma_{w_e}}{\lambda_{w_a}(\lambda_{w_l} + \gamma_{w_l})(\lambda_{w_p} + \gamma_{w_p})} W_e^*$$

(v). By solving

$$\frac{\Lambda_{w_e} W_{f_1}^* W_{a_1}^*}{T} - \lambda_{w_e} W_{e_1}^* - \gamma_{w_e} W_{e_1}^* = 0$$

We get the value of W_e^* as,

$$W_e^* = \frac{\Lambda_{w_e}}{T(\lambda_{w_e} + \gamma_{w_e})} W_f^* W_a^*$$

Substitute the value of W_f^* and W_a^* from (iii) and (iv),

$$W_e^* = \frac{T \lambda_{w_f} \lambda_{w_a} (\lambda_{w_e} + \gamma_{w_e})(\lambda_{w_l} + \gamma_{w_l})^2 (\lambda_{w_p} + \gamma_{w_p})^2}{\rho(1-\rho)\Lambda_{w_e} \gamma_{w_p}^2 \gamma_{w_e}^2 \gamma_{w_l}^2}$$

Appendix A.2. Wild Mosquitoes Free Equilibrium

Suppose a successful release of Wolbachia infected mosquitoes replaces the wild mosquitoes by Wolbachia infected mosquitoes. Then the possible equilibrium points can be found by substituting $W_{e_2}^* = 0$, $W_{l_2}^* = 0$, $W_{p_2}^* = 0$, $W_{f_2}^* = 0$ and $W_{a_2}^* = 0$ in the following system of equations

$$\begin{aligned}
0 &= \frac{\Lambda_{i_e} I_{f_2}^* (W_{a_2}^* + I_{a_2}^*)}{T} - \lambda_{i_e} I_{e_2}^* - \alpha \gamma_{i_e} I_{e_2}^* \\
0 &= \alpha \gamma_{i_e} I_{e_2}^* - \lambda_{i_l} I_{l_2}^* - \beta \gamma_{i_l} I_{l_2}^* \\
0 &= \beta \gamma_{i_l} I_{l_2}^* - \lambda_{i_p} I_{p_2}^* - \epsilon \gamma_{i_p} I_{p_2}^* \\
0 &= \rho_i \epsilon \gamma_{i_p} I_{p_2}^* - \lambda_{i_f} I_{f_2}^* \\
0 &= (1-\rho_i)\epsilon \gamma_{i_p} I_{p_2}^* - \lambda_{i_a} I_{a_2}^*.
\end{aligned}$$

(i) By solving

$$0 = (1-\rho_i)\epsilon \gamma_{i_p} I_{p_2}^* - \lambda_{i_a} I_{a_2}^*$$

We get,

$$I_{a_2}^* = \frac{(1-\rho_i)\epsilon\gamma_{i_p}}{\lambda_{i_a}} I_{p_2}^*$$

(ii) By solving

$$0 = \rho_i\epsilon\gamma_{i_p} I_{p_2}^* - \lambda_{i_f} I_{f_2}^*$$

We get,

$$I_{f_2}^* = \frac{\rho_i\epsilon\gamma_{i_p}}{\lambda_{i_f}} I_{p_2}^*$$

(iii) By solving

$$\beta\gamma_{i_l} I_{l_2}^* - \lambda_{i_p} I_{p_2}^* - \epsilon\gamma_{i_p} I_{p_2}^*$$

We get,

$$I_{l_2}^* = \frac{(\lambda_{i_p} + \epsilon\gamma_{i_p})}{\beta\gamma_{i_l}} I_{p_2}^*$$

(iv) By solving

$$0 = \alpha\gamma_{i_e} I_{e_2}^* - \lambda_{i_l} I_{l_2}^* - \beta\gamma_{i_l} I_{l_2}^*$$

We get,

$$I_{e_2}^* = \frac{(\lambda_{i_l} + \beta\gamma_{i_l})}{\alpha\gamma_{i_e}} I_{l_2}^*$$

Substitute the value of $I_{l_2}^*$ from (iii),

$$I_{e_2}^* = \frac{(\lambda_{i_l} + \beta\gamma_{i_l})(\lambda_{i_p} + \epsilon\gamma_{i_p})}{\alpha\beta\gamma_{i_e}\gamma_{i_l}} I_{p_2}^*$$

(v) By solving,

$$0 = \frac{\Lambda_{i_e} I_{f_2}^* (W_{a_2}^* + I_{a_2}^*)}{T} - \lambda_{i_e} I_{e_2}^* - \alpha\gamma_{i_e} I_{e_2}^*$$

Put $W_{a_2}^* = 0$,

$$\begin{aligned}
\frac{\Lambda_{i_e} I_{f_2}^* I_{a_2}^*}{T} - \lambda_{i_e} I_{e_2}^* - \alpha\gamma_{i_e} I_{e_2}^* &= 0 \\
\left(\frac{\Lambda_{i_e}}{T}\right)\left(\frac{\rho_i\epsilon\gamma_{i_p}}{\lambda_{i_f}} I_{p_2}^*\right)\left(\frac{(1-\rho_i)\epsilon\gamma_{i_p}}{\lambda_{i_a}} I_{p_2}^*\right) &= (\lambda_{i_e} + \alpha\gamma_{i_e}) I_{e_2}^* \\
I_{p_2}^* &= \frac{T\lambda_{i_f}\lambda_{i_a}(\lambda_{i_e} + \alpha\gamma_{i_e})(\lambda_{i_l} + \beta\gamma_{i_l})(\lambda_{i_p} + \epsilon\gamma_{i_p})}{\Lambda_{i_e}\alpha\beta\rho_i(1-\rho_i)\epsilon^2\gamma_{i_p}^2\gamma_{i_e}\gamma_{i_l}}
\end{aligned}$$

From (i)–(v) we have the following equilibrium point,

$$P_3 = (0,0,0,0,0, I_{e_2}^*, I_{l_2}^*, I_{p_2}^*, I_{f_2}^*, I_{a_2}^*)$$

where,

$$
\begin{aligned}
I^*_{e_2} &= \frac{(\lambda_{i_l}+\beta\gamma_{i_l})(\lambda_{i_p}+\epsilon\gamma_{i_p})}{\alpha\beta\gamma_{i_e}\gamma_{i_l}} I^*_{p2} \\
I^*_{l_2} &= \frac{(\lambda_{i_p}+\epsilon\gamma_{i_p})}{\beta\gamma_{i_l}} I^*_{p2} \\
I^*_{p_2} &= \frac{T\lambda_{i_f}\lambda_{i_a}(\lambda_{i_e}+\alpha\gamma_{i_e})(\lambda_{i_l}+\beta\gamma_{i_l})(\lambda_{i_p}+\epsilon\gamma_{i_p})}{\Lambda_{i_e}\alpha\beta\rho_i(1-\rho_i)\epsilon^2\gamma^2_{i_p}\gamma_{i_e}\gamma_{i_l}} \\
I^*_{f_2} &= \frac{\rho_i\epsilon\gamma_{i_p}}{\lambda_{i_f}} I^*_{p2} \\
I^*_{a_2} &= \frac{(1-\rho_i)\epsilon\gamma_{i_p}}{\lambda_{i_a}} I^*_{p2}
\end{aligned}
$$

Appendix A.3. Both Wolbachia and Non-Wolbachia Mosquitoes Co-Existence Equilibrium

The equilibrium point for the co-existence state can be found by solving the following systems of equations

$$
\begin{cases}
\dfrac{\Lambda_{w_{en}} W^*_{f_n} W^*_{a_n}}{T} - \lambda_{w_e} W^*_{e_n} - \gamma_{w_e} W^*_{e_n} & = 0 \\
\gamma_{w_e} W^*_{e_n} - \lambda_{w_l} W^*_{l_n} - \gamma_{w_l} W^*_{l_n} + (1-\alpha)\gamma_{i_e} I^*_{e_n} & = 0 \\
\gamma_{w_l} W^*_{l_n} - \lambda_{w_p} W^*_{p_n} - \gamma_{w_p} W^*_{p_n} + (1-\beta)\gamma_{i_l} I^*_{l_n} & = 0 \\
\rho\gamma_{w_p} W^*_{p_n} - \lambda_{w_f} W^*_{f_n} + (1-\epsilon)\gamma_{i_p}\rho_{i_w} I^*_{p_n} & = 0 \\
(1-\rho)\gamma_{w_p} W^*_{p_n} - \lambda_{w_a} W^*_{a_n} + (1-\epsilon)\gamma_{i_p}(1-\rho_{i_w}) I^*_{p_n} & = 0 \\
\dfrac{\Lambda_{i_e} I^*_{f_n}(W^*_{a_n}+I^*_{a_n})}{T} - \lambda_{i_e} I^*_{e_n} - \alpha\gamma_{i_e} I^*_{e_n} & = 0 \\
\alpha\gamma_{i_e} I^*_{e_n} - \lambda_{i_l} I^*_{l_n} - \beta\gamma_{i_l} I^*_{l_n} & = 0 \\
\beta\gamma_{i_l} I^*_{l_n} - \lambda_{i_p} I^*_{p_n} - \epsilon\gamma_{i_p} I^*_{p_n} & = 0 \\
\rho_i\epsilon\gamma_{i_p} I^*_{p_n} - \lambda_{i_f} I^*_{f_n} & = 0 \\
(1-\rho_i)\epsilon\gamma_{i_p} I^*_{p_n} - \lambda_{i_a} I^*_{a_n} & = 0.
\end{cases}
$$

(i)

$$
\begin{aligned}
(1-\rho_i)\varepsilon\gamma_{i_p} I^*_{p_n} - \lambda_{i_a} I^*_{a_n} &= 0 \\
I^*_{p_n} &= \frac{\lambda_{i_a}}{(1-\rho_i)\varepsilon\gamma_{i_p}} I^*_{a_n}
\end{aligned}
$$

(ii)

$$
\begin{aligned}
\rho_i\varepsilon\gamma_{i_p} I^*_{p_n} - \lambda_{i_f} I^*_{f_n} &= 0 \\
I^*_{f_n} &= \frac{\rho_i\varepsilon\gamma_{i_p}}{\lambda_{i_f}} I^*_{p_n} \\
I^*_{f_n} &= \frac{\rho_i\lambda_{i_a}}{(1-\rho_i)\lambda_{i_f}} I^*_{a_n}
\end{aligned}
$$

(iii)

$$\begin{aligned}\beta\gamma_{i_l} I_{l_n}^* - \lambda_{i_p} I_{p_n}^* - \varepsilon\gamma_{i_p} I_{p_n}^* &= 0\\ I_{l_n}^* &= \frac{\lambda_{i_a}}{\beta\gamma_{i_l}(1-\rho_i)}\left[1+\frac{\lambda_{i_p}}{\varepsilon\gamma_{i_p}}\right] I_{a_n}^*\end{aligned}$$

Let $B_1 = 1 + \frac{\lambda_{ip}}{\varepsilon\gamma_{ip}}$

$$I_{l_n}^* = \frac{\lambda_{i_a} B_1}{\beta\gamma_{i_l}(1-\rho_i)} I_{a_n}^*$$

(iv)

$$\begin{aligned}\alpha\gamma_{i_e} I_{e_n}^* - \lambda_{i_l} I_{l_n}^* - \beta\gamma_{i_l} I_{l_n}^* &= 0\\ I_{e_n}^* &= \frac{(\lambda_{i_l}+\beta\gamma_{i_l})}{\alpha\gamma_{i_e}} I_{l_n}^*\\ I_{e_n}^* &= \frac{B_1 B_2 \lambda_{i_a}}{\alpha\gamma_{i_e}(1-\rho_i)} I_{a_n}^*\end{aligned}$$

Where, $B_1 = \left[1+\frac{\lambda_{ip}}{\varepsilon\gamma_{ip}}\right]$; $B_2 = \left[1+\frac{\lambda_{i_l}}{\beta\gamma_{i_l}}\right]$

(v)

$$\begin{aligned}\frac{\Lambda_{I_{en}} I_{f_n}^* W_{a_n}^* + \Lambda_{I_{en}} I_{f_n}^* I_{a_n}^*}{T} - \lambda_{i_e} I_{e_n}^* - \alpha\gamma_{i_e} I_{e_n}^* &- 0\\ W_{a_n}^* &= \frac{T(\lambda_{i_e}+\alpha\gamma_{i_e})}{\Lambda_{i_e} I_{f_n}^*} I_{e_n}^* - I_{a_n}^*\\ W_{a_n}^* &= \frac{TB_1B_2\lambda_{i_f}(\lambda_{i_e}+\alpha\gamma_{i_e})}{\alpha\Lambda_{i_e}\rho_i\gamma_{i_e}} - I_{a_n}^*\\ W_{a_n}^* &= \frac{TB_1B_2B_3\lambda_{i_f}}{\rho_i\Lambda_{i_e}} - I_{a_n}^*\end{aligned}$$

Where, $B_3 = 1 + \frac{\lambda_{ie}}{\alpha\gamma_{ie}}$

(vi)

$$\begin{aligned}(1-\rho)\gamma_{w_p} W_{p_n}^* - \lambda_{w_a} W_{a_n}^* + (1-\varepsilon)\gamma_{i_p}(1-\rho_{iw}) I_{p_n}^* &= 0\\ W_{p_n}^* &= \frac{\lambda_{w_a}}{(1-\rho)\gamma_{w_p}} W_{a_n}^* - \frac{(1-\varepsilon)\gamma_{i_p}(1-\rho_{iw})}{(1-\rho)\gamma_{w_p}} I_{p_n}^*\\ W_{p_n}^* &= \frac{TB_1B_2B_3\lambda_{i_f}\lambda_{w_a}}{\Lambda_{i_e}(1-\rho)\rho_i\gamma_{w_p}} - B_4 I_{a_n}^*\end{aligned}$$

where, $B_4 = 1 + \frac{(1-\varepsilon)(1-\rho_{iw})\lambda_{ia}}{(1-\rho)(1-\rho_i)\varepsilon\gamma_{wp}}$

(vii)

$$\begin{aligned}\rho\gamma_{w_p} W_{p_n}^* - \lambda_{w_f} W_{f_n}^* + (1-\varepsilon)\gamma_{i_p}\rho_{i_w} I_{p_n}^* &= 0\\ W_f^* &= \frac{\rho\gamma_{w_p}}{\lambda_{w_f}}\left[\frac{TB_1B_2B_3\lambda_{i_f}\lambda_{w_a}}{\Lambda_{i_e}(1-\rho)\rho_i\gamma_{w_p}} - B_4 I_{a_n}^*\right] + \frac{(1-\varepsilon)\rho_{iw}\lambda_{i_a}}{\varepsilon\lambda_{w_f}(1-\rho_i)} I_{a_n}^*\end{aligned}$$

(viii)

$$\begin{aligned}
\gamma_{w_l} W^*_{l_n} - \lambda_{w_p} W^*_{p_n} - \gamma_{w_p} W^*_{p_n} + (1-\beta)\gamma_{i_l} I^*_{l_n} &= 0 \\
W^*_{l_n} &= \left[\frac{\lambda_{w_p} + \gamma_{w_p}}{\gamma_{w_l}}\right] W^*_{p_n} - \left[\frac{(1-\beta)\gamma_{i_l}}{\gamma_{w_l}}\right] I^*_{l_n} \\
W^*_{l_n} &= \frac{\lambda_{w_p} + \gamma_{w_p}}{\gamma_{w_l}} \left[\frac{TB_1B_2B_3\lambda_{i_f}\lambda_{w_a}}{\Lambda_{i_e}(1-\rho)\rho_i\gamma_{w_p}} - B_4 I^*_{a_n}\right] \\
&\quad - \frac{(1-\beta)\gamma_{i_l}}{\gamma_{w_l}} \left[\frac{\lambda_{i_a} B_1}{\beta\gamma_{i_l}(1-\rho_i)} I^*_{a_n}\right]
\end{aligned}$$

(ix)

$$\begin{aligned}
\gamma_{w_e} W^*_{e_n} - \lambda_{w_l} W^*_{l_n} - \gamma_{w_l} W^*_{l_n} + (1-\alpha)\gamma_{i_e} I^*_{e_n} &= 0 \\
W^*_{e_n} &= \left[\frac{\lambda_{w_l} + \gamma_{w_l}}{\gamma_{w_e}}\right] W^*_{l_n} - \left[\frac{(1-\alpha)\gamma_{i_e}}{\gamma_{w_e}}\right] I^*_{e_n} \\
W^*_{e_n} &= \left(\frac{\lambda_{w_l} + \gamma_{w_l}}{\gamma_{w_e}}\right)\left(\frac{\lambda_{w_p} + \gamma_{w_p}}{\gamma_{w_l}}\right)\left[\frac{TB_1B_2B_3\lambda_{i_f}\lambda_{w_a}}{\Lambda_{i_e}(1-\rho)\rho_i\gamma_{w_p}}\right] \\
&\quad - \frac{I^*_{a_n}}{\gamma_{w_e}}\left[B_4(\lambda_{w_l} + \gamma_{w_l})\left(\frac{\lambda_{w_p} + \gamma_{w_p}}{\gamma_{w_l}}\right)\right. \\
&\quad + (\lambda_{w_l} + \gamma_{w_l})\left(\frac{(1-\beta)\gamma_{i_l}}{\gamma_{w_l}}\right)\left(\frac{\lambda_{i_a} B_1}{\beta\gamma_{i_l}(1-\rho_i)}\right) \\
&\quad \left. + \frac{(1-\alpha)\gamma_{i_e}\lambda_{i_a} B_1 B_2}{\alpha\gamma_{i_e}(1-\rho_i)}\right]
\end{aligned}$$

(x)

$$
\begin{aligned}
\Lambda_{w_e}\frac{W_f^* W_a^*}{T} - \lambda_{w_e} W_e^* - \gamma_{w_e} W_e^* &= 0 \\
\frac{W_f^* W_a^* \Lambda_{w_e}}{T} &= \left(\frac{\Lambda_{w_e}\rho B_4 \gamma_{w_p}}{T\lambda_{w_f}}\right)(I_a^*)^2 - \left(\frac{\Lambda_{w_e}\rho B_1 B_2 B_3 \lambda_{i_f}}{\rho_i \Lambda_{I_e}\lambda_{w_f}}\right) \\
&\quad \times \left(\frac{\lambda_{w_a}}{(1-\rho)} + B_4\gamma_{w_p}\right) I_a^* + \left(\frac{\Lambda_{w_e}\rho B_1^2 B_2^2 B_3^2 \lambda_{i_f}^2 \lambda_{w_e}}{\Lambda_{I_e}^2 (1-\rho)\rho_i^2 \lambda_{w_f}}\right) \\
(\lambda_{w_e} + \gamma_{w_e}) W_e^* &= \frac{(\lambda_{w_e}+\gamma_{w_e})\left(\lambda_{w_p}+\gamma_{w_p}\right)(\lambda_{w_l}+\gamma_{w_l})}{\gamma_{w_e}\gamma_{w_l}} - \frac{(\lambda_{w_e}+\gamma_{w_e})}{\gamma_{w_e}} \\
&\quad \times \left[\frac{\left(\lambda_{w_p}+\gamma_{w_p}\right)(\lambda_{w_l}+\gamma_{w_l}) B_4}{\gamma_{w_l}} + \frac{(\lambda_{w_l}+\gamma_{w_l})(1-\beta)\lambda_{i_a} B_1}{\gamma_{w_l}\beta(1-\rho_i)}\right. \\
&\quad \left. + \frac{(1-\alpha)\lambda_{i_a} B_1 B_2}{\alpha(1-\rho_i)}\right] I_a^* \\
\Lambda_{w_e}\frac{W_f^* W_a^*}{T} - \lambda_{w_e} W_e^* - \gamma_{w_e} W_e^* &= 0 \\
\left(\frac{\Lambda_{w_e}\rho B_4 \gamma_{w_p}}{T\lambda_{w_f}}\right)(I_a^*)^2 &- \left(\frac{(\lambda_{w_e}+\gamma_{w_e})}{\gamma_{w_e}}\right)\left(\frac{\Lambda_{w_e}\rho B_1 B_2 B_3 \lambda_f^*}{\rho_i \Lambda_{I_e}\lambda_{w_f}}\right)\left(\frac{\lambda_{w_a}}{(1-\rho)} + B_4\gamma_{w_p}\right) \\
&\quad \times \left[\frac{\left(\lambda_{w_p}+\gamma_{w_p}\right)(\lambda_{w_l}+\gamma_{w_l}) B_4}{\gamma_{w_l}} + \frac{(\lambda_{w_l}+\gamma_{w_l})(1-\beta)\lambda_{i_a} B_1}{\gamma_{w_l}\beta(1-\rho_i)}\right. \\
&\quad \left. + \frac{(1-\alpha)\lambda_{i_a} B_1 B_2}{\alpha(1-\rho_i)}\right] I_a^* + \left(\frac{\Lambda_{w_e}\rho B_1^2 B_2^2 B_3^2 \lambda_{i_f}^2 \lambda_{w_e}}{\Lambda_{I_e}^2 (1-\rho)\rho_i^2 \lambda_{w_f}}\right) = 0
\end{aligned}
$$

The above equation is a quadratic equation on $I_{a_n}^*$. That is,

$$a_1 I_{a_n}^{*^2} + a_2 I_{a_n}^* + a_3 = 0,$$

where,

$$
\begin{aligned}
a_1 &= \frac{\Lambda_{w_e}\rho B_4 \gamma_{w_p}}{T\lambda_{w_f}}; \\
a_2 &= \left(\frac{\lambda_{w_e}+\gamma_{w_e}}{T\lambda_{w_f}}\right)\left(\frac{\lambda_{w_e}\lambda_{i_f}\rho B_1 B_2 B_3}{\rho_i \Lambda_{i_e}\lambda_{w_f}}\right)\left(\frac{\lambda_{w_a}}{(1-\rho)} + B_4\gamma_{w_p}\right) \\
&\quad \left(\frac{(\lambda_{w_l}+\gamma_{w_l})(\lambda_{w_p}+\gamma_{w_p}) B_4}{\gamma_{w_l}} + \frac{(\lambda_{w_l}+\gamma_{w_l})(1-\beta)\lambda_{i_a} B_1}{\gamma_{w_l}\beta(1-\rho_i)} + \frac{(1-\alpha)\lambda_{i_a} B_1 B_2}{\alpha(1-\rho_i)}\right); \\
a_3 &= \frac{\Lambda_{w_e}\rho T B_1^2 B_2^2 B_3^2 \lambda_{w_a}}{\Lambda_{i_e}^2 (1-\rho)\rho_i^2 \lambda_{w_f}}.
\end{aligned}
$$

These are the equilibrium points presented in Section 4.4.

References

1. kilbas, A.A.; Sirvastava, H.M.; Trujillo, J.J. *Theory and Applications of Fractional Differential Equations*; Elsevier: Amsterdam, The Netherlands, 2006.
2. Rahimy, M. Applications of fractional differential equations. *Appl. Math. Sci.* **2010**, *4*, 2453–2461.
3. Samko, S.G.; Kilbas, A.A.; Marichev, O.I. *Fractional Integrals and Derivatives: Theory and Applications*; Gordon & Breach, Science Publications: London, UK; New York, NY, USA, 1993.

4. Podlubny, I. *Fractional Differential Equations: An Introduction to Fractional Derivatives, Fractional Differential Equations, to Methods of Their Solution and Some of Their Applications*; Academic Press: Cambridge, MA, USA, 1999.
5. Podlubny, I. Geometric and physical interpretation of fractional integration and fractional differentiation. *Fract. Calc. Appl. Anal.* **2002**, *5*, 367–386.
6. Oldham, K.B.; Spanier, J. *The Fractional Calculus*; Academic Press: New York, NY, USA; London, UK, 1974.
7. Li, Y.; Chen, Y.; Podlubny, I. Stability of fractional-order nonlinear dynamic systems: Lyapunov direct method and generalized Mittag-Leffler stability. *Comput. Math. Appl.* **2010**, *59*, 1810–1821. [CrossRef]
8. Gibbons, R.; Vaughn, D. Dengue: An escalating problem. *BMJ* **2002**, *324*, 1563–1566. [CrossRef] [PubMed]
9. Bhatt, S.; Gething, P.W.; Brady, O.J.; Messina, J.P.; Farlow, A.W.; Moyes, C.L.; Drake, J.M.; Brownstein, J.S.; Hoen, A.G.; Sankoh, O.; et al. The global distribution and burden of dengue. *Nature* **2013**, *496*, 504–507. [CrossRef] [PubMed]
10. Chye, J.K.; Lim, C.T.; Ng, K.B. Vertical transmission of dengue. *Clin. Infect. Dis.* **1997**, *25*, 1374–1377. [CrossRef] [PubMed]
11. Kraemer, M.; Sinka, M.; Duda, K.; Mylne, A.; Shearer, F.; Barker, C. The global distribution of the arbovirus vectors Aedes aegypti and *Ae. albopictus*. *eLife* **2015**, *4*, 1–18. [CrossRef]
12. Gubler, D.J. Dengue and dengue hemorrhagic fever. *Clin. Microbiol. Rev.* **1998**, *11*, 480–496. [CrossRef]
13. Gubler, D.J. Epidemic dengue/dengue hemorrhagic fever as a public health, social and economic problem in the 21st century. *Trends Microbiol.* **2002**, *10*, 100–103. [CrossRef]
14. Ong, A.; Sandar, M.; Chen, M.I.; Sin, L.Y. Fatal dengue hemorrhagic fever in adults during a dengue epidemic in Singapore. *Int. J. Infect. Dis.* **2007**, *11*, 263–267. [CrossRef]
15. World Health Organization. Vector-Borne Diseases. 2020. Available online: http://www.who.int/mediacentre/factsheets/fs387/en/ (accessed on 25 January 2021).
16. Alphey, L.; Benedict, M.; Bellini, R.; Clark, G.G.; Dame, D.A.; Service, M.W.; Dobson, S.L. Sterile-insect methods for control of mosquito-borne diseases: An analysis. *Vector-Borne Zoonotic Dis.* **2010**, *10*, 295–311. [CrossRef] [PubMed]
17. Bouyer, J.; Lefrancois, T. Boosting the sterile insect technique to control mosquitoes. *Trends Parasitol.* **2014**, *30*, 271–273. [CrossRef] [PubMed]
18. Fu, G.; Lees, R.S.; Nimmo, D.; Aw, D.; Jin, L.; Gray, P.; Berendonk, T.U. Femalespecific flightless phenotype for mosquito control. *Proc. Natl. Acad. Sci. USA* **2010**, *107*, 4550–4554. [CrossRef]
19. James, A.A. Gene drive systems in mosquitoes: rules of the road. *Trends Parasitol.* **2005**, *21*, 64–67. [CrossRef]
20. Scott, T.W.; Takken, W.; Knols, B.G.J.; Boëte, C. The ecology of genetically modified mosquitoes. *Science* **2002**, *298*, 117–119. [CrossRef] [PubMed]
21. Masud, M.A.; Kim, B.N.; Kim, Y. Optimal control problems of mosquito-borne disease subject to changes in feeding behaviour of Aedes mosquitoes. *Biosystems* **2017**, *156–157*, 23–39. [CrossRef]
22. Momoh, A.A.; Fugenschuh, A. Optimal control of intervention strategies and cost effectiveness analysis for a zika virus model. *Oper. Res. Health Care* **2018**, *18*, 99–111. [CrossRef]
23. Segoli, M.; Hoffmann, A.A.; Lloyd, J.; Omodei, G.J.; Ritchie, S.A. The effect of virus-blocking Wolbachia on male competitiveness of the dengue vector mosquito, Aedes aegypti. *PLOS Negl. Trop. Dis.* **2014**, *8*, e3294. [CrossRef] [PubMed]
24. Walker, T.; Johnson, P.H.; Moreira, L.A.; Iturbe-Ormaetxe, I.; Frentiu, F.D.; McMeniman, C.J.; Leong, Y.S.; Dong, Y.; Axford, J.; Kriesner, P.; et al. The WMel Wolbachia strain blocks dengue and invades caged Aedes aegypti populations. *Nature* **2011**, *476*, 450–453. [CrossRef]
25. Xi, Z.; Khoo, C.C.; Dobson, S.L. Wolbachia establishment and invasion in an Aedes aegypti laboratory population. *Science* **2005**, *310*, 326–328. [CrossRef]
26. Ormaetxe, I.; Walker, T.; Neill, S.L.O. Wolbachia and the biological control of mosquito-borne disease. *Embo Rep.* **2011**, *12*, 508–518. [CrossRef]
27. World Mosquito Program. Available online: https://www.worldmosquitoprogram.org (accessed on 25 January 2021).
28. Dutra, H.L.C.; Rocha, M.N.; Dias, F.B.S.; Mansur, S.B.; Caragata, E.P.; Moreira, L.A. Wolbachia blocks currently circulating Zika virus isolates in Brazilian Aedes aegypti mosquitoes. *Cell Host Microbe* **2016**, *19*, 771–774. [CrossRef] [PubMed]
29. Hancock, P.; Sinkins, S.; Godfray, H. Population dynamic models of the spread of Wolbachia. *Am. Nat.* **2011**, *177*, 323–333. [CrossRef]
30. Hughes, H.; Britton, N. Modelling the use of Wolbachia to control dengue fever transmission. *Bull. Math. Biol.* **2013**, *75*, 796–818. [CrossRef]
31. McMeniman, C.J.; Lane, R.V.; Cass, B.N.; Fong, A.W.; Sidhu, M.; Wang, Y.F.; Neill, S.L.O. Stable introduction of a life-shortening Wolbachia infection into the mosquito *Aedes aegypti*. *Science* **2009**, *323*, 141–144. [CrossRef]
32. Jiggins, F. The spread of Wolbachia through mosquito populations. *PLoS Biol.* **2017**, *15*, e2002780. [CrossRef]
33. Ndii, M.Z.; Hickson, R.I.; Allingham, D.; Mercer, G.N. Modelling the transmission dynamics of dengue in the presence of Wolbachia. *Math. Biosci.* **2015**, *262*, 157–166. [CrossRef]
34. Koiller, J.; da Silva, M.A.; Souza, M.O.; Codeco, C.; Iggidr, A.; Sallet, G. *Aedes, Wolbachia and Dengue*; Inria Nancy-Grand Est: Villers-lès-Nancy, France, 2014; pp. 1–47.
35. Adekunle, A.I.; Michael, M.T.; McBryde, E.S. Mathematical analysis of a Wolbachia invasive model with imperfect maternal transmission and loss of Wolbachia infection. *Infect. Dis. Model.* **2019**, *4*, 265–285. [CrossRef]

36. Xue, L.; Manore, C.; Thongsripong, P.; Hyman, J. Two-sex mosquito model for the persistence of Wolbachia. *J. Biol. Dyn.* **2017**, *11*, 216–237. [CrossRef]
37. Rock, K.S.; Wooda, D.A.; Keeling, M.J. Age- and bite- structured models for vector-borne diseases. *Epidemics* **2015**, *12*, 20–29. [CrossRef] [PubMed]
38. Rafikov, M.; Meza, M.E.M.; Correa, D.P.F.; Wyse, A.P. Controlling Aedes aegypti populations by limited Wolbachia-based strategies in a seasonal environment. *Math. Methods Appl. Sci.* **2019**, *42*, 5736–5745. [CrossRef]
39. Supriatna, A.K.; Anggriani, N.; Melanie; Husniah, H. The optimal strategy of Wolbachia- infected mosquitoes release program an application of control theory in controlling Dengue disease. In Proceedings of the 2016 International Conference on Instrumentation, Control and Automation(ICA), Bandung, Indonesia, 29–31 August 2016; pp. 38–43.
40. Dianavinnarasi, J.; Cao, Y.; Raja, R.; Rajchakit, G.; Lim, C.P. Delay-dependent stability criteria of delayed positive systems with uncertain control inputs: Application in mosquito-borne morbidities control. *Appl. Math. Comput.* **2020**, *382*, 125210. [CrossRef]
41. Dianavinnarasi, J.; Raja, R.; Alzabut, J.; Cao, J.; Niezabitowski, M.; Bagdasar, O. Application of Caputo—Fabrizio operator to suppress the Aedes Aegypti mosquitoes via Wolbachia: An LMI approach. *Math. Comput. Simul.* **2021**. [CrossRef]
42. Nisar, K.S.; Ahmad, S.; Ullah, A.; Shah, K.; Alrabaiah, H.; Arfan, M. Mathematical analysis of SIRD model of COVID-19 with Caputo fractional derivative based on real data. *Results Phys.* **2021**, *21*, 103772. [CrossRef]
43. Boyd, S.; Ghaoui, L.; Feron, E.; Balakrishnan, V. *Linear Matrix Inequalities in System and Control Theory*; SIAM Philadelphia: Philadelphia, PA, USA, 1994.
44. Wu, H.; Zhang, X.; Xue, S.; Wang, L.; Wang, Y. LMI conditions to global Mittag-Leffler stability of fractional-order neural networks with impulses. *Neurocomputing* **2016**, *193*, 148–154. [CrossRef]
45. Stamova, I. Global stability of impulsive fractional differential equations. *Appl. Math. Comput.* **2014**, 237, 605–612. [CrossRef]
46. Agarwal, R.P.; Meehan, M.; O'Regan, D. *Fixed Point Theory and Applications*; Cambridge University Press: Cambridge, UK, 2001.
47. Iswarya, M.; Raja, R.; Rajchakit, G.; Alzabut, J.; Lim, C.P. A perspective on graph theory based stability analysis of impulsive stochastic recurrent neural networks with time-varying delays. *Adv. Differ. Equ.* **2019**, *502*, 1–21. [CrossRef]
48. Stamov, G.; Stamova, I.; Alzabut, J. Global exponential stability for a class of impulsive BAM neural networks with distributed delays. *Appl. Math. Inf. Sci.* **2013**, *7*, 1539–1546. [CrossRef]
49. Stamov, G.T.; Alzabut, J.O.; Atanasov, P.; Stamov, A.G. Almost periodic solutions for impulsive delay model of price fluctuations in commodity markets. *Nonlinear Anal. Real World Appl.* **2011**, *12*, 3170–3176. [CrossRef]
50. Zada, A.; Waheed, H.; Alzabut, J.; Wang, X. Existence and stability of impulsive coupled system of fractional integrodifferential equations. *Demonstr. Math.* **2019**, *52*, 296–335. [CrossRef]
51. Zada, A.; Alam, L.; Kumam, P.; Kumam, W.; Ali, G.; Alzabut, J. Controllability of impulsive non-linear delay dynamic systems on time scale. *IEEE Access* **2020**, *8*, 93830–93839. [CrossRef]
52. Ndii, M.Z.; Hickson, R.I.; Mercer, G.N. Modelling the introduction of Wolbachia into Aedes aegypti to reduce dengue transmission. *Anziam J.* **2012**, *53*, 213–227.
53. Yang, H.M.; Macoris, M.L.G.; Galvani, K.C.; Andrighetti, M.T.M.; Wanderley, D.M.V. Assessing the effects of temperature on the population of Aedes aegypti, the vector of dengue. *Epidemiol. Infect.* **2009**, *137*, 1188–1202. [CrossRef] [PubMed]
54. Maidana, N.A.; Yang, H.M. Describing the geographic spread of dengue disease by traveling waves. *Math. Biosci.* **2008**, *215*, 64–77. [CrossRef]

symmetry

MDPI

Article

New Command Mechanism of Flaps and Wings of a Light Sport Aircraft

Ion-Marius Ghiţescu [1], Maria Luminita Scutaru [1,*], Marilena Ghiţescu [1], Paul Nicolae Borza [2] and Marin Marin [3]

1 Department of Mechanical Engineering, Transilvania University of Brasov, B-dul Eroilor 29, 500036 Brasov, Romania; marius.ghitescu@unitbv.ro (I.-M.G.); marilenaradu@unitbv.ro (M.G.)
2 Department of Computers and Electronics, Transilvania University of Brasov, B-dul Eroilor 29, 500036 Brasov, Romania; borzapn@unitbv.ro
3 Department of Mathematics and Computers Science, Transilvania University of Brasov, B-dul Eroilor 29, 500036 Brasov, Romania; m.marin@unitbv.ro
* Correspondence: lscutaru@unitbv.ro or luminitascutaru@yahoo.com

Abstract: Commercial aircraft have well-designed and optimized systems, the result of a huge experience in the field, due to the large fleet of aircraft in operation. For light, utility, or sports aircraft, with a multitude of shapes, tasks, and construction types, there are different solutions that seek to best meet the requirements of the designed aircraft. In this sense, for a sport plane, an increased maneuverability is desired, and the system that controls flaps and wing must be properly designed. A new flap mechanism command solution is proposed and justified in the paper, for use in sports and recreational aviation, in order to achieve angles of braking greater than 40°, take-off and landing in a shorter time and over a shorter distance, as well as the gliding of the aircraft in critical flight conditions or when fuel economy is needed. A finite element model is used to verify the optimized command system for the flap and wing and to check if the strength structure of the aircraft is properly designed. The main result consists of the new design command system for flaps and wings and in verifying, by calculation, the acceptability of the new mechanism proposed from the point of view of the strength of the materials.

Keywords: Light Sport Aircraft; conceptual aircraft design; wing; flap; aileron; weight estimation; symmetric profile

Citation: Ghiţescu, I.-M.; Scutaru, M.L.; Ghiţescu, M.; Borza, P.N.; Marin, M. New Command Mechanism of Flaps and Wings of a Light Sport Aircraft. *Symmetry* **2021**, *13*, 221. https://doi.org/sym13020221

Academic Editors: Victor A. Eremeyev and Sergei D. Odintsov

Received: 30 December 2020
Accepted: 26 January 2021
Published: 29 January 2021

1. Introduction

Aircraft engineering represents an intensive process full of evaluation and decision-making [1]. Much as 80% of the life cycle costs in aeronautical engineering are determined in the conceptual design phase [2]. In this, the phase will define the aircraft characteristics, such as type of being propulsion used; the purpose of aircraft—Light Sport Aircraft; technology of aircraft—materials, engine; occupant comfort requirements—fuselage cavity, pressurization; ergonomics of both crew and passenger; aircraft's aerodynamics and auxiliary lifting surfaces; certification basic of aircraft—Light Sport Aircraft FAR 23 [3]; ways of manufacturing the aircraft; aircraft's maintainability in the future; cost estimation [4].

Flight mechanics is a field studied in numerous works, and the results obtained are already classic [5,6]. The preliminary design phase is carried out after all characteristics presented in the conceptual phase had been made, and a rough aircraft sketch had been made. This phase is critical where normally it is a go situation where the negative outcome will result in a further continuation of the project until the aircraft is being manufactured [7].

In discussing the structural layout of an aircraft and its strength, it is important that we know about the material that is being used so that we can know the material load limit. There are several factors that influence the selection of materials that are being used in building the structure of an aircraft, mainly material fatigue, toughness, stiffness, and resistance to corrosion, but the overall lightness of the material is commonly looked upon by the structure and material engineer [8–11]. The main groups of materials that are usually selected to be the part of aircraft are duralumin or aluminum.

All aircraft in the world, including Light Structure Aircraft (LSA), are strictly regulated and bounded by certification based on the system of aircraft certification it possesses for each country. Several variants of light aircraft are presented below. The light aircraft Extra 300 LT, at which the force load of the direction reaches 90 Kg [12], is a one- or two-seater acrobat aircraft produced by German Extra Flugzeugproduktions- und Vertriebs-GmbH.

The technical characteristics of some known aircraft are shown in Table 1.

Table 1. Technical features.

Technical Features	Airplane Extra 300 LT	Airplane Zlin 142	Festival R40 Light Aircraft	F2 Light Aircraft	I.-M.G. Light Aircraft LSA
Crew	1	1	1	1	1
Passengers	1	1	1	1	1
Engine model	Lycoming AEIO-540-L1B5	Walter/LOM M-337AK	Rotax 912ULS	Rotax 912iS	-
Engine power	224 kW/300 CP	156 kW/209 CP	75 kW/101 CP	100 CP	200 CP
Speed	407 km/h	272 km/h	175 km/h	252 km/h	470 km/h
Flight ceiling	4.877 m	4.250 m	4.000 m		5.000 m
Flight range	769 km	941 km	800 km		700 km
Empty weight	667 g	730 kg	354 kg	380 kg	670 kg
Take-off weight	950 kg	1090 kg	1900 kg	600 kg	970 kg
Wing span	8 m	9.16 m	9.17 m	9.872m	9.7 m
Wing surface area	10.7 m^2	13.2 m^2	13.97 m^2	11.1 m^2	11.92 m^2
Height	2.62 m	2.75 m	2.47 m	2.34 m	2.57 m
Length	6.94 m	7.33 m	6.74 m	6.86 m	7 m

In the last column, the characteristics of the sport I.-M.G. light aircraft LSA studied in the paper are presented. It is known as the Zlin 142 light aircraft, in which the load with forces reaches 200–250 N [13]. Zlin Z 142 is a single-engine aircraft with two seats for tourism and produced by Czechoslovak manufacturer Moravan Otrokovice (now ZLIN Aircraft Otrokovice, Czech Republic). The Festival R40 is a single-engine aircraft with two seats for tourism produced by SC. Aerostar Bacau S.A. [14]. The F2 light aircraft is produced by Flight Design Hoerselberg [15]. On the light aircrafts Extra 300 LT, Zlin 142, Festival R40, and F2, the flaps poach up to a maximum of 45°. Before a design layout can be started in order to obtain a new type of light aircraft with greater angles of braking, a number of parameters must be chosen. There is a light aircraft wing with a rectangular form and asymmetric profile, mounted at the top, which has been modeled in 3D in Solid Works and analyzed with the finite element method (FEM) in Abaqus CAE. The wing is made of aluminum and steel. The wing is loaded with distributed pressures, and the stress and strain fields in the wing structure are studied [16].

A light aircraft wing has been modeled as CAD in Solid Works and imported into ANSYS Workbench [17]. The aircraft wings are attached to both sides of the fuselage to produce lift force [18]. A fuselage wing is used in the light acrobatic aircraft, which has been subjected to the extended finite element method (XFEM) for determining stress intensity factors (FIS) [19]. The presence of a crack through the attachment opening is not allowed, as its growth due to high dynamic loads is usually rapid and can lead to catastrophic consequences. In order to demonstrate how dangerous the appearance of the crack could be, as well as to estimate the residual strength and life of fatigue of the cracked component, the analyses are carried out using the maximum load that may occur during the flight. The expected number of cycles to complete the failure is decreased, confirming that the fastening ears must be designed using the safety approach [20]. A reconfigurable active vibration control (AVC) system is known [21,22].

The S-LSA aircraft, for which the design methodology, design requirements, weight sizing, performance sizing, weight and balance, aerodynamics, and stability and control are established, is presented in [23]. This aircraft is a tandem aircraft with sufficient visibility for both occupants. There are studies on the material from which an airplane wing is made. Tests are performed on a wing made of three types of materials: aluminum alloy, conventional carbon fiber reinforced composite, and towed carbon fiber reinforced composite. Significant performance is found in terms of the wing made of the conventional aluminum composite [24].

A high wing of an ultra-light aircraft is known, which is followed by the structural design and analysis of the wing. Wing design involves its initial considerations, such as planning shape selection, aircraft location, and structural design involves design calculations for air profile selection, wing area, wing load characteristics, and wing weight. The design is done according to the values calculated using the ANSYS FLUENT software design [25].

There is a study of the design and development of a new light aircraft, which meets the standards of the European regulations on ultra-light aircraft and the US regulations for light sports aircraft. The aircraft is a two-seater model [26]. To this end, the development of the wings, propeller, and fuselage is performed with extra caution to get the best possible results.

In connection with an aerodynamic profile used for an ultralight aircraft sailing at low speed, the invention relates to an aerodynamic profile for an ultralight aircraft, which maintains the aerodynamic performance of a main aerial profile made of plastic. Air wing pressure is reduced by bending a curved surface from a leading edge to the bottom surface and improves manufacturing comfort and structural strength by increasing the thickness of a leading-edge [27].

The invention relates to the weight-shift-controlled airplane that is characterized by a novel design of airplanes with closed passenger compartment and that combines the advantages of conventional aerodynamically controlled airplanes with those of weight-shift-controlled ultra-light airplanes. The inventive weight-shift-controlled airplane comprises a passenger compartment with a fitted propeller as the pressure drive and a chassis, an airfoil 1, and an adjustable support system between the passenger compartment. The airfoil 1 allows that the airfoil is tilted for controlling the airplane [28,29].

The utility model discloses and provides a flap control mechanism of a Light Sport Aircraft. The flap control mechanism is simple in structure, reduces resistance, and allows each flap to be independently controlled and completely sealed in a flap structure. The flap control mechanism of the Light Sport Aircraft comprises flaps and a flap driving mechanism, wherein the flap driving mechanism is arranged in the flaps and is used for enabling each flap to be independently operated and unfolded; the utility model is applied to the technical field of airplane structures [30]. All these results give an image of the effort made to obtain optimized solutions for the control mechanisms of a light aircraft.

The presented literature summarizes the main researches performed for the study of some types of related airplanes, with constructive solutions close to the structure studied in the paper. There are two aspects that are studied in the paper: first, the proposed solution for the control mechanism, and then a study of the strength of the aircraft structure to see if the proposed solution for the mechanism is compatible with the requirements appearing in the aircraft structure. The mechanism introduced in the control system must not cause stresses that exceed the strength of the material.

The paper is in line with this research trend, proposing a new system optimized for the control of flaps and wings of I.-M.G. Light Sport Aircraft.

The new Light Sport Aircraft has a maximum take-off weight of 900 Kg, and the cruising speed is 470 km per hour, and the maximum flight ceiling is 5000 m. In the design of the aircraft, it has been considered that the parts of the aircraft must be simple to manufacture and install and accessible to repair.

In determining the aerodynamic shape and design dimensions of the light aircraft, the optimal design of the aircraft is considered, easily taking into account the parameters influencing the aerodynamic shape of the fuselage and the wing, as well as the systems and mechanisms of the flap and aileron that are mounted in these areas. The technical characteristics of a new proposed I.-M.G. light aircraft are shown in Table 1 (column 5). In the new I.-M.G. light aircraft, the constructive solution of the flaps control mechanism allows the flaps to be braced up to a maximum of 55°. The fuselage is to accommodate the engine, people, fuel, and baggage. The two seats are arranged in a cote-a-cote configuration.

The new model of light aircraft proposed has a certain aerodynamic shape, symmetry geometry, and good stability being made up of middle wings of rectangular shape and having in section a symmetrical profile, from the fuselage with a certain aerodynamic shape, propeller helmet, ailerons, flaps, cockpit, vertical tail, horizontal tail, rudder with profiled shapes. The choice of the optimal solution, from an engineering and maneuverability point of view but also of the simplicity of the solution, is made following the analysis of some functional models, varying the parameters of the proposed model until obtaining an advantageous solution from several points of view.

The purpose of this article is to create the virtual model of the I.-M.G. light aircraft, the flap, and the wing that will be built and analyzed, optimizing the shape of these using CAD-CAE systems with CATIA V5R21 software. The subject of the paper is the proposal of a new flap and wing control mechanism. It is obvious that this mechanism is incorporated in a complex mechanical structure, with multiple functions and requirements, to which it must respond accordingly. Therefore, in order to see if this new type of mechanism satisfies the needs, a calculation of the strength of the structure of the entire plane is made in order to determine the deformations, strains, and stresses that appear in the components of the structure. The new type of mechanism changes the stresses that can occur in the elements of the aircraft structure, and it must be verified if, in these conditions, safety is ensured from the point of view of strength. To achieve this, the FEM is applied.

2. A New Model of I.-M.G. Light Sport Aircraft

The light aircraft and its flap control mechanism improve the dynamic behavior of the aircraft easily through the aircraft's constructive shape and the vole control mechanism that allows braking angles greater than 40° required for take-offs and landings over a short distance and in a shorter and greater load time, as well as the operation of the aircraft in critical flight conditions or when fuel economy is required, the airplane being able to hover in these situations, under conditions of reduced manufacturing costs. The proposed Light Sport Aircraft designed has as its areas of use sports and recreational aviation with a maximum capacity of two seats and having a certain aerodynamic shape and has on its wings mounted steering voles without a run-away, in order to achieve angles of braking greater than 40°, take-off and landing in a shorter time and over a shorter distance, as well as the gliding of the aircraft in critical flight conditions or when fuel economy is needed.

The main components of this light aircraft variant are the propeller helmet, propeller, cockpit, plane fuselage, a left wing with symmetrical profile, a right wing with symmetrical profile, ailerons (one on each wing) with symmetrical profile, flaps (one on each wing) with symmetrical profile, the depth on the horizontal tail, the left horizontal tail, the right horizontal tail, the direction on the vertical tail, the vertical tail. The newly designed light aircraft, named I.-M.G., shown in Figure 1, has a maximum take-off weight of 900 Kg, and the cruising speed is 470 km per hour, and the maximum flight ceiling is 5.000 m.

In determining the aerodynamic shape and design dimensions of the light aircraft, the optimal design of the aircraft is considered, easily taking into account the parameters influencing the aerodynamic shape of the fuselage and the wing, as well as the systems and mechanisms of the flap and aileron that are mounted in these areas (Figure 1).

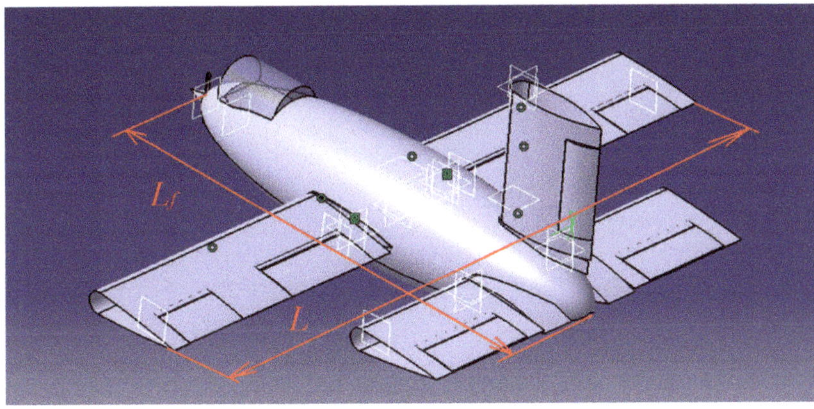

Figure 1. A mew model of I.-M.G. Light Sport Aircraft with a symmetric profile of wing–flaps–aileron.

Figure 2 shows a wing with a symmetric profile equipped with a flap and an aileron with a symmetrical profile each. They can be made of duralumin or aluminum.

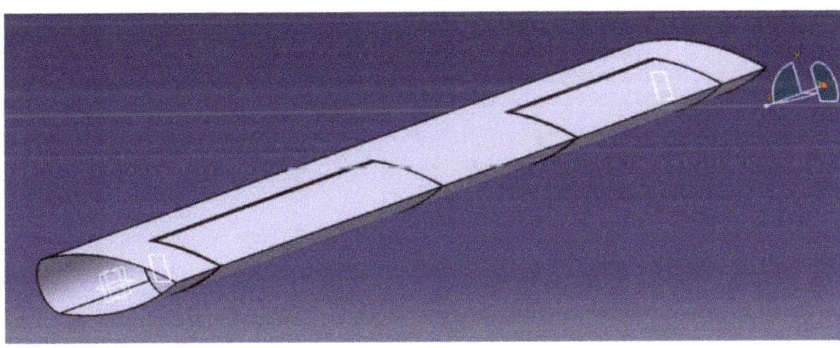

Figure 2. Left wing with symmetrical profile equipped with flap and aileron of I.-M.G. light aircraft.

Figure 3 presents the structure of the wing with a symmetrical profile unequipped with flap and aileron.

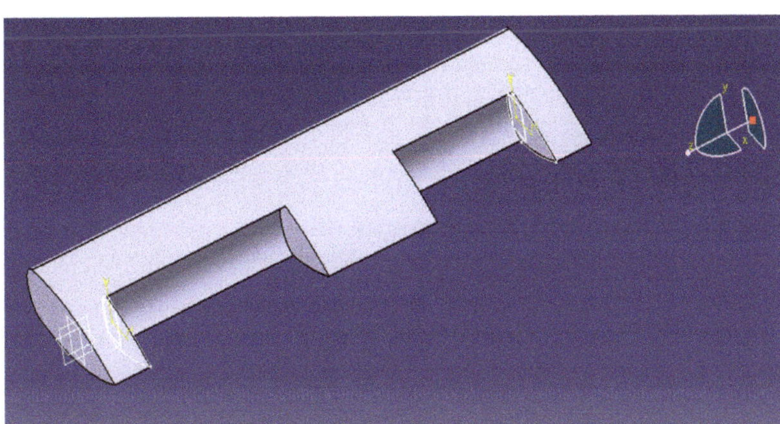

Figure 3. Wing with symmetrical profile unequipped with flap and aileron.

The control mechanism of the voles of a light aircraft, shown in Figure 4, transmits the rotational motion from a valve lever 1 (1—leading element being the input into the mechanism), to a torsion tube 2, to the kinematic drive elements 3, to the kinematic actuators 4, to the complex connecting bodies 5, which transmit the rotational motion, on the one hand, to the non-snake-board curvature flaps 6, and, on the other hand, the kinematic connecting elements 9 transmit the movement between the complex connecting bodies 5 and the oscillating arms 10.

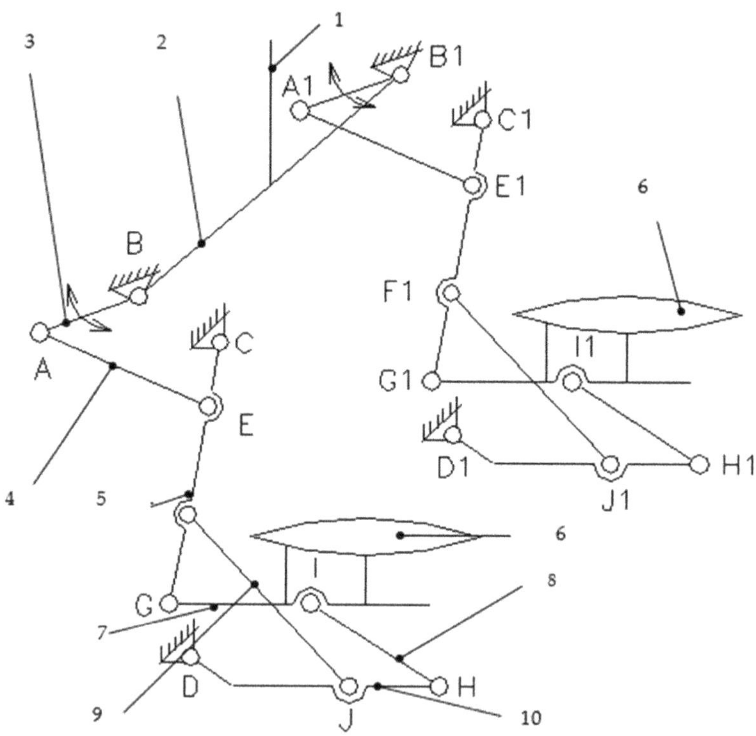

Figure 4. Kinematic structural scheme of the flaps command mechanism.

Elements 7 finally transmit the movement from the oscillating arms 10 to the 6th of the frictionless curvature for the escape board, thus transmitting the rotational movement, amplifying the force, and transmitting the mechanical power from lever 1 to the non-snake-board curvature flaps 6 (final driven elements set in motion with elements 1–10), with the aim of achieving larger gears corresponding to a high-performance flight regime characteristic of these aircraft and allowing take-off and landing over a short distance and in a short time.

The kinematic connecting elements (9, 8) are arranged at an angle with values between (0° … 90°) to the complex connecting bodies 5 and the oscillating arms 10, respectively, to the non-snake-board curvature flaps 6 and the oscillating arms 10. When operating the 1 control valve, the rotational movement is transmitted to a torsion tube 2, in solidarity with the kinematic drive elements 3 articulated at the base by the rotational couplings of A and A1 (which is the base, A and A1) and further to the kinematic actuators 4 by the rotation couplings of B and B1.

From the kinematic actuators 4, 16, the movement is transmitted to the kinematic actuator 5 through the rotational couplings in E and E1, the kinematic actuator 5 elements being connected to the bases by the rotational couplings in C and C1. These five bodies have three links (rotation couplings E, F, G, E1, F1, G1, respectively) and bases C and C1, respectively. The rotational couplings in F and F1 provide the connection between bodies 5 and bodies 9, and the rotation couplings in G and G1 provide the connection with the output bodies 6 (the endless curvature of the escape board, the asymmetric profile flap).

From bodies 5 via rotational couplings F and F1, the rotational movement is transmitted to bodies 10 and 20 by means of the rotational couplings J and J1, and from the 10 in rotational bodies of H and H1, the movement is transmitted to bodies 8, and then through the rotational couplings I and I1 to the non-snake-board curvature flaps 6 (6—output body = final driven element). Bodies 10 have a base connection through the rotational couplings in D and D1. This flap control mechanism of the new I.-M.G. light aircraft allows the volts to be braced up to a maximum of 55°.

In Figures 5 and 6, the proposed optimized solution for the command system and the kinematical model of the mechanism are presented.

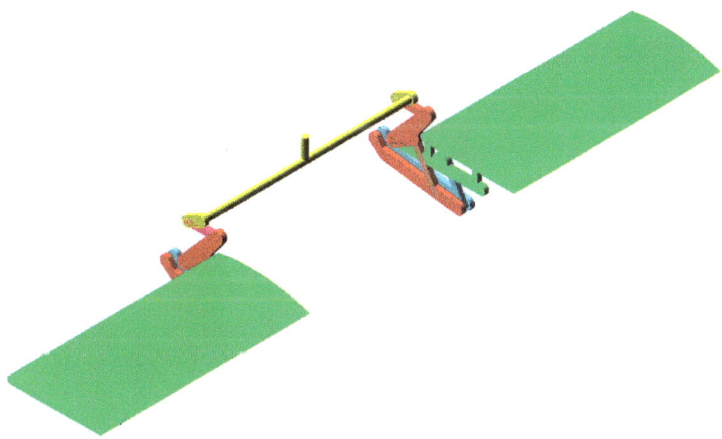

Figure 5. The proposed optimized solution.

Figure 6. Kinematical model of the mechanism.

3. Significant Loads on I.-M.G. Aircraft

The fundamental factors to be taken into account in the design of this type of aircraft, class LSA, are the strength, weight, and reliability of the aircraft material. The aircraft housing shall be of low weight, but the strength shall make the total mass of the empty aircraft as small as possible and be able to support an additional load, crew, and fuel to be operated for as long as possible. The materials to be used in the manufacture of the aircraft must be reliable so as to minimize the likelihood of material failure and sudden failure during operation. In this section, we have tried to get more ideas on the strength of the aircraft structure, as it is a part of the field of the research study.

Before going deeper into understanding the loads and stresses acting on an aircraft that will lead to the idea of aircraft strength, we first must recognize the major forces on aircraft. This is illustrated in Figure 7, where these major forces occur when the aircraft is in steady unaccelerated flight or called straight and level flight. The forces acting on the aircraft during a flight are thrust, drag force, weight, and lift [31–33]. This is illustrated in Figure 7 in horizontal rectilinear and uniform flight. Thrust is the forward force produced by the propeller.

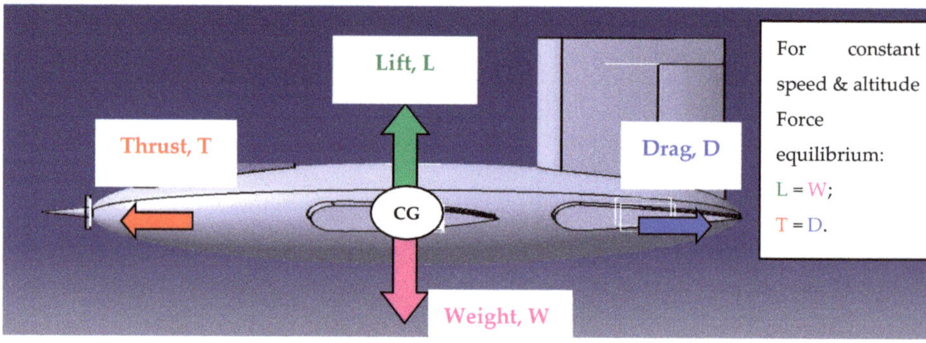

Figure 7. Major equivalent forces acting on the aircraft during straight and level flight.

The total weight consists of the weight of the empty aircraft plus the weight of the two crew members plus the weight of the fuel plus the weight of the aircraft installations plus the weight of the crew baggage.

4. Optimization, Using Finite Element Model, of Wing and Flap

The method of finite elements is used as a need to study the state of deformations and stresses in the structure of the light aircraft named I.-M.G. The wing structure is analyzed using the CATIA computer-aided engineering (CAE) FEA software [34].

The properties of the material are diversified according to the type of element and the type of problem to analyze. The mechanical characteristics of the material/materials used are defined. For the analysis of elastic structures, restrictions are necessary on the movements of translation and rotation with known values. To analyze the elastic mechanical structures, the loading of the model is carried out by inserting concentrated forces in knots and/or normal pressures in the centers of the faces of the finished elements or moments. Figure 8 shows the 3D model of the wing with a symmetrical profile realized from duralumin. The introduction of the values of the material characteristics necessary for the analysis with finite elements is done using the material library of the CATIA software, from which the metallic material from duralumin is chosen, with the following characteristics: Young's modulus = 7.31×10^{10} N/m^2, Poisson ratio = 0.33, density = 2790 kg/m^3.

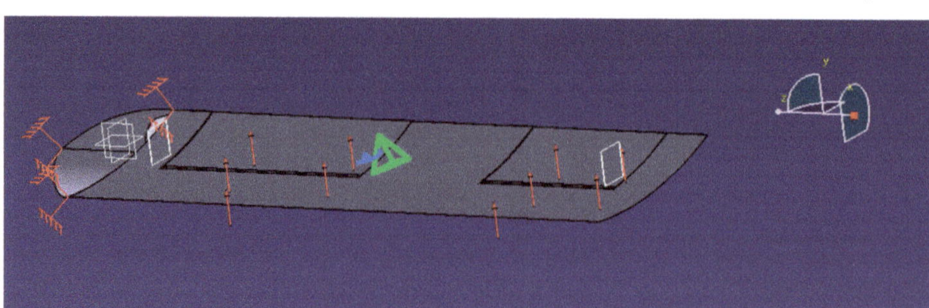

Figure 8. Boundary conditions and load with the force of wing with symmetrical profile from duralumin.

To generate the model, the finite element method is performed, which involves the static analysis of the structure under conditions of imposed constraints and independent time loads. The link with the base imposed on the model is defined by canceling the 6 degrees of possible freedom associated with the lateral surface of the wing (Figure 9) [22].

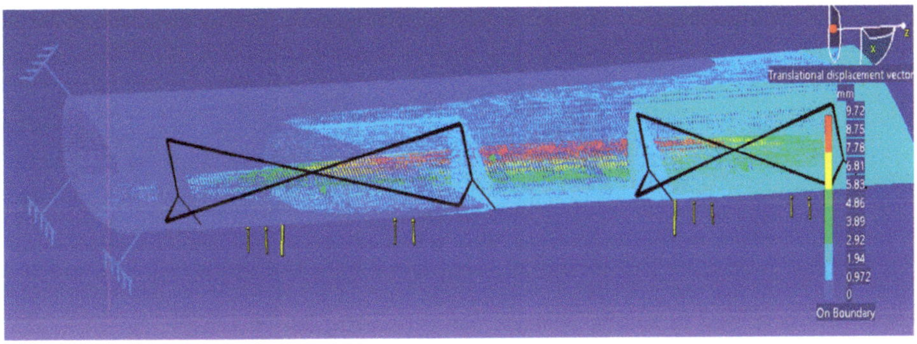

Figure 9. The displacements in the wing with symmetric duralumin profile.

The loading of the model is a uniform pressure after the Y-axis on the face of the wing. The model solution is offered by soft. The displacement field of the wing with a symmetrical profile from duralumin is presented in Figure 9, the maxim displacement being 9.72 mm.

The introduction of the values of the material characteristics necessary for the analysis with finite elements is done using the material library of the CATIA software, from which the metallic material from aluminum is chosen, with the following characteristics: Young's modulus = 7.0×10^{10} N/m^2, Poisson ratio = 0.346, and density = 2710 kg/m^3.

The finite element method is performed to generate the model, involving the static analysis of the structure under conditions of imposed constraints and independent time loads. The liaisons with the base imposed on the model are defined by canceling the 6 degrees of possible freedom associated with the lateral surface of the wing (Figure 10).

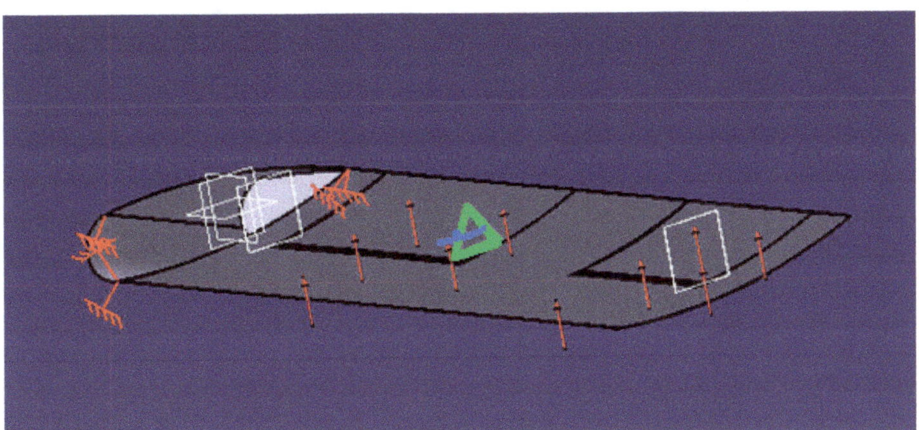

Figure 10. Boundary conditions and load with the force of wing with symmetrical profile from aluminum.

For the flap, the material characteristics of duralumin and aluminum are defined, and the structure is analyzed with finite elements to obtain the Von Misses field of deformations and stresses. The aim is to see the influence of the material on the deformations appearing in the structure of the flap. The mechanical characteristics of aluminum and duralumin are presented in Table 2. It can be observed that the elasticity module for duralumin is higher than the elasticity module for aluminum, and the Poisson's coefficient for duralumin is lower than the Poisson's coefficient for aluminum.

Table 2. Mechanical characteristics of aluminum and duralumin.

Mechanical Characteristics	Material	
	Duralumin	**Aluminum**
Young's elasticity modulus, E (N/m^2)	7.31×10^{10}	7×10^{10}
Poisson coefficient ν	0.33	0.346
Density, ρ (kg/m^3)	2790	2.710

Figure 11 shows the symmetric profile, which can be made by aluminum and which has relief holes that also have a functional role, through which the rods of the control mechanisms and the 3D model of the flap with symmetric profile from duralumin are passed. The introduction of the values of the material characteristics necessary for the analysis with finite elements is done using the material library of the CATIA software. Boundary conditions are presented in Figure 12. To generate the model, the finite element method is performed, which involves the static analysis of the structure under conditions of imposed constraints and independent time loads.

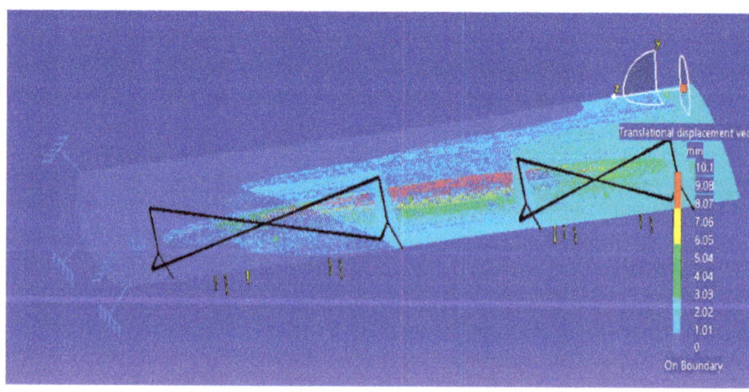

Figure 11. The displacements in the wing with the symmetric aluminum profile.

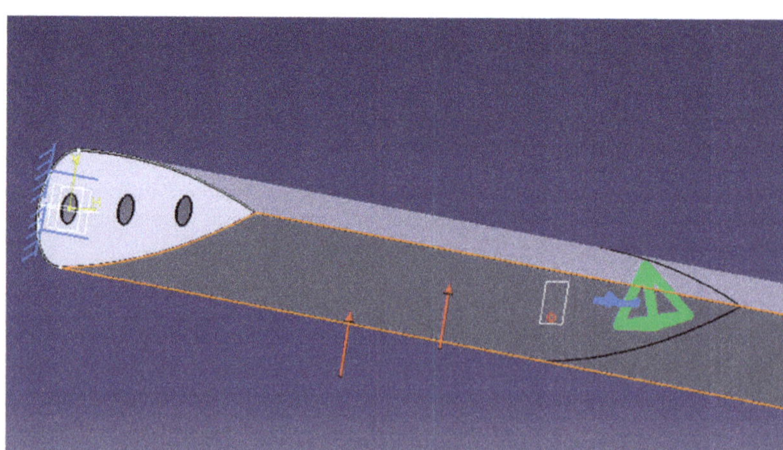

Figure 12. Boundary conditions and load with the force of flap with symmetric profile from aluminum.

The displacement field is calculated, the maxim displacement being 0.106 mm. The results of Equivalent Von Misses stress is visualized in Figure 13, the maxim value being 2.63×10^6 N/m^2.

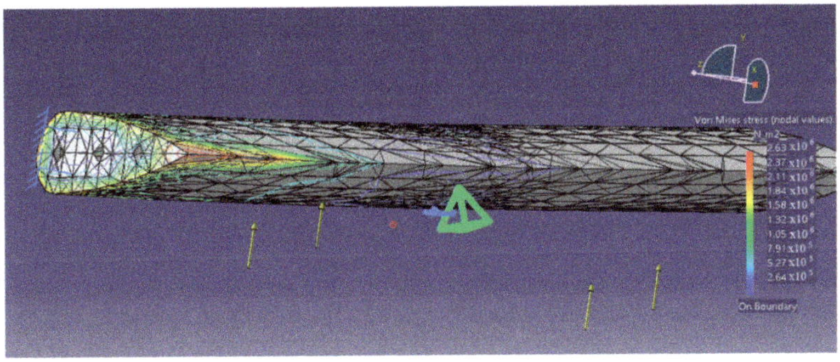

Figure 13. Von Misses stress in the flap with symmetric duralumin profile.

The link with the base imposed by the model is defined by canceling the 6 degrees of possible freedom associated with the lateral surface of the flap (Figure 14). The loading of the model is a uniform pressure, after Y-direction, on the face of a wing made from aluminum.

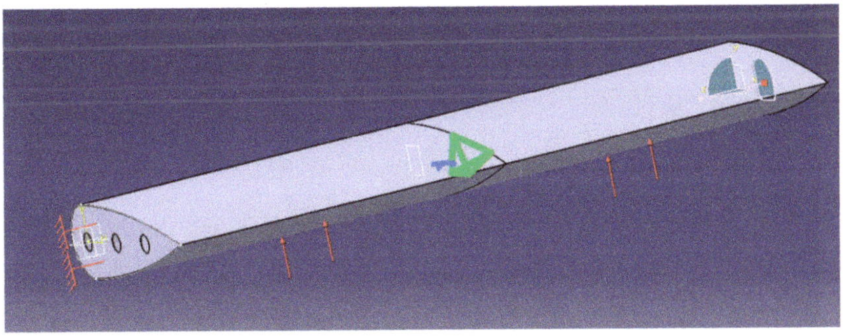

Figure 14. Boundary conditions and load with the force of flap.

The displacement field in the flap with the symmetric aluminum profile is presented in Figure 15. The maxim displacement is 0.011 mm to the extremity free of the flap from aluminum.

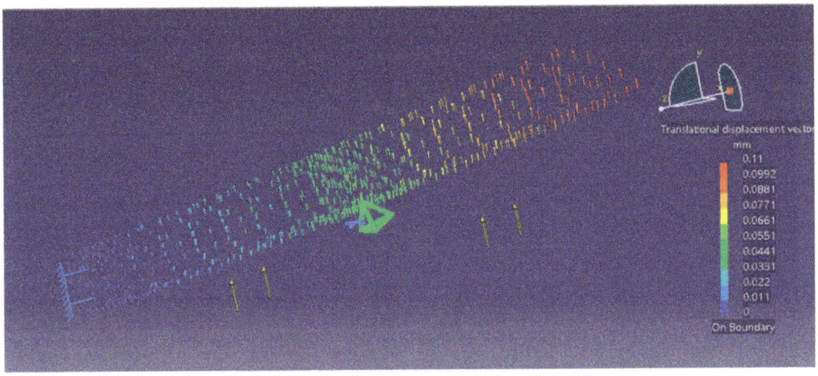

Figure 15. The displacements in the flap.

Table 3 presents the results of finite element analysis for the wing with a symmetric profile. Table 4 shows the results of finite element analysis for the flap with a symmetric profile.

Table 3. The results of finite element analysis for the wing with a symmetric profile.

Obtained Results	Wing with Symmetric Profile	
	Material	
	Duralumin	Aluminum
Maximum deformation obtained, (mm)	1.45×10^3	1.45×10^3

Table 4. The results of finite element analysis for the flap with a symmetric profile.

F = 270 N	Flap with Symmetric Profile	
Obtained Results	Material	
	Duralumin	Aluminum
Maximum deformation obtained, (mm)	0.106	0.11
Maximum von Misses stress, (N/m^2)	2.63×10^6	2.63×10^6

Equivalent Von Misses stress in the flap with the symmetric aluminum profile is presented in Figure 16, the maxim value being 2.63×10^6 N/m^2.

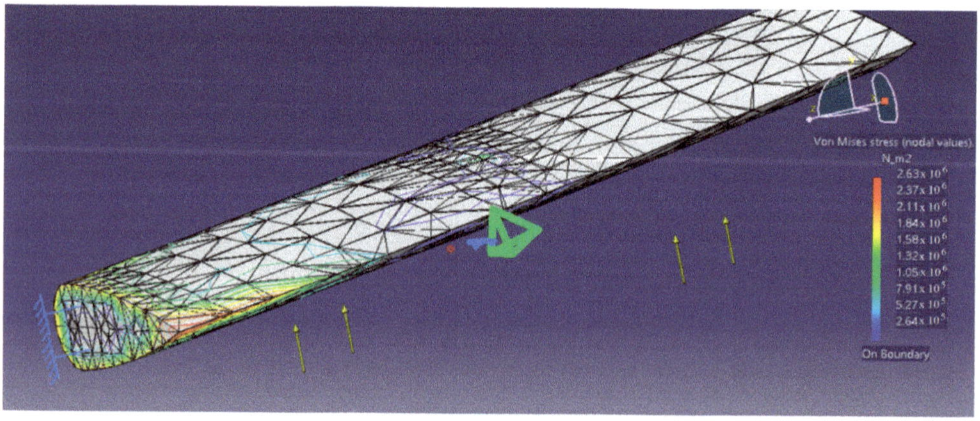

Figure 16. Von Misses stress that appears in the flap with the symmetric aluminum profile.

5. Discussions

The modern design process is a complex process that involves, among other things, the use of modeling methods and analysis based on the use of high-performance software. The finite element method offers solutions to the problems to identify the fields of variation in stresses and deformations. Field identification problems provide a spatial distribution of dependent variables, mathematical modeling of distributions, and dependencies being achieved by differential equations and expressions that contain integrals. CAD modeling of the I.-M.G. light aircraft and finite element analysis are done with the Catia V5 software using the Part Design module that allows the creation of the 3D model of pieces and the application of material characteristics specific to the material chosen for the creation of the pieces, features existing in the material library of the program or by introducing in the program the material characteristics specific to the material of the piece created.

After solid modeling in the CATIA Part Design module, the piece is considered to have a material (aluminum, duralumin), with physical properties, important during the analysis. These values are indicated by default by the CATIA program after selecting the piece in the specification tree and choosing the material "Aluminum/Duralumin" in the "Library" dialog window.

For the finite element analysis of the I.-M.G. light aircraft, the Catia V5 R21 software is used with the CATIA Generative Structural Analysis Module.

From the FEM analysis of the symmetrical profile wing of I.-M.G. aircraft loaded with a force, uniformly distributed and analyzed in two situations (duralumin, aluminum), it is apparent that the maximum deformation occurs in the duralumin wing. From the FEM analysis of the symmetrical profile wing and for the asymmetric profile wing loaded each with force, uniformly distributed, and analyzed in two situations (duralumin, aluminum), it is apparent that the deformation is minimal in the case of the asymmetric profile wing.

The advantages of this mechanism consist of a simple realization, easy maintenance, low-cost price but also the achievement of excellent performances for the studied plane. For this reason, it becomes important to study the compatibility of this type of mechanism with the structure of the studied aircraft. From all the results presented above, this compatibility, from the point of view of strength, is ensured.

6. Conclusions

The ever-increasing complexity of the products leads to some difficulties in design and manufacture. There are several solutions to this feature of modern production—the most used being the realization of new tools and technologies to support the approach of the project without significantly affecting the time of realization or the quality obtained.

Thus, improvements are imposed in the processes of design, calculation and optimization, simulation of manufacturing, or in the way of information management. Between all this, the assisted design presents a decisive link.

Reducing the duration of the product production cycle is possible when the design, manufacture, and finite element method are increasingly integrated.

CATIA allows the design of parts and assemblies directly in three dimensions, without first drawing the plates in representation (two-dimensional). Note that calculations can be made, which can be of great help to the engineer who wants to hold a single platform for CAD-CAE-CAM domains.

On the I.-M.G. light aircraft, a static analysis is carried out, with uniformly distributed forces, after the state of tensions and deformations in the analyzed parts is obtained. The I.-M.G. Light Sport Aircraft is designed with the median wing with a rectangular section, with a symmetrical profile. The wing is equipped with a flap and an aileron. The median wing is advantageous in terms of interaction with the fuselage. By increasing the length of the wings, the length of the flaps is also increased, which increases the lift during the take-off and landing of the aircraft slightly. The proposed I.-M.G. lightweight aircraft is designed by using original component elements like shape and size to improve the aerodynamics of the aircraft and its stability as follows: rectangular-shaped median wings and with an asymmetrical profile in its section so that the profile string forms an angle with the forward direction, of a range to a value greater than the fuselage length, $L = (1.1...1.3) * L_f$, and implicitly increasing the surface of the flaps, which may allow the airplane to glide in critical flight situations when the engine is no longer running or when needed for fuel economy, allowing the aircraft to behave easily and like a plane, the central plane of the fuselage, the propeller helmet, the previous fuselage (hoods), the propellers and the handles of curvature without a run-out board (one on each wing), the cabin in which the crew sits, the depth, drift, direction, rear fuselage, stabilizer, and the 2nd section of the central fuselage. By mounting the flaps on the wings, the landings are smooth, and the plane does not hit the ground. The designed constructive solution of the flaps control mechanism for the new I.-M.G. light aircraft allows the steering to be braced up to a maximum of 55°, this value being superior to the classic solutions of mechanisms with flaps that poach up to a maximum of 45°. The finite element analysis of the flap and wing offers us a convenient solution with real maneuverability advantages.

A comparison of the stress and strain fields that occur in the aircraft structure when equipped with the new type of control mechanism, with the stress field that occurs when equipped with current mechanisms [35], shows that these fields are admissible from the point of view of strength. So the proposed mechanism is compatible with the structure analyzed in the paper, and the calculated stress is lower than the limit stress.

Author Contributions: Conceptualization, I.-M.G., M.L.S., M.M.; methodology, I.-M.G., M.G., M.L.S.; software, I.-M.G., M.G., M.M.; validation, I.-M.G., M.G., M.L.S.; formal analysis, I.-M.G., M.G., P.N.B.; investigation, I.-M.G.; resources, I.-M.G., M.L.S., M.M.; data curation, I.-M.G., M.G., P.N.B.; writing—original draft preparation, M.G.; writing—review and editing, M.G., I.-M.G., M.L.S.; visualization, M.G., M.L.S.; supervision, I.-M.G., M.G., M.L.S., M.M. All authors have read and agreed to the published version of the manuscript.

Funding: This research received no external funding.

Institutional Review Board Statement: Not applicable.

Informed Consent Statement: Not applicable.

Data Availability Statement: Not applicable.

Acknowledgments: We want to thank the reviewers who have read the manuscript carefully and have proposed pertinent corrections that have led to an improvement in our manuscript.

Conflicts of Interest: The authors declare no conflict of interest.

References

1. Zhang, R.Q.; Noon, C.; Winer, E.; Oliver, J.H.; Gilmore, B.; Duncan, J. Immersive product configurator for conceptual design. In Proceedings of the ASME 2007 International Design Engineering Technical Conferences and Computers and Information in Engineering Conference, Las Vegas, NV, USA, 4–7 September 2007. AIAA-2007-193.
2. Manning, V.M. Large-Scale Design of Supersonic Aircraft via Collaborative Optimization. Ph.D. Thesis, Department of Aeronautics and Astronautics, Stanford University, Stanford, CA, USA, 1999.
3. EASA. *Certification Specification and Acceptable Means of Compliance for Light Sport Aeroplanes, CS-LSA*; European Aviation Safety Agency: Cologne, Germany, 2013.
4. Gudmundsson, S. *General Aviation Aircraft Design—Methods and Procedures*, 1st ed.; Elsevier Inc.: Oxford, UK, 2014.
5. Warren, F.P. *Mechanics of Flight*; Wiley: Hoboken, NJ, USA, 2009.
6. Kermode, A.C. *Mechanics of Flight*; Longman Pub Group: London, UK, 1996.
7. Roskam, J. *Airplane Design, Part I. Preliminary Sizing of Airplanes*; University of Kansas: Lawrence, KS, USA, 1985.
8. Zipfel, P.H. *Modeling and Simulation of Aerospace Vehicle Dynamics, Revised Edition*; AIAA: Reston, VA, USA, 2014.
9. Kroes, M.J.; Rardon, J.R.; Nolan, M.S. *Aircraft Basic Science*, 8th ed.; Mac Graw Hill: New York, NY, USA, 2013.
10. Abzug, M.J. *Airplane Stability and Control*, 2nd ed.; Cambridge University Press: Cambridge, UK, 2008.
11. Postelnicu, A. *Aerodynamic Profiles*; Transilvania University of Brasov: Brasov, Romania, 1997; pp. 27–54.
12. Maintenance Manual Extra 300 LT Aircraft. Available online: https://www.extraaircraft.com/docs/tech-manuals/MM300LT/300LT_MM_20150612.pdf (accessed on 22 September 2020).
13. Maintenance Manual Zlin 142 Aircraft. Available online: http://aerodromclinceni.ro/wp-content/uploads/2012/07/Man-de-zbor-Z142-rev8-lb-eng.pdf (accessed on 22 September 2020).
14. Maintenance Manual Festival Aircraft. Available online: http://aerodromclinceni.ro/wp-content/uploads/2016/01/Manual-de-utilizare-avioane-Festival.pdf (accessed on 22 September 2020).
15. Technical Data Flightdesignf2. Available online: https://flightdesign.com/flightdesignf2 (accessed on 23 October 2020).
16. Bourchak, M.; Dobah, Y. Effect of Finite Element Mesh and Load Location on the Stress and Deflection of a Light Aircraft Metal Wing Structure. *Trans. Jpn. Soc. Aeronaut. Space Sci.* **2013**, *56*, 70–74. [CrossRef]
17. Itu, C.; Ochsner, A.; Vlase, S.; Marin, M.I. Improved rigidity of composite circular plates through radial ribs. *Proc. Inst. Mech. Eng. Part L J. Mater.-Des. Appl.* **2019**, *233*, 1585–1593. [CrossRef]
18. Vlase, S. Elimination of Lagrangian-multipliers. *Mech. Res. Commun.* **1987**, *14*, 17–22. [CrossRef]
19. Vlase, S.; Marin, M.; Öechsner, A.; Scutaru, M.L. Motion equation for a flexible one-dimensional element used in the dynamical analysis of a multibody system. *Contin. Mech. Thermodyn.* **2019**, *31*, 715–724. [CrossRef]

20. Gharibi, A.; Ovesy, H.R.; Khaki, R. Development of wing deflection assessment methods through experimental ground tests and finite element analysis. *Thin-Walled Struct.* **2016**, *108*, 215–224. [CrossRef]
21. Prakash, S.; Renjith, T.G.; Kumar, S.; Dwarakanatha, D.; Subramani, H.; Karthikeyan, C. Active vibration control of a full scale aircraft wing using a reconfigurable controller. *J. Sound Vib.* **2016**, *361*, 32–49. [CrossRef]
22. Scutaru, M.L.; Vlase, S.; Marin, M.; Modrea, A. New analytical method based on dynamic response of planar mechanical elastic systems. *Bound. Value Probl.* **2020**, *2020*, 104. [CrossRef]
23. Anemaat, W.; Kaushik, B.; Wiedenmann, K. Preliminary Design of a Tandem Seater Light Sport Airplane. *SAE Trans. Sect. 1 J. Aerosp.* **2006**, *115*, 568–579.
24. Brooks, T.; Martins, J.; Kennedy, G. Aerostructural Tradeoffs for Tow-Steered Composite Wings. Available online: https://arc.aiaa.org/doi/pdf/10.2514/1.C035699 (accessed on 8 November 2020). [CrossRef]
25. Sarath Raj, N.S.; Chithirai Pon Selvan, M.; Bseliss, M.G. Design and analysis of wing of an ultralight aircraft. *Int. J. Eng. Sci. Res. Technol.* **2017**, *6*, 799–821. Available online: https://zenodo.org/record/569965/files/Pon%20Selvan.docx (accessed on 9 November 2020).
26. Design of Ultra Light Aircraft. Available online: https://lisafea.com/pdf/Design_of_Ultralight_Aircraft.pdf (accessed on 9 November 2020).
27. Cheol Wan, K.; Kwang Min, L. Patent KR20140098600(A)—2014-08-08. Airfoil for Ultra Light Plane. Available online: https://worldwide.espacenet.com/publicationDetails/biblio?II=3&ND=3&adjacent=true&locale=en_EP&FT=D&date=20140808&CC=KR&NR=20140098600A&KC=A# (accessed on 9 November 2020).
28. Wacker, M.; Soler, B. Patent WO0119671(A1)—2001-03-22, Weight-Shift-Controlled Airplane. Available online: https://worldwide.espacenet.com/publicationDetails/biblio?II=11&ND=3&adjacent=true&locale=en_EP&FT=D&date=20010322&CC=WO&NR=0119671A1&KC=A1# (accessed on 9 November 2020).
29. Chunmao, G. Patent CN101214858(A)—2008-07-09. Method for Producing Light Aircraft. Available online: https://worldwide.espacenet.com/publicationDetails/biblio?II=9&ND=3&adjacent=true&locale=en_EP&FT=D&date=20080709&CC=CN&NR=101214858A&KC=A# (accessed on 9 November 2020).
30. Jinye, P. Patent CN209795809(U)—2019-12-17. Flap Control Mechanism of Light Sport Aircraft. Available online: https://worldwide.espacenet.com/publicationDetails/biblio?CC=CN&NR=209795809U&KC=U&FT=D&ND=4&date=20191217&DB=&locale=en_EP# (accessed on 9 November 2020).
31. Crane, D. *Aviation Mechanic Handbook: The Aviation*; Aviation Supplies & Academics, Inc.: Newcastle, WA, USA, 2017.
32. Yechout, T.R. *Introduction to Aircraft Flight Mechanics*; American Institute of Aeronautics & Astronautics: Reston, VA, USA, 2014.
33. Peery, D.J. *Aircraft Structures*; Dover Publications: Mineola, NY, USA, 2011.
34. Catia V5 Release 21 Software. Available online: https://books.google.com.hk/books?hl=zh-CN&lr=&id=NvP2pThCPzIC&oi=fnd&pg=PA1&dq=Catia+V5+Release+21+Software.&ots=zpsRGDTzQP&sig=-3QEbqUXnoYxEyN1xR4XOF8P4k0&redir_esc=y&hl=zh-CN&sourceid=cndr#v=onepage&q=Catia%20V5%20Release%2021%20Software.&f=false (accessed on 22 December 2020).
35. Radu, M. Experimental and Theoretical Researches on the Flexible Coupling. Ph.D. Thesis, Transylvania University of Brasov, Brasov, Romania, 2005.

symmetry

MDPI

Article

A New Optimized Solution for a Flexible Coupling with Bolts Used in the Mechanical Transmissions

Marilena Ghiţescu [1], Ion-Marius Ghiţescu [1,*], Paul Nicolae Borza [2] and Sorin Vlase [1,*]

1 Department of Mechanical Engineering, Transilvania University of Brasov, B-dul Eroilor 29, 500036 Brasov, Romania; marilenaradu@unitbv.ro
2 Department of Computer Science, Transilvania University of Brasov, B-dul Eroilor 29, 500036 Brasov, Romania; borzapn@unitbv.ro
* Correspondence: marius.ghitescu@unitbv.ro (I.-M.G.); svlase@unitbv.ro (S.V.)

Abstract: The paper presents a new constructive and functional solution of flexible coupling with bolts and nonmetallic intermediary elements. The bolts have a special shape in the sense that they have a milled area on a certain length equal to the width of the non-metallic element, a cylindrical area that reduces the stress concentrators at the diameter passages. The novelty of this coupling consists of the existence of one intermediary disk between two semi couplings (driver and driven semi couplings). This is fixed by the semi-coupling on the right side, the intermediary disc having milled four locations in which to mount the left, as the right has a metal plate with a special shape of eight metallic plates that are mounted with screws in the four milled places processed on the intermediate disc. The nonmetallic elements have various forms and can be made from different qualities of rubber mounted on the bolts. Eight milled bolts (four on each semi-coupling) allow the transmission of the torsion moment in both directions of rotation, in one direction becoming rigid and thus behaving like a safety coupling. Finite element method is used to obtain the mechanical response of this new type of coupling.

Keywords: flexible coupling; bolt; non-metallic element; finite element method; elastic characteristic

Citation: Ghiţescu, M.; Ghiţescu, I.-M.; Borza, P.N.; Vlase, S. A New Optimized Solution for a Flexible Coupling with Bolts Used in the Mechanical Transmissions. *Symmetry* **2021**, *13*, 171. https://doi.org/10.3390/sym13020171

Received: 27 December 2020
Accepted: 20 January 2021
Published: 22 January 2021

1. Introduction

Rotating machines in which the drive system and the working machine are different require an adapting element in order to transmit power between the two sides. For this, a drive shaft can be used with universal joints or a particular coupling system. Ideally, the two shafts should be properly connected by a rigid coupling so that the two shafts work as one. This offers advantages such as not allowing relative movement between the two shafts, the motor, and the conduced shaft. Vertically positioned applications, such as vertical pumps, use this solution. More used, however, are flexible couplings, which aim to transmit torque or rotary motion without slipping and to compensate for axial, parallel, or angular misalignment. Additionally, the non-linear flexibility of the couplings makes the dynamics of the rotary movement smoother and improves the dynamic response for both the electric motor and the work machine. Examples of such couplings can be pumps, air pumps, engine generator sets, conveyors, crushers, vibrating screens, etc. In practice, flexible coupling can be a major contributor to performance problems of the two machines [1–8], which is why an extensive study is required for each particular application.

Flexible couplings with bolts used in industrial application are presented in a wide variety of constructive solutions, in accordance with their functional role and the demands they must respond to. These demands differ not only by dimensions for each constructive solution but also by the shape of the elastic element. Two of these variants are standardized: variant N, normal, and variant B, with spacer bush [9]. In general, couplings are frequently used in the composition of mechanical transmissions. Flexible couplings with non-metallic elements are of great importance due to the numerous and very diverse fields of activity of equipment and installations they equip (compressors, pumps, generators, pulleys, cranes, conveyors, mixers, piston motors, general industrial applications, as well as in the metallurgical industry, the mining industry, the paper industry, and the pulp industry). In applications, the couplings are made in a great diversity of constructive solutions, a unitary and generally accepted classification of them being difficult. Specific to flexible couplings with non-metallic elements is the fact that the transmission of rotational motion and torque is achieved through the non-metallic element, which dampens shocks and vibrations and can take axial, radial, angular, or combined deviations. The stress characteristic of non-metallic coupling elements together with their size and shape of these elements represent a key design factor for mitigating the transmission of power between the sides of coupling.

Herein are presented some constructive coupling solutions for flexible coupling. These emphasize the advantages of the proposed solution, its applicability, and the way in which this solution meets the requirements of the industry. Known from literature [10], constructive solutions that include flexible bolt coupling consist of two identical semi-couplings. In this case, the torsion moment is transmitted through the rubber sleeves mounted on the bolts, which are rigidly fixed in the other half of the coupling. The profiled sleeves mounted on the bolts are fixed at one end. This type of coupling transmits small and medium.

There are flexible couplings with rubber pins whose half-couplings are machined differently from each other. Thus, one has holes for pins, and the other has the holes machined so as to allow the mounting of rubber bushes. The holes are arranged in a circle on the circumference of each coupling half [11,12].

In [13,14] is presented flexible coupling with studs and washers or rubber bushes. The rubber elements are mounted in one coupling half on the studs, and on the other, the studs are fixed with flanges. Flexible coupling allows, within certain limits, the relative rotation of semi-couples and the variation of shafts alignment.

The Guardian Superflex coupling has two hubs: a super flexible rubber element and a bolts with self-locking nuts [15]. The Superflex Coupling has been used world-wide on a variety of applications, including light towers, air pumps, welding sets, and other machinery with large driven inertia. The coupling consists of two cast iron hubs, a super-elastic rubber element, and a locking hardware. In a rigid coupling, the torque is transmitted from one half of the coupling to the other through the bolts, and in this arrangement, shafts need to be aligned [16].

A flexible Rupex coupling consists of two semi-couplings, bolts and non-metallic rubber element. The non-metallic element is mounted on bolts fixed only on one side of the coupling [17,18]. This coupling absorbs shocks and cushions vibrations. The movement is transmitted from one half-coupling to the other through non-metallic elements fixed with bolts.

Flexible couplings with bolts and bushings are made up of two semi-coupled panels that are mounted on the shafts of the machine [19]. These pieces allow additional torsional flexibility compared to simple cylindrical pieces.

Bolt and barrel bushings couplings are widespread on engineering applications. A flange coupling and non-metallic barrel element is used to connect trees that have a small parallel misalignment, angular misalignment, or axial misalignment. This coupling works with coupling surge. In general, this is used to assemble electric machine with working machines [20–22].

A flexible coupling with elastic rubber bushings and pins consists of two semi-couplings with different flanges having conical bore inside and elastic rubber bushings that are mounted on notched head pins [23]. The flexible coupling with elastic rubber bushings allows an increased load-bearing capacity and durability. The resulted stiffness characteristics are non-linear and asymmetrical. These couplings are used by the transport systems.

The rubber-cushioned sleeve bearing coupling is fast to replace, and this comprises a right half coupling body, a pin, a flexible sleeve, and a nut [24]. The flexible sleeve can be quickly replaced, and it does not need a coaxially correction after replacement. Due to this, an improved production efficiency is obtained.

It is well known that the flexible coupling consists of fixing screws, transition sleeves, elastic rubber rings, a connecting pressure disc, a brake wheel, and studs [25]. The multiple holes for screws and sleeves are evenly distributed alternately on the same circle in the left brake wheel. The connecting pressure disc and the brake wheel are fixedly connected by the fixed screws. The flexible coupling cannot be worn completely and has a long service life. The coupling can be easily removed.

In case of maritime couplings, it is known that these consist of two different half-couplings, a tapered sleeve, screws, drive couplings, connecting pins, elastic rubber rings, nuts, transition sleeves, and connecting pressure plates [26]. Depending on the maritime coupling, the service life of the coupling can be extended by the transition sleeves and the rate of replacement of rubber spring rings, and the frequency of maintenance can be reduced, thus maintenance is convenient. The marine coupling especially shows its superiority when it is used in places where the maintenance is difficult.

A constructive solution of flexible coupling with bolts can be made up by two half-couplings. These have distributed holes on their circumference on a certain circle diameter where cylindrical bolts are placed. These are threaded at one end, being alternately mounted on the two semi-couplings [27]. The movement is transmitted from one half-coupling to another by bolts and non-metallic elements.

Another constructive solution for flexible coupling with bolts and non-metallic elements consists of a coupling part placed radially and having non-metallic elastic link elements. As these elements use materials with composite structures as armatures, the different shape and the used materials generate the desired anisotropy. Additionally, the deformation characteristics of link elements will be different. Thus, the coupling is able to be adapted for a wide set of applications without major changes. In mounting processes, different solutions can be adopted by usage of specific mounting pieces having different kinds of shape, size, and screws. By usage of anysotropic link elements in correlation with disc shape, bolt placement, spacer elements, and screws, all of these generate cheaper and simpler solutions for transfer of movement by coupling. Shape, size, and materials used for link elements represent key factors for adapting to application by the amount of energy that can be absorbed inside coupling. Additionally, the coupling's elements design influences the time response of the system at variation of torque from both sides: motor and load. Moreover, by using elastomer based link elements, these provide a lubrication effect that has beneficial effects for coupling lifespan and its usage [28].

The study of the specialized literature in the field of flexible couplings with non-metallic elements revealed the existence of flexible couplings with non-metallic elements with relatively simple functional and constructive principles. These couplings are made in a wide range of types and sizes, being produced by specialized companies with international reputation. Because the literature in the field of flexible couplings with non-metallic elements is very poor, new research is welcome in the field.

Published articles are very rare, and the books (courses, monographs, etc. [1–8]) present general references regarding design and construction. In modern transmissions, the couplings are mounted immediately after the drive motor, which can be direct current motor with series or parallel excitation or with permanent magnets or could be asynchronous or synchronous motors. The motor or the load torque inherently vary in time function as a result of machine functioning and also as a result of functioning of the control system. The designer is able to design a coupling that smooth movement harmonics in time (speed angular acceleration in time). The dynamic response of the acting system can by optimized by the designed coupling for target work-torque, speed, and for the transient regimes. Thus, the designed coupling is able to perform oscillation dumping during transitory and stationary regimes.

The problems that appear with the couplings listed above would be related to the following: a technological achievement, generally difficult assembly of the component elements, and generally high loads on the coupling elements.

Our paper proposes a coupling system that has several advantages: easy disassembly of non-metallic elements; easy disassembly of bolts; by mounting the non-metallic element between the plates and not in bores processed in semi-couplings, it allows the non-metallic element to relax, this being required besides crushing, shearing, and traction; transmits the torque in both directions, in a direction of rotation that becomes rigid and thus behaves like a safety coupling (which means that it fulfills a new function with this coupling compared to the classic solutions of bolt couplings and non-metallic elements); fulfills the function of limiting the load in order to avoid breaking the non-metallic elements.

2. Description of the Proposed Flexible Coupling with Milled Bolts and Non-Metallic Elements

In the literature, there is no flexible coupling solution with milled bolts and non-metallic elements by type of coupling designed (Figure 1). The novelty of this coupling consists of the existence of one intermediary disk which has four processed holes, the intermediary disk being mounted between two semi couplings. The disk is and fixed to the right half-coupling with screws and centered with the semi-coupling with one cylindrical pin of eight metallic plates fixed with one left and one right to each processed hole on the intermediate disc (there are four disk stakes arranged at 90 degrees). The nonmetallic elements from different qualities of rubber are mounted with bolts and with eight milled bolts, four on each semi-coupling. By using the milled bolts, four mounted on each half-coupling, the transmission of the movement and the torque in both directions are obtained. Each metal plate attaches to the intermediate disc by means of two screws.

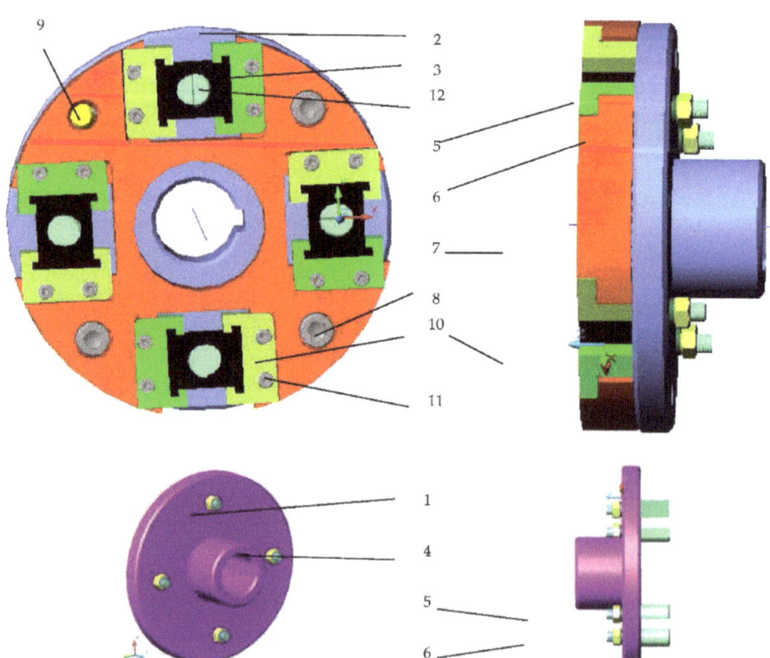

Figure 1. The flexible coupling with milled bolts and non-metallic elements. 1—driver semi-coupling; 2—driven semi-coupling; 3—non-metallic elements; 4—milled bolts; 5—Grower washers; 6—M8 nuts 6; 7—intermediary disk; 8—screws; 9—cylindrical pin; 10—metal plates; 11—screws; 12—milled bolts.

In existing locations between two plates (there are four locations), a non-metallic element is mounted. Non-metallic rubber elements of different qualities are mounted in the space between two plates, thus there are four non-metallic elements mounted. At the same time, the non-metallic element is mounted on the bolts.

This is a new flexible coupling with bolts that differs from the classic solutions of flexible couplings with milled bolts and non-metallic elements by the existence of the intermediate disc, the existence and the form of non-metallic plates and of non-metallic elements that have various constructive forms that can be changed, and by the existence of the eight milled bolts, which are mounted/fixed with four on each half-coupling (Figure 1). On each semi-coupling, four milled bolts are mounted, fixing each semi-coupling bolt by means of a nut and a Grower washer.

In this constructive variant, the bolts 4 and 12 are milled, and their number is eight, four on each semi-coupling. The contact between the milled faces of the bolts is about the length equal to the width of the non-metallic element 3 [9].

The number of bolts mounted on each semi-coupling is four and equals the number of non-metallic elements 3 mounted between the metal plates 10 fixed to the intermediate disk 7, and this number is equal to the maximum number of milled holes on the intermediate disk so as not to endanger the strength of the milled disk and the allocators. The milled part of the bolts allows the transmission of the movement and the torque from one half-coupling to another. The two milled areas of two bolts (one on each half-coupling) come into contact (Figure 1 on the width of the non-metallic element. There are four mounted milled bolts from a total of eight on each half-coupling; only four milled bolts work at a chosen direction of rotation.

The semi-couplings are approximately identical, the constructive differences consisting of the presence on the driven semi-coupling 2 at the level of diameter D_s, of four equidistant holes, three of them being threaded, for fixing by means of screws 8 the intermediate disk 7 of the driven semi-coupling 2, and of the fourth orifice for centering by means of the cylindrical pin 9 of the intermediate disc on the driven semi-coupling. The subassembly consists of the half-coupling (2), the milled bolts 12 fixed by the semi-coupling by means of the Grower washers 5 and the M8 nuts 6, the metal plates 10 fixed to the intermediate disc 7 by means of the screws 11 in the space between the plates and the non-metallic elements 3, which come into contact with the milled bolts that are fitted. The intermediate disk 7 is fixed to the semi-coupling driven 2 by means of the screws 8, the centering of the disk in relation to the intermediate disk being achieved by means of the cylindrical pin 9. In the coupling component, in the milled locations on the intermediate disc 7 are mounted four subassemblies plates 10, a non-metallic element 3, and screws 11, one in each milled location of the type shown in Figure 1.

The non-metallic element has various shapes presented below in the form of the letter H. The foot of the letter H ensures fastening and centering in the places of the plates.

The non-metallic element 3 that is mounted between the plates is made of the same rubber quality. Non-metallic element 3 can be made of three rubber qualities (natural rubber, NR; butadiene rubber acrylonitrile, NBR; ethylene propylene diene rubber, EPDM. When designing the coupling, the influence of the shape of the non-metallic element and the quality of the rubber on the coupling performance were taken into account.

In the optimal design process of the flexible coupling with milled bolts and non-metallic elements, the following aspects were taken into account:

- Driver machine (electric motor): characterized by engine power P(kW) and speed n (rot/min);
- Maximum diameter of d (mm) shafts on which the semi-couplings are mounted;
- Coupling gauge: L (mm)—length of coupling, De (mm)—exterior diameter of semi coupling;
- Limits (minimum, maximum) of the torsion moment transmitted by coupling: M_{tmin}, M_{tmax} (Nmm);

- Performing the functions characteristic of flexible couplings with non-metallic elements: transmission of rotational motion and torsion moment; taking over the position albeit deviations of the shafts; shock and vibration damping;
- Safety in operation: superior mechanical strength of the fastening screws of the two semi-couples of the cylindrical centering stud;
- Economic requirements: execution processes and means of economic processing;
- Establishment of technology for semi-manufactured: non-metallic elements made by vulcanization; semi-couplings are obtained by free forging or molding; the intermediate disc is obtained by free forging; bolts are made of round laminated steel; metal plates are made of square laminated steel;
- Maintenance conditions: simple maintenance; visual control at each use;
- Ergonomic costs and requirements: simple constructive form; minimum material consumption, cost per minimum product.

The main parameter resulting from the characteristic of the flexible couplings is the torque. Another important parameter is the rigidity, which represents the dependence of the relative rotation angle φ of the half-couplings, depending on the value of the torque M_t.

The flexible coupling with milled bolts and non-metallic elements is characterized by damping capacity and rigidity. The existence of the flexible coupling in the mechanical systems favorably influences their behavior at oscillating loads, frequently encountered in operation, high values of the degree of damping leading to a quieter operation of the mechanical systems equipped with such a coupling. In this constructive version, by applying at the input of a torque in the direction indicated by the continuous arrow (I) (see Figure 1), the torque is transmitted from entry driver semi-coupling 1, by shape, through the milled bolts 4 to the milled bolts 12, fixed rigidly by the semi-coupling 2, in this sense, the coupling becoming rigid. In the direction of rotation indicated by the interrupted arrow (II), the milled bolts 4 act on the elastic element 3 of different constructive forms, compressing it in the direction of movement.

Figure 2 presents the 3D model of the constructive form of the milled bolt realized in Autodesk Mechanical Desktop 6 Power Pack [29] and Catia V5 [30]. The milled bolts 4 and 12 are fixed to the semi-coupled leading 1 and driven 2 by means of Grower washers and nuts that prevent the bolts from breaking. The cylindrical area of the bolt with a diameter greater than the diameter of the milled area is intended to reduce the stresses concentrators and break the bolt at the passage of diameter from the diameter of the milled area to the diameter of the cylindrical zone to the right of the threaded area. The cylindrical area with the largest diameter in the bolt component represents a shoulder, stopping on the milled end of the bolt, which is mounted in the opposite direction. The installation of the milled bolts 4 of the driver semi-coupling 1 is done easily. From right to left, the bolts are inserted into the bores processed in the driver semi-coupling 1, then Grower washers 5 and nuts 6 are mounted, and the nuts are tightened. The installation of the milled bolts 12 of the driven semi-coupling 2 (the right half-coupling, Figure 1) is easy as well. From left to right, the bolts are inserted into the handles processed in the semi-coupled driven 2, and then the Grower washers 5 and the nuts 6 are mounted, and the nuts are squeezed against the unraveling of the bolts.

Figure 2. 3D model of the constructive shape of milled bolt.

Figure 3 presents the 3D model of forms of plates on the left side and on the right side. Each plate has a milled channel, in which one end of the non-metallic element is mounted, while the other end of the non-metallic element is mounted in the processed channel in the plate fixed on the opposite side of each milled slot located on the intermediate disk. Figure 4 shows the subassembly consisting of plates 10, non-metallic element 3 mounted between those two plates, and screws 11, the subassembly being part of the constructive variant of the prototype. For fixing the metal plate 10 of the intermediate disc 7, two screws 11 are sufficient for low execution costs for the plate and the intermediate disc, saving material and reducing the weight of the coupling. Figure 5 shows the subassembly consisting of the semi-coupling of the driver 1, the milled bolts 4, the Grower 5 washers, and the nuts 6, the subassembly being part of the constructive variant of the prototype. Semi-coupling drive 1 input transmits the torsion moment from the electric motor to the milled bolts 4 fixed by this semi-coupling.

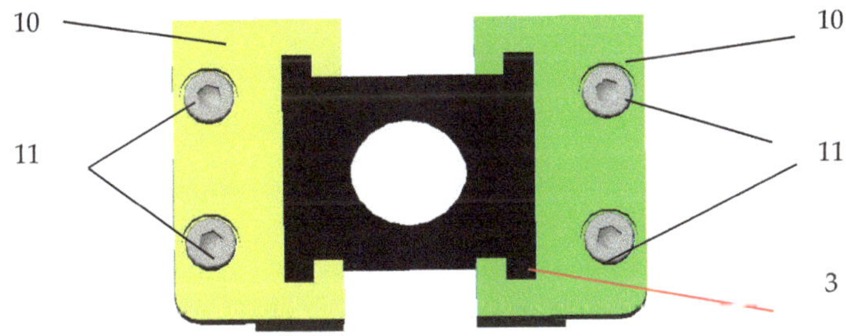

Figure 3. Subassembly plates 10, non-metallic element 3, screws 11.

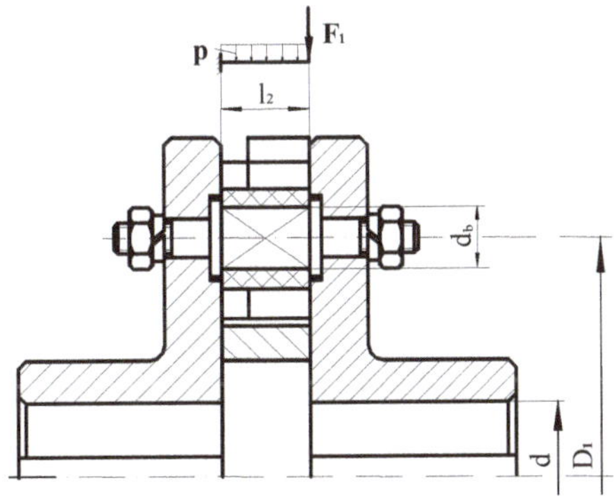

(a)

Figure 4. *Cont.*

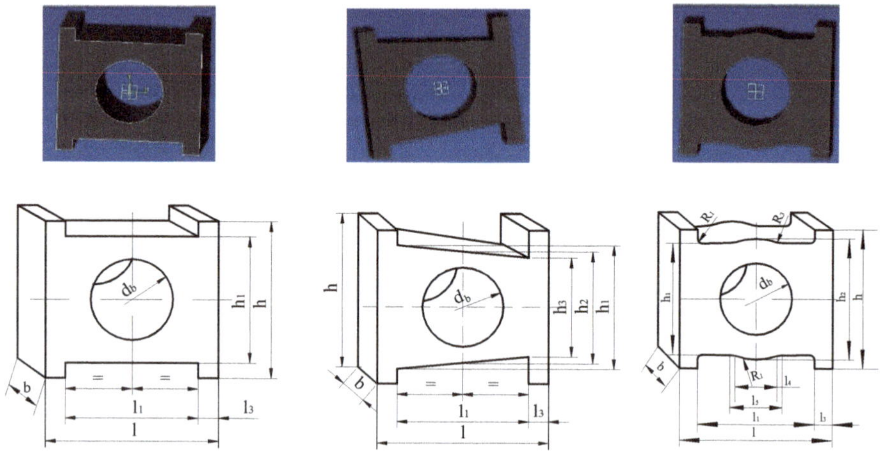

shape 1 shape 2 shape 3

(**b**)

Figure 4. (**a**) Scheme of coupling; (**b**) nonmetallic elements.

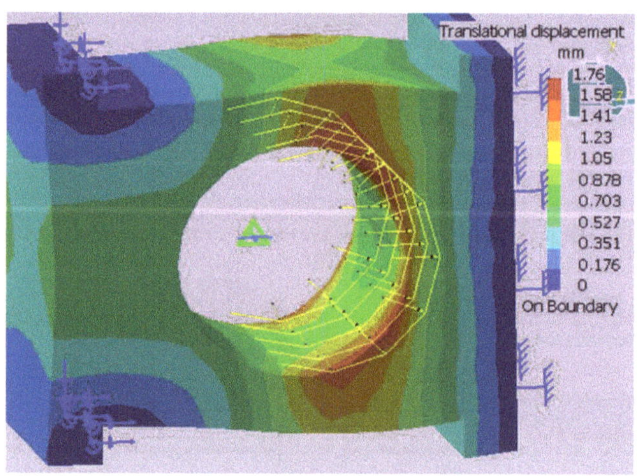

Figure 5. The maximum deformation of nonmetallic element from natural rubber.

The milled bolts 4 and 12, the metallic elements 10, the intermediary disc 7, and the semi couplings 1 and 2 can be made from improved OLC 45 steel (improved quality rolled steel brand with 0.45% carbon content). The mechanical characteristics of the improved OLC 45 (equivalent steel is C 45) steel are: density = 8.31 g/cm^3, Young's modulus = 200 GPa, Poisson's ratio = 0.287.

The coupling was modeled with Computer Aid Design (CAD) in the Autodesk Mechanical Desktop 6 Power Pack and Catia V5.

The non-metallic element is required for crushing and traction. The bolt is required for bending.

3. Calculus Element, Execution and Testing of Flexible Coupling with Milled Bolts and Non-Metallic Elements

The numerical modeling of the performances of the designed coupling is performed on the basis of a calculation scheme, this being presented in Figure 4 [9].

According to the assembly of Figure 4a, the bolt is an integral part of the semi-coupling 1. Non-metallic elastic elements are incompressible, as they can change their shape, but they do not change their volume to a large extent.

Two hypotheses are used in the calculus:

- the bolts are considered uniformly loaded;
- the pressures are considered uniformly distributed along the bolt.

During operation, other additional loads act on the coupling elements, such as: inertia loads, which appear in the non-stationary operating mode of the transmission equipped with coupling; shock and vibratory loads, which occur in both non-stationary and stationary operating modes; loads due to the elements of the couplings. The magnitude of the load acting on the couplings depends on: the characteristic of the motor car, the operating mode of the driven car, the influence of the coupling on the moment of inertia, the rigidity, and the vibration behavior of the kinematic chain.

Due to the fact that it is not possible to know exactly the variation of the torque over the entire operating time, the coupling calculation is performed at an imposed value of calculation torque M_{tc}. Taking into account the additional loads that may occur at installation or during operation, the so-called torsion moment of calculation M_{tc}, is calculated with the formula:

$$M_{tc} = K_s M_{tn} \leq M_{tcn} \quad (1)$$

where K_S is a safety factor experimentally obtained and offered by literature [1], M_{tn} is nominal torque calculated using the power of the electric motor P and its corresponding rotation speed n, with the relation $M_{tn} = 9.55 \cdot 10^6 \cdot \frac{P\ [\text{kW}]}{n\ [\text{rot/min}]}$ [Nmm] [9].

Assuming that all bolts are loaded evenly and their number is z, the force loading a bolt is

$$F = \frac{2M_{tc}}{zD_1} \quad (2)$$

The crush check is performed at the bolt level. Assuming the uniform distribution of pressures along the generator and in section, the crushing between the bolt and the elastic element is determined by the relation [12].

$$\sigma_s = \frac{F}{A_s} = \frac{2M_{tc}}{zD_1 d_b l_2} \leq \sigma_{as} \quad (3)$$

where: A_s is the crushing surface, d_b is the bolt diameter, l_2 is the length on which the crushing occurs ($l_2 = b$, b being the thickness of the non-metallic element), σ_{as} is the permissible crushing resistance ($\sigma_{as} = (5\ldots7)$MPa), and M_{tc} is the calculation torque [9,11].

Applying a traction, the stress is given by [12]:

$$\sigma_t = \frac{F_t}{A_t} = \frac{2M_{tc}}{zD_1 A_t} \leq \sigma_{at} \quad (4)$$

where: A_t is the section area ($A_t = (h_2 - d_b)b$—for the shape of Figure 2), σ_{at} = 1.5 MPa from [9].

The bending stress is

$$\sigma_i = \frac{M_i}{W_z} = \frac{32Fl_2}{\pi d_b^3} = \frac{64M_{tcapi}l_2}{\pi z D_1 d_b^3 l_2} \leq \sigma_{ai} \quad (5)$$

where $\sigma_{ai} = (0.25\ldots0.4)\sigma_{02}$, $\sigma_{02} = 400$ MPa—for bolt realized from improved OLC 45 steel. The values obtained applying Equation (5) is $\sigma_i = 9.54$ MPa $\langle\langle\ \sigma_{ai} = 160$MPa .

Angle of relative rotation of those semi couplings is offered by:

$$\Delta\varphi = \frac{4M_{tc}}{zD_1^2}\frac{1}{Eb}\left[\frac{l-d_b}{h} + 4\sqrt{\frac{h+d_b}{h-d_b}}arctg\sqrt{\frac{h+d_b}{h-d_b}} - \frac{4d_b}{\sqrt{h^2-d_b^2}}arctg\sqrt{\frac{h+d_b}{h-d_b}} - \pi\right] \quad (6)$$

The rigidity is obtained by

$$k_r = \frac{M_t(\Delta\varphi)}{\Delta\varphi} = \frac{zD_1^2}{4}\frac{bE}{\left[\frac{l-d_b}{h} + 4\sqrt{\frac{h+d_b}{h-d_b}}arctg\sqrt{\frac{h+d_b}{h-d_b}} - \frac{4d_b}{\sqrt{h^2-d_b^2}}arctg\sqrt{\frac{h+d_b}{h-d_b}} - \pi\right]} \quad (7)$$

The non-metallic element was studied with the finite element method(FEM) [30,31]. The deformation of the non-metallic element is presented in Figure 5. Von Mises stresses appearing in the non-metallic element are presented in Figure 6. This numerical has been used to determine the elastic response of the components of the proposed coupling. A total of 13,482 octree (parabolic tetrahedron) discretization elements were used. The FEM analysis shows a non-metallic element deformation of 1.76 mm. The experimental determinations performed in static regime gave a maximum deformation of the non-metallic element of 1.77 mm.

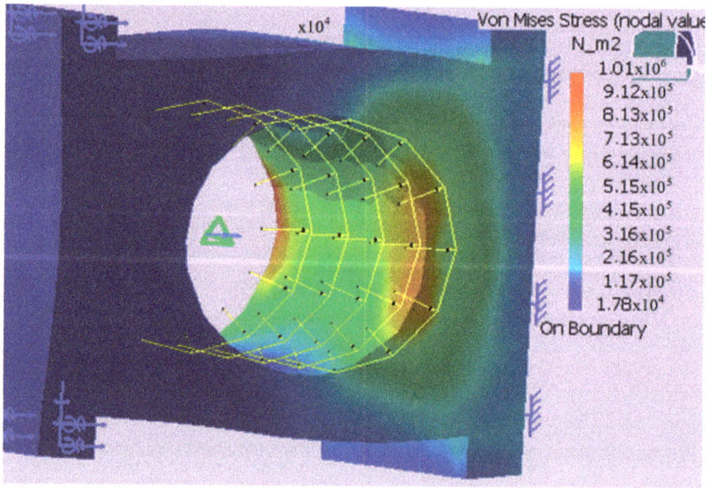

Figure 6. The maximum Von Mises stress of nonmetallic element from natural rubber.

The maximum Von Mises stresses are presented in Figure 7. The results of the experimental tests of the coupling containing the non-metallic element 3 made of natural rubber are presented in the following figure. The static characteristic can be seen (when the input-output shafts are collinear) with a hysteresis loop of the non-metallic element. The nonlinear character of the characteristic of the non-metallic element exists both for loading and for unloading. The experimental setup made to perform the measurement is presented in Figure 8.

Curve 1 from Figure 9 presents variation of torque of input shaft in time. Curve 2 in the same figure presents variation of output torque of output shaft during unloading. M_{ti} represents the input torque and M_{to} the output torque.

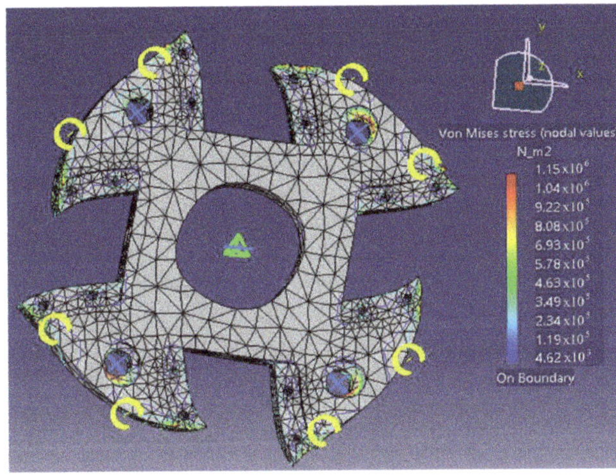

Figure 7. The maximum Von Mises stress of intermediary disc.

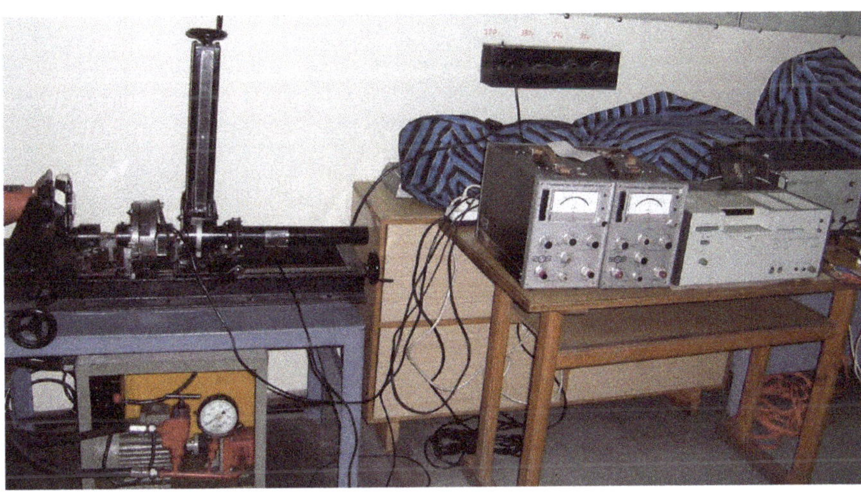

(**a**)

(**b**)

Figure 8. The measurement stand (**a**) experimental stand for static and dynamic testing; (**b**) non-metallic elements from natural rubbers.

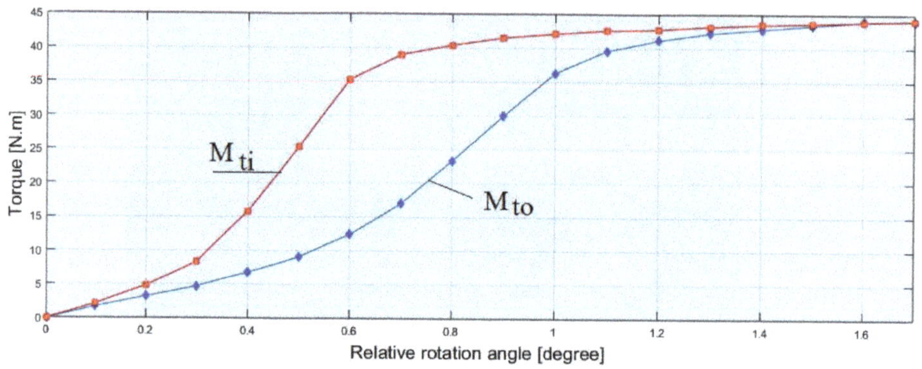

Figure 9. Measured torque in static mode.

The shock takes place by deforming the non-metallic elastic intermediate element, which transforms the shock energy into mechanical deformation work L_e (the area under the load characteristic). Part of this energy (the area between the charging and the discharging characteristics) is transformed into heat, representing the mechanical friction work L_f in the coupling. In non-metallic intermediate elements, friction takes place inside the elastic element (internal friction). From calculus results, mechanical deformation work L_e = 211.94 J. The high values of the shock absorption capacity lead to the quiet operation of the transmission equipped with such a coupling, even to oscillating loads. The flexible coupling with damping also improves the behavior in vibrational mode.

Figure 10 shows the experimental torque at the input and the output shaft as well as for the attenuation factor of the constructive variant of the prototype tested in dynamic mode with the input shafts (collinearly output). The damping capacity of torsion shocks is the characteristic of flexible couplings to convert some of their energy into heat, the rest being converted into deformation energy, which will be returned to the system after the shock action ceases.

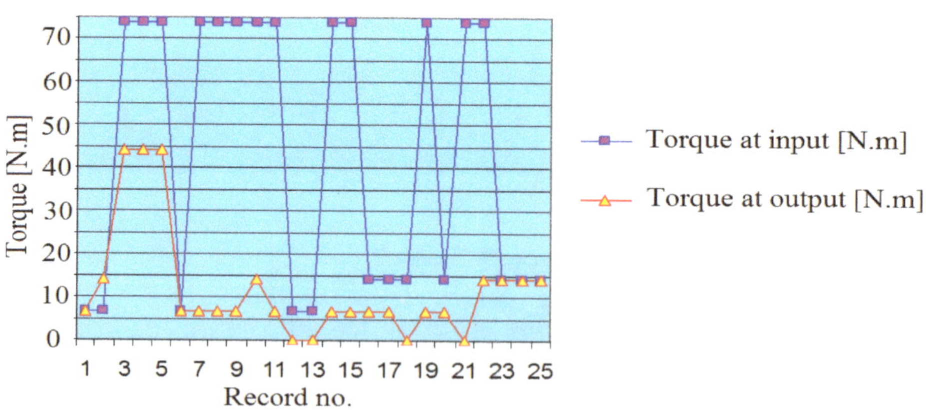

Figure 10. Variation of moments to the input and the output shaft in dynamic approach; input and output shafts are collinear.

Figure 10 shows the variation of the relative rotation angle between the semi-coupled and the torsion moment at the dynamic output shaft, the data recording speed at the oscillograph being 100 mm/s in time of 1 s. In this figure can be seen the dynamic characteristic with a hysteresis loop of the non-metallic element. The hysteresis loop is smaller and more attenuated in dynamic mode. In dynamic behavior, the shocks and the vibrations are higher, and the coupling damps these shocks and vibrations very well through the non-metallic elements. The relative rotation angle between the dynamic couplings is 2.25°, and the input the output shafts have collinearity. In dynamic mode, the coupling must also take over the overloads that appear during starting as well as the shocks that may occur during the operation of the system equipped with coupling. For Figure 11, from calculus results, a mechanical deformation work L_e = 394.11 J.

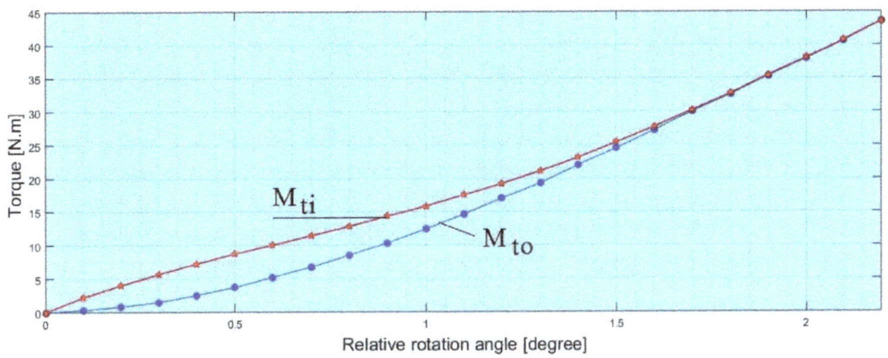

Figure 11. Measured torque in dynamic mode.

4. Discussion

The main parameter resulting from the characteristic of the flexible couplings is the torque. Another important parameter is the rigidity, which represents the dependence of the relative rotation angle φ of the half-couplings, depending on the value of the torque M_t. The verification of the proposed mathematical model for determining the elastic characteristic of the flexible couplings, the validation of the constructive solution, and the adopted technology are performed by comparing the theoretical diagrams with the experimental ones, determined both in static and dynamic mode.

Following the calculation using FEM for the three shapes presented in Figure 4, we obtained results presented in Table 1.

Table 1. Results of verification calculations at the solicitation of traction and crushing for non-metallic element (theoretical values).

Form Shape	Obtain Results
Shape 1	$\sigma_{tF1} = 0.43$ MPa $\langle$ $\sigma_{at} = 1.5$ MPa; $\sigma_{sF1} = 0.24$ MPa $\langle$ $\sigma_{as} = 7$ MPa.
Shape 2	$\sigma_{tF2} = 0.55$ MPa $\langle$ $\sigma_{at} = 1.5$ MPa; $\sigma_{sF1} = 0.24$ MPa $\langle$ $\sigma_{as} = 7$ MPa.
Shape 3	$\sigma_{tF2} = 0.35$ MPa $\langle$ $\sigma_{at} = 1.5$ MPa; $\sigma_{sF1} = 0.24$ MPa $\langle$ $\sigma_{as} = 7$ MPa.

From the results presented for the effective stresses to traction/compression and crushing, it follows that the three constructive forms of the non-metallic element resist crush and traction, which indicates a correct design.

The torque that can be transmitted and that resulted in the three cases is presented in Figure 12. The maximum torque that can be transmitted by the coupling is determined by the traction conditions.

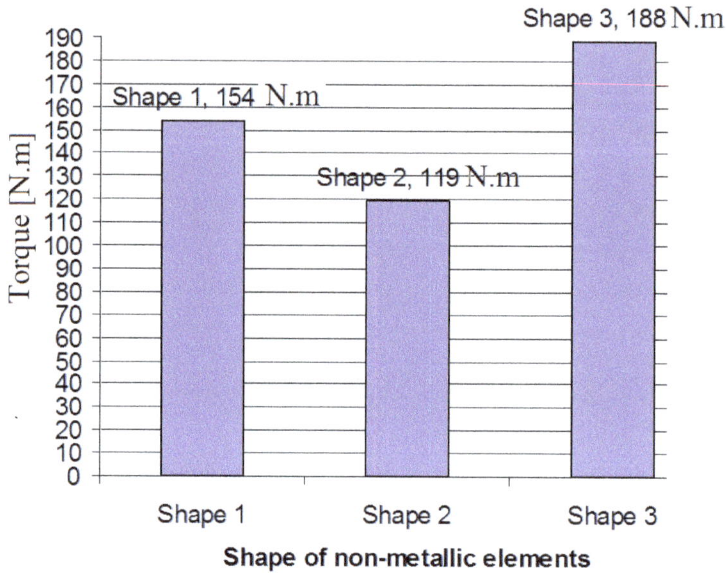

Figure 12. Torque transmitted by coupling.

The third case, presented in Figure 4, allows the coupling to transmit a maximum torque of 188.232 Nmm.

5. Conclusions

The following criteria must be taken into account when designing such a coupling: the coupling must take axial, radial, and/or angular deviations; the relative movement between the semi-couplings to be done without shocks; the coupling must have a low rigidity; feature to have an increasing slope and high damping capacity; changing the non-metallic elastic elements to achieve the modification of the coupling elasticity; in case of damage to a non-metallic element, the coupling can continue to operate; destroyed elastic elements can be easily replaced; the component elements of the coupling must not have protrusions/roughness, thus increasing the safety in operation.

The study permits us to draw the following conclusions:

- the coupling has the possibility to transmit the torque in any direction;
- the proposed solution of coupling ensures compensation of radial and angular deviations;
- in the case of small deformations, $E \approx$ constant, and taking into account the other constants in the expression of rigidity, it appears that the characteristic of the coupling is linear;
- the calculation algorithm of the flexible coupling with milled bolts and non-metallic elements presented is original and based on the hypothesis made in the paper;
- the maximum moment that can be transmitted by the coupling is 188.232 N·m for the third form of the non-metallic element;
- when exceeding the torsion moment determined from the tensile strength condition of the non-metallic element, the elastic element will break, thus the coupling can also perform the safety function (load limitation);
- from the calculations performed, it results that the rigidity of the coupling is small when the non-metallic elements are made of natural rubber (NR);
- the rigidity of the coupling is high when the non-metallic elements are made of ethylene propylene diene rubber (EPDM);

- this type of coupling can be used in mechanical engineering, lifting installations, or in electromechanical systems;
- the coupling designed and executed has a simple construction, small size, and a low cost compared to the classic ones with non-metallic elements and bolts.

The main objective was the level of performance of the coupling, namely the torque of being transmitted by the coupling. Experimental determinations were made in different cases by loading and unloading. Another objective was to carry out tests in different modes (static and dynamic, respectively) taking into account the situation that the two input and output shafts are collinear. From the point of view of functional performance (the torque transmitted by coupling to a load-discharge cycle and the torque to input shaft I and output shaft II, respectively), the coupling performed well, the values obtained experimentally being close to the theoretical ones.

The tested construction variant has a very good shock and vibration attenuation factor of up to 9%.

Figures 9 and 11 show the non-linear characteristic of non-metallic elements at loading and unloading, which means that the flexible coupling with milled bolts and non-metallic elements has the non-linear characteristic, and the coupling is with variable rigidity. Damping occurs if there is a difference between the elastic loading characteristic and the elastic unloading characteristic of the coupling. The hysteresis loop is smaller and more attenuated in dynamic mode. In dynamic mode, the shocks and the vibrations are higher, and the coupling damps these shocks and vibrations very well through the non-metallic elements.

From the dynamic variation of the torsion moment at the output shaft according to the relative angle of rotation between the two semi-couples it can be observed that the non-metallic element has a negligible hysteretic loss, on a small portion of the discharge characteristic practically coinciding with the loading. Note the non-linear nature of the non-metallic element feature when loading and unloading.

Author Contributions: Conceptualization, M.G.; methodology, M.G.; software, I.-M.G. and M.G.; validation, M.G., I.-M.G.; formal analysis, S.V., I.-M.G., M.G., P.N.B.; investigation, I.-M.G.; resources, M.G. and I.-M.G.; data curation, M.G. and I.-M.G.; writing—original draft preparation, M.G. and I.-M.G.; writing—review and editing, S.V., M.G. and I.-M.G.; visualization, S.V., M.G. and I.-M.G.; supervision, S.V., M.G., I.-M.G., P.N.B. All authors have read and agreed to the published version of the manuscript.

Funding: This research received no external funding.

Institutional Review Board Statement: Not applicable.

Informed Consent Statement: Not applicable.

Data Availability Statement: Data is contained within the article.

Conflicts of Interest: The authors declare no conflict of interest.

References

1. Spotts, M.; Shoup, T.; Hornberger, L. *Design of Machine Elements*, 8th ed.; Pearson: New Delhi, India, 2003.
2. Oberg, E. *Machinery's Handbook Toolbox*, 31st ed.; Industrial Press Inc.: Norwalk, CT, USA, 2020.
3. Budynas, R.; Nisbett, K. *Mechanical Engineering Design*, 10th ed.; McGraw-Hill: New York, NY, USA, 2014.
4. Parmlez, R. *Mechanical Components*, 1st ed.; McGraw-Hill: New York, NY, USA, 2000.
5. Mzsyka, D. *Machines & Mechanisms: Applied Kinematic Analysis*, 4th ed.; Pearson: London, UK, 2010.
6. Mott, R.; Vavrek, E.; Wang, J. *Machine Elements in Mechanical Design*, 6th ed.; Pearson: London, UK, 2017.
7. Dieter, G.; Schmidt, L. *Engineering Design*, 5th ed.; McGraw-Hill Education: New York, NY, USA, 2012.
8. Norton, R. *Design of Machinery*, 5th ed.; McGraw-Hill: New York, NY, USA, 2011.
9. Radu, M. Theoretical and Experimental Studies as Concerns Couplings with Nonmetallic Elastic Elements. Ph.D Thesis, Transylvania University of Brasov, Brasov, Romania, 2005.
10. Standard STAS 5982. Flexible Coupling with Bolts. Available online: https://www.etansari-mecanice-pompe.ro/cat/cuplaje-elastice-mecanice-permanente/cuplaje-mobile-elastice-cu-bolturi-element-elastic-din-cauciuc/ (accessed on 20 October 2020).

11. Radu, M.; Chişu, E. Some aspects concerning calculus loads determination at the elastic couplings with nonmetallic elements. In Proceedings of the International Conference TEHNONAV 2004, In Ovidius University Annals of Mechanical Engineering, Constanta, Romania, 27–29 May 2004; Volume 6, pp. 347–352.
12. Flexible Couplings with Rubber Studs. Available online: http://www.velhightech.com/Documents/ME8593%20Design%20of%20Machine%20Elements.pdf (accessed on 20 October 2020).
13. A Flexible Coupling with Stifts and Washers or Rubber Bushes. Available online: https://www.kdkce.edu.in/upload//Machine%20elements,%20Power%20Transmission%20Devices.pdf (accessed on 20 October 2020).
14. Flexible-Coupling. Available online: https://www.comintec.com/en/project/flexible-coupling (accessed on 20 October 2020).
15. Guardian-Couplings. Available online: https://www.guardiancouplings.com/-/media/Files/Literature/Brand/guardian-couplings/related/brochures/p-7733-gc.ashx (accessed on 20 October 2020).
16. Module 5 Couplings. A Typical Flexible Coupling with Rubber Bushings. Available online: https://nptel.ac.in/content/storage2/courses/112105125/pdf/module-5%20lesson-2.pdf (accessed on 20 October 2020).
17. Rupex-Claw-Couplings. Available online: https://www.geartech.no/en/products/transmission/shaft-couplings/rupex-claw-couplings/ (accessed on 26 October 2020).
18. RUPEX®Couplings. Available online: https://www.flender.com/en/Products/Couplings/RUPEX-Pin-and-Bush-Coupling/p/ATN03101 (accessed on 26 October 2020).
19. Construction-Installation-and-Maintenance-of-Flexible-Couplings.pdf. Available online: https://practicalmaintenance.net/wp-content/uploads/Construction-Installation-and-Maintenance-of-Flexible-Couplings.pdf (accessed on 26 October 2020).
20. A Flange Coupling and Non-Metallic Barrel Elements. Available online: https://www.researchgate.net/publication/331154663_Design_and_Analysis_of_Bushed_Pin_Flexible_Co.upling?enrichId=rgreq-10a027bed8d6a339a6fd0b9e3c43d0b4-XXX&enrichSource=Y292ZXJQYWdlOzMzMTE1NDY2MztBUzo3MjY5OTMwMTY0Nzk3NDRAMTU1MDM0MDA1MDY4Ng%3D%3D&el=1_x_2&_esc=publicationCoverPdf (accessed on 25 October 2020).
21. Timur, C.K. Design and Analysis of Bushed Pin Flexible Coupling. *IJARESM* **2017**, *5*, 38–44.
22. Elaziz, A.E.M.; Marin, M.; Othman, M.I.A. On the effect of Thomson and initial stress in a thermo-porous elastic solid under GN electromagnetic theory. *Symmetry* **2019**, *11*, 413. [CrossRef]
23. Patent SU 1262417 A1. Elastic Coupling with Rubber-Bushed Studs. Available online: https://worldwide.espacenet.com/publicationDetails/biblio?II=15&ND=3&adjacent=true&locale=en_EP&FT=D&date=19861007&CC=SU&NR=1262147A1&KC=A1 (accessed on 10 November 2020).
24. Patent CN 103925303/16.07.2014. Rubber-Cushioned Sleeve Bearing Coupling Fast to Replace. Patent CN 106321676A·2017-01-11. Elastic Coupling. Available online: https://worldwide.espacenet.com/patent/search/family/057733218/publication/CN106321676A?q=elastic%20coupling%20rubber%20bolt (accessed on 11 November 2020).
25. Hangzhou, Xin Kai Transmission Machinery Co Ltd. Maritime Coupling. Utility Model CN 203847581U, 24 September 2014.
26. Herwarth Reich. Elastic Coupling. U.S. Patent 2924082A, 9 February 1960.
27. Prodănescu, G.S.; Iliuc, I. Elastic Coupling, with Elastic, Non-Metallic, Radial, Discrete Elements. RO 116430 (B1), 30 January 2001.
28. Autodesk Mechanical Desktop 6 Power Pack. Available online: https://softadvice.informer.com/Autocad_Mechanical_6_Power_Pack.html (accessed on 20 November 2020).
29. Catia®Software Corporation. Catia® Version 8; Copyright © Dassault Systemes: New York, NY, USA, 2016.
30. Hughes, T.J.R. *Finite Element Method: Linear Static and Dynamic Finite Element Analysis*; Dover Publication: Garden City, NY, USA, 2000.
31. Draghicescu, T.H.; Stanciu, A.; Vlase, S.; Scutaru, L.; Calin, M.R.; Serbina, L. Finite element method analysis of some fibre-reinforced composite laminates. *Optoelectron. Adv. Mater. Rapid Commun.* **2011**, *5*, 782–785.

symmetry

MDPI

Article

Design and Analysis of Inertial Platform Insulation of the ELI-NP Project of Laser and Gamma Beam Systems

Calin Itu [1], Polidor Bratu [2,3], Paul Nicolae Borza [1], Sorin Vlase [1,2,*] and Dorin Lixandroiu [1]

1 Department of Mechanical Engineering, Transilvania University of Brasov, B-dul Eroilor, 20, 500036 Brasov, Romania; calinitu@unitbv.ro (C.I.); borzapn@unitbv.ro (P.B.); lixi.d@unitbv.ro (D.L.)
2 Romanian Academy of Technical Sciences, B-dul Dacia 26, 030167 Bucharest, Romania; icecon@icecon.ro
3 Institute ICECON S.A., Soseaua Pantelimon, 266, Sect.2, 021652 Bucharest, Romania
* Correspondence: svlase@unitbv.ro; Tel.: +40-722-643020

Received: 3 November 2020; Accepted: 25 November 2020; Published: 28 November 2020

Abstract: All the installations, devices, and annexes within the laser and the gamma ray production system within the ELI-NP project from Magurele are installed on an inertial platform that weighs over 54,000 tons. The platform is made of concrete, is insulated from the outside environment, and is supported by spring batteries and shock absorbers. The flatness of this platform respects some very strict standards, and, taking into account the processes that take place on the platform, the transmission of the different trepidations of the environment to the inertial mass must be extremely low. For this reason, a static study and a vibration analysis of the platform, performed in this paper, are required. The static analysis verifies if the flatness of the platform can be observed in operating conditions, and the dynamic analysis verifies how excitations coming from the external environment can be transmitted to the measuring equipment. The finite element method is used both to determine the deformability of the concrete platform for different loads, placed at different points and to determine its eigenvalues and its eigenmodes of vibration. The obtained results are analyzed and constructive solutions are proposed to improve the realized system, through a judicious placement of the installations and the distribution of the masses on the platform.

Keywords: stability; laser; nuclear installation; insulation; Extreme Light Infrastructure; gamma ray

1. Introduction

ELI-NP abbreviation represents Extreme Light Infrastructure-Nuclear Physics facility and is a project incorporating 13 European countries. The researches presented in the paper concern a study of the inertial platform of the project developed in Magurele-Bucharest. This project has two components: a very high intensity laser system (10–30 PW) and a very brilliant, intense γ beam of up to 19 MeV, 0.1 % band width, and 1013 γ/s, both positioned on a concrete insulating platform. The ELI-NP facility has the following purposes: to include new experiments and to upgrade the laser power and gamma beam intensity and energy and to include the most ambitious and far-reaching projects as well as the ones that are yet to be discovered. It may include an upgrade of the γ beam facility, using a superconducting energy recovery linac reaching to higher intensities of 1015 γ/s and improved bandwidth. The realization of the project was possible due to the support provided by the EU and the management and scientific supervision was carried out by an international committee of renowned specialists. The gamma beam system being under construction at ELI-NP will provide a very bright photon beam with unprecedented bandwidth and tunability. To fully take advantage of this performance, a number of interesting physics cases were identified in the WhiteBook [1] of the

ELI-NP project and then detailed in the Technical Design Reports [2]. Two of the major experimental directions are the study of Nuclear Resonance Fluorescence (NRF) and the study of states above the neutron emission threshold [3,4]. Both these experimental set-ups state in their technical design reports the desire to use isotopically enriched targets. The gamma source under construction at ELI-NP is especially well suited for these studies because of the small beam size at the target position which allows matching the full beam flux with a very small target. Nevertheless, this advantage can quickly turn into an experimental quagmire if the overlap between the beam spot at the target position and the target is not very good or at least measurable. To achieve this is necessary to realize a good insulation and to reduce significantly the vibration of the ground transmitted to the platform of the laser and gamma beam system. The researcher must realize a very precise target alignment with the gamma beam and this is not possible if the transmissibility of the vibration is not satisfactory controlled. The design of the inertial platform imposes the implementation of a high precision of positioning the devices, the pieces, the alignment system for the experiments due, in the more general view, to the huge impact on the experiments. This engineering achievement is unique in its size and the results obtained in the design and study of this structure should be known for future projects.

The complexity of the study comes from the fact that the ELI-NP will deliver a beam of high energy photons of unprecedented intensity within a very small opening angle, and for this, the precision become an important thing. Insulation of the base platform is a means of reducing the transmission of vibrations in buildings, and the first attempt to address the problem in this way was in the 60 s (last century) [5,6]. Since then, many types of springs have been fitted to many constructions in order to reduce the effects of anthropogenic source vibrations. In most cases, the building rests on steel springs or rubber elements. The aim is to reduce the transmission of vibrations by at least 10 dB for frequencies above 10 Hz, but this goal is rarely achieved [7]. An important element in insulation is the type of spring used but also important are the cost of operation and the difficulty of implementing the system. A more accurate modeling allows the choice of parameters for the elastic and dissipative elements that correspond as well as are possible to the final purpose of vibration isolation.

The design of inertial platform used to ELI-NP laser project represents a unique achievement realizing the insulation of the laser and gamma beam installations from the excitations that may come from the environment. The solutions found for this project can be useful for similar isolation conditions imposed by other project. Such projects and their design are very rare, and the results obtained within them are even more important, through the information they bring, to the scientific community.

If the platform is considered as a rigid with six degrees of freedom (Steward platform), the isolation methods have been intensively studied. There are several parameters that influence the ability of such a system to be isolated. A study method using a genetic sorting algorithm is developed in [8]. In these cases, genetic algorithms have been shown to have higher optimization efficiency, better computational accuracy, and reduced optimization time.

A dynamic model for a 6-axis hybrid vibration isolation system based on a cubic configuration of Stewart platform is proposed. The displacement excitations and the loads are simultaneously considered using a Newton-Euler formulation. A force feedback control principle is proposed and studied. Numerical simulation strengthens confidence in the proposed method [9].

An analysis of a six-degree-of-freedom (DOF) active vibration isolation system is presented in [10]. The system is applied to isolate a Stewart platform, for a better and precise control and for a wide range of earth-based systems. The paper presents the design and analysis of the platform as a performing and robust vibration isolation system. A prototype has been realized and tested.

Anthropogenic sources of vibration are currently complex due to large flexible structures, activities of all kinds, daily and transport. The inertial platform of a laser system is, in the first approximation, a system with 6 degrees of freedom and can be studied with classical means of analysis. A presentation of these methodologies is made in [11] in which the latest generation technology and the ways of active vibration isolation for the precise realization of the experiments on the platform are reviewed.

An inerter-based vibration isolation system (IVIS) using an analytical model is established and validated by comparing the experimental and numerical results and is presented in [12].

A platform to prevent the vibration caused by the low-frequency seismic excitation is presented in [13]. This design isolates the horizontal and vertical vibrations simultaneously. For this, a combination of a rolling isolation system and four three-parameter isolator with active damping is used. The dynamic model offers the numerical results that show an effective suppression of the vibrations.

In [14], a new type of vibration isolation to reduce the dynamic response in case of an excitation with wide frequency range is proposed. For installations used in nuclear physics, there are researches presented in [15–18].

2. Methodology and Model of the Inertial Platform

The inertial platform is suspended on a number of almost 1000 spring and shock absorber batteries. All installations in the ELI-NP project are located on a platform (Figure 1) that should be isolated from the influence of any vibrations of the environment. The mass of the entire concrete platform, without partition walls, roofs, and equipment is about 25,000 tons. If you add the partition walls, then it reaches about 35,000 tons, and if you add all the equipment expected and roofs, you get a mass of 54,000 tons. The concrete used for the building is steel reinforced.

(a) (b)

Figure 1. (**a**) Platform under its own weight. (**b**) Platform with partition walls.

The platform is placed on a series of springs and dampers (Figures 2 and 3) that have the role of isolating the equipment on the platform from any excitations that could come from the environment. Due to the dimensions of the platform and the particularly large mass, the concrete can deform under the action of its own weight or the weight of the walls and equipment placed over the platform. The numerous springs located at a short distance from each other have the role of uniformizing the loads that appear in the elastic supports of the platform. In this way, the loads that press on the platform as well as its own weight will not give large deformations of the platform plate. The fact that it is possible to act on the springs so that some are active and others do not allow an adjustment and control of the uniformity of the platform support on the base table.

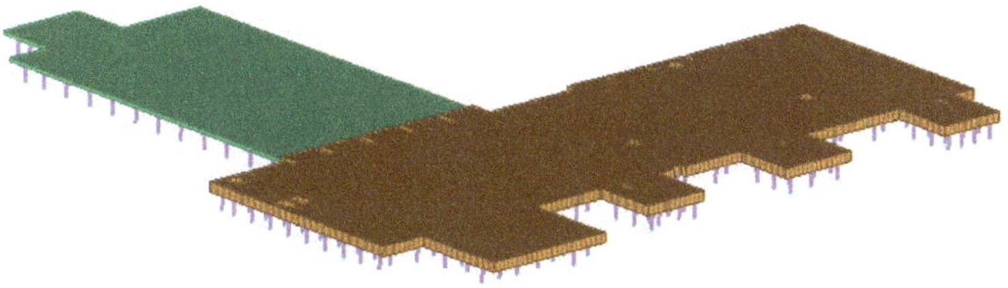

Figure 2. Platform supported on batteries of springs.

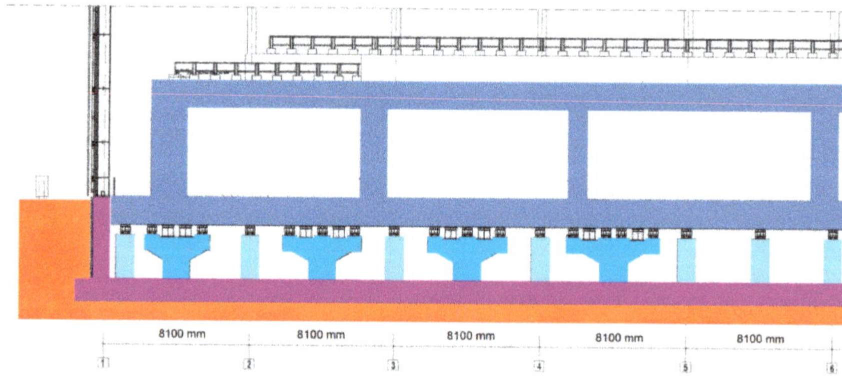

Figure 3. Platform with partition walls and roofs (detail).

Essentially, the platform consists of two concrete blocks, rigidly connected to each other, placed on spring batteries that evenly distribute the weight on the floor. On the concrete platform are placed massive concrete blocks, with thicknesses of 1600 mm or more, over which are placed concrete slabs as a roof. On this whole structure are placed the equipment and different control and measurement systems. Experiments with laser beam and gamma beam ray are conducted on the concrete platform. The platform must ensure a very good isolation from the anthropic activities around the building, isolation imposed by the experiments that are performed on the platform. At the same time, the platform's response to an earthquake must be within certain parameters that ensure the integrity of the structure in a high-intensity earthquake.

For the analysis of this structure and of the static and dynamic response to the action of some external (of natural or anthropic source) excitations, a model was made using the finite element method. This method was the main tool used in this paper. Based on the numerical results obtained, a study was made of the deformability of the massive concrete block in static case. It was studied to what extent the deformations of the concrete block can influence the flatness of the entire platform and what would be the influence of additional loads placed on this block, in different positions. Then, the modal numerical analysis of the system was made, obtaining the eigenfrequencies of the elastic structure suspended on the spring batteries. The first three eigenfrequencies represent the frequencies of the rigid movement of the platform mass. In the other modes of vibration, the effect of the deformability of the concrete slabs can be observed.

For the analysis of this structure and of the static and dynamic response to the action of some external, natural, or anthropic excitation, a calculation model was made using the finite element method. This method was the main tool used in this paper. Based on the numerical results obtained, a study was made of the deformability of the massive concrete block in static case. It was studied to what extent the deformations of the concrete block can influence the flatness of the entire platform and what would be the influence of additional loads placed on this block. Then, the modal numerical analysis of the system was made, obtaining the eigenfrequencies of the elastic structure suspended on the spring batteries. The first three eigenfrequencies represent the frequencies of the rigid movement of the platform mass. In the other modes of vibration, the effect of the deformability of the concrete slabs can be observed.

The types of analyzes performed are in the linear domain (linear static analysis). Under these conditions, the equations used follow Hooke's law. Software used to prepare the model: Altair HyperMesh. Solver for running analysis: OPTISTRUCT. The type of elements used has 4 corner nodes with 24 degrees of freedom (3×4 translations + 3×4 rotations).

3. Static Analysis

The insulation platform consists of a solid concrete block, made up of two parts, i.e., one of them (which we call the main part) having a thickness of 1600 mm and the other (which we call secondary) having 600 mm. At the dimensions of the construction and at the masses to be placed on top, the deformability of the platform becomes important for obtaining precise (accurate) experiments. For this reason, it is necessary to study the deformability of the concrete platform both under the action of its own weight and as a result of additional loads. The spring suspensions of the platform must be distributed as evenly as possible, and thus, a support system has been designed on about 1000 spring batteries. The inactive elements in the spring batteries can be activated in case of need and can rebalance portions of the platform, leveling its deformations and the tensions that appear in it. Each battery consists of 3–7 springs, and between the springs are located vibration dampers. The springs can be operated (i.e., support the platform) or are not in service, do not support the platform at this time, but can be released and become active, if necessary. Thus, a study was made of the stresses and deformability of the platform placed on the spring batteries, without additional masses placed on top. The finite element method was used to determine the stress and stress that occurs in the platform block. Figure 4 shows the set of two blocks consisting of the main body (shown in yellow, 1600 mm thick) and the secondary body (shown in green, 600 mm thick). The positions where 1000 tons of loads will be applied successively are shown in purple. We put in a limited area, equivalent with the basic surface of a wall (approx. 20 m^2), the considered supplementary mass.

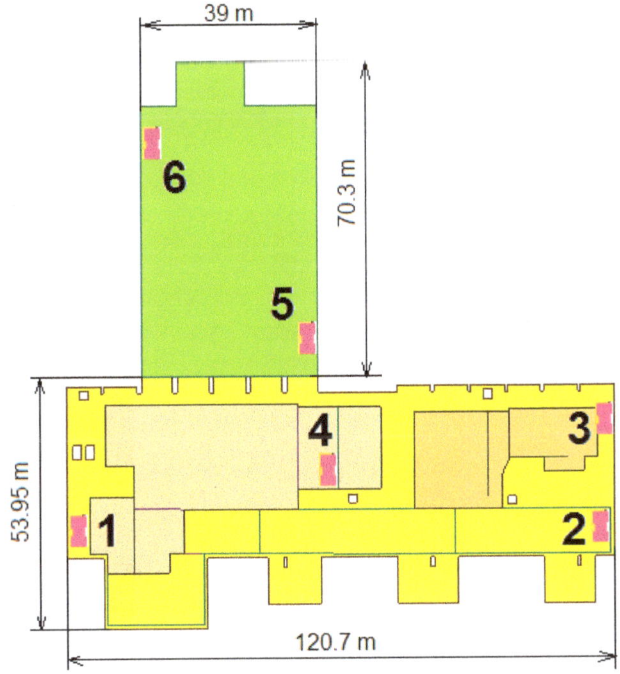

Figure 4. The platform made by two solid blocks.

In Figure 5, Figure 6, Figure 7, Figure 8, Figure 9, Figure 10, and Figure 11, two cases are presented in parallel: the case of the unloaded platform (total mass 25,000 tons) and the case of the platform with vertically mounted walls (total mass 35,000 tons). The deformation of the concrete platform loaded with its own weight; placed on the spring batteries, is presented in Figure 5, i.e., on the left is the 25,000 ton variant, while on the right is the 35,000 ton variant. Figure 6 shows the deformability of the platform if the load of 1000 tons is applied in point 1. A logarithmic scale is used for a better visualization of the deformation situation. In Appendix A, the deformations using a linear scale are represented (Figures A1–A4).

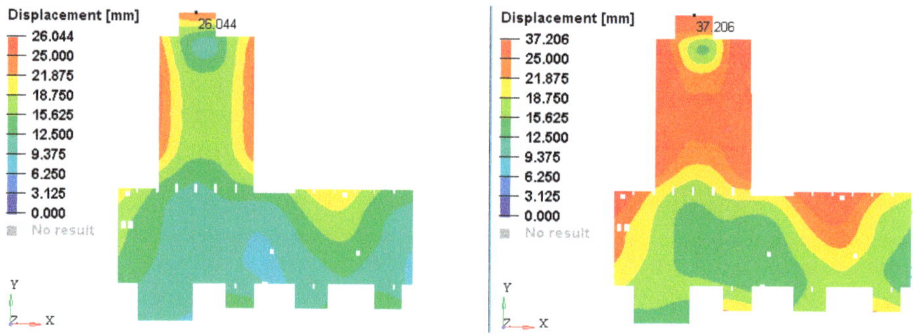

Figure 5. Case (a) plate under its own weight.

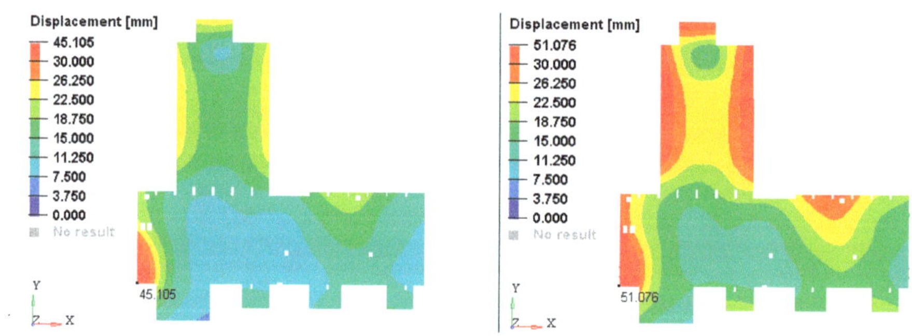

Figure 6. Case (b) plate loaded with 1.000 tons, placed in pct.1.

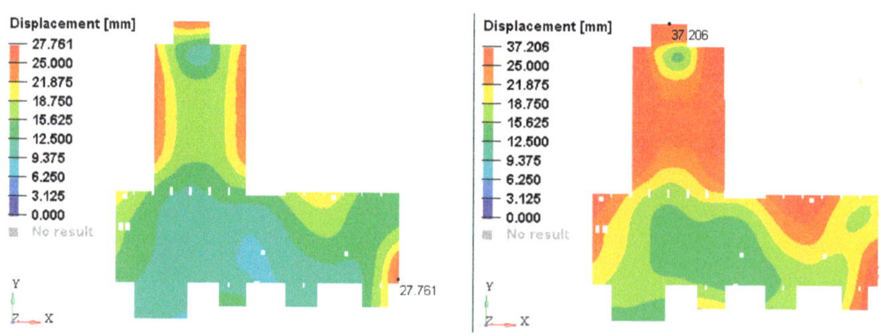

Figure 7. Case (c) plate loaded with 1.000 tons, placed in pct.2.

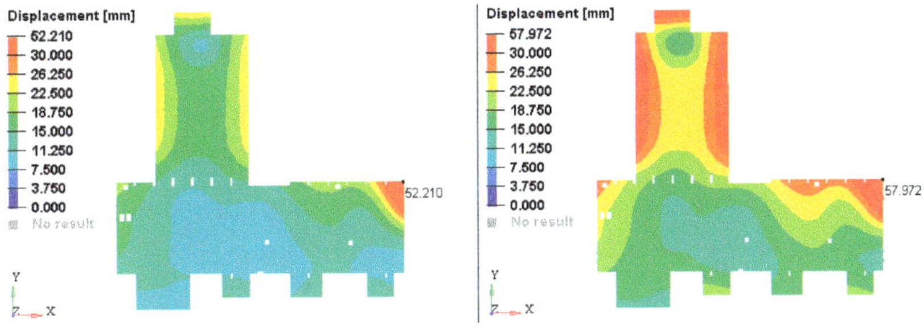

Figure 8. Case (d) plate loaded with 1.000 tons, placed in pct.3.

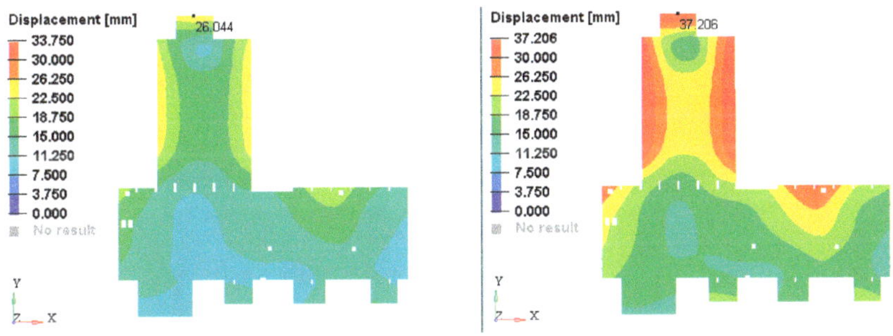

Figure 9. Case (e) plate loaded with 1.000 tons, placed in pct.4.

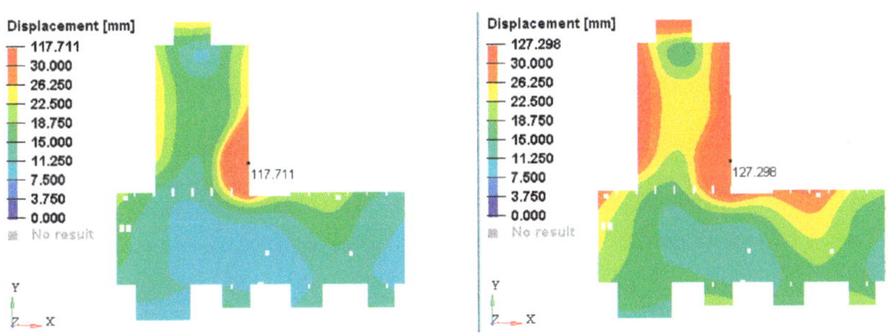

Figure 10. Case (f) plate loaded with 1.000 tons, placed in pct.5.

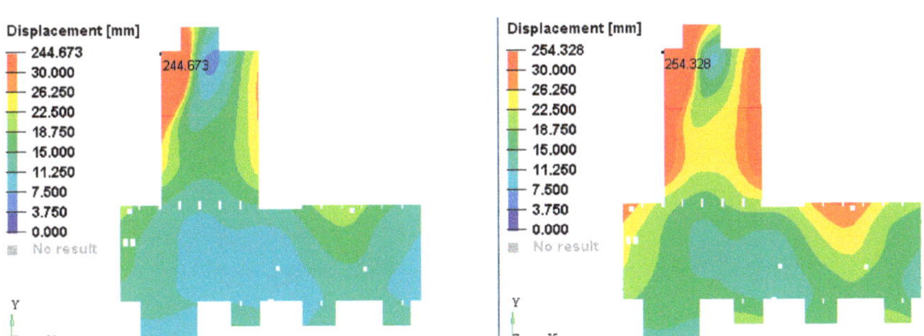

Figure 11. Case (f) plate loaded with 1.000 tons, placed in pct.6.

The finite element model used has a number of 50,898 nodes and a number of elements of 50,053. CBUSH type elements (12 degrees of freedom) were used for modeling the spring batteries, and QUAD4 plate elements (24 degrees of freedom per element) were used for the discretization of the element plates [19–22]. The mechanical properties of the concrete, used in this paper, are obtained from literature [23]. We mention that the concrete is reinforced with steel. In this case, it is expected that the Young's Modulus can increase. Some studies [24,25] show that, in reality, the increase in the Young's modulus is not significant. For a fiber volume ratio of 0.025, the Young's modulus is 30.000 N/mm^2, the value considered in our study. In [26] has been presented the values of reinforced concrete offered by Eurocode (Structural design calculations according to Eurocodes). The Poisson's ration used is 0.2, as recommended by [26]. The connection with the ground, after the three directions, is made by means of suspension springs. The springs are fixed to the ground and are connected to the inertial mass (boundary conditions). The rigidity of a single spring is $k_z = 1.85e + 6$N/m after the vertical direction and kx = ky = 0.55e + 6 N/m for the other two directions. The dampers have the coefficient $c_x = c_y = cz = 2000$ kNs/m.

In Figure 5, Figure 6, Figure 7, Figure 8, Figure 9, Figure 10, and Figure 11, the graphical representations are made in case the loads are applied in points 1–6. The stresses that appear in the plate, for the studied cases, are presented in Appendix B (Figure A5).

In order to see how the placement of additional weights or the removal of weights from the platform can influence the deformability of the platform, it was considered an operation that was performed in the modification of a concrete wall placed in position 4 in Figure 12. The wall that was removed weighed 250 tons. The deformation field in the two cases was presented in Figure 12. If the figure is analyzed, it is found that there are no significant differences between the two situations, i.e., the removal of a wall of 250 tons will not significantly influence the placement of the platform.

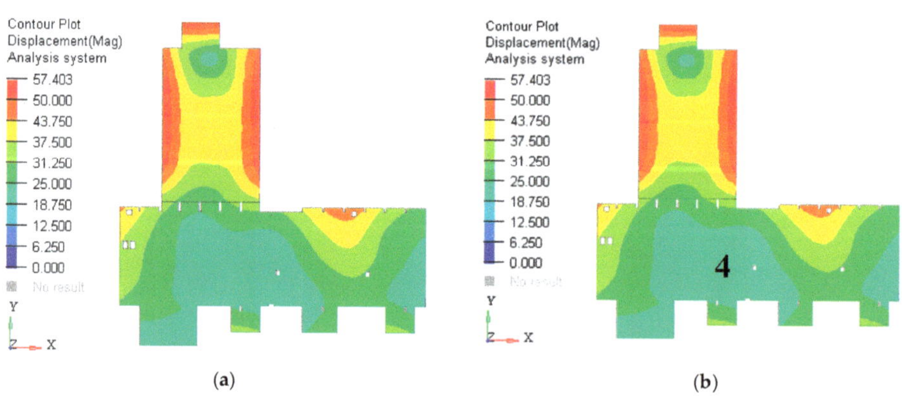

Figure 12. Comparison between the deformations of the standard plate and the deformations after removing a 250 tons wall: (**a**) deformation field for the initial platform and (**b**) deformation field if a wall is removed. There are no significant differences between in the two cases.

4. Modal Analysis

The elastic elements that influence the vibration behavior of the platform are the spring batteries on which the platform is placed and the deformability of the platform under the action of its own weight, the weight of the walls, and the weight of the equipment.

Analysis of the vibrations of the concrete slab placed on the spring batteries is made using the model with finite elements presented in Section 3 [27–31]. We are interested in the vibrations of the concrete platform caused by the elastic suspension based on placement the equations of motion of the platform are given by:

$$[\,m\,]\{\,\ddot{x}\,\}+[\,c\,]\{\,\dot{x}\,\}+[\,k\,]\{\,x\,\}=\{f(t)\,\}, \tag{1}$$

where $[m]$ is the inertial matrix, $[c]$ is the *damping matrix*, $[k]$ is the stiffness matrix (these three matrix are real, symmetric and positive definite), $\{\,x\}$ is the vector of generalized displacements, $\{\,\dot{x}\,\}$ is the column vector of velocities, and $\{\,\ddot{x}\,\}$ is the vector of the generalized accelerations.

Let us consider the form free vibrations of this system, a particular expression of (1):

$$[\,m\,]\{\,\ddot{x}\,\}+[\,k\,]\{\,x\,\}=0, \tag{2}$$

then, the time dependence of $\{\,x\,\}$ is harmonic

$$\{\,x(t)\,\}=C\{u\}\cos(\omega t-\varphi), \tag{3}$$

where C is constant, ω is the circular frequency, and φ is the initial phase shift.

Substitution of (3) into (2) yields

$$[\,k\,]\{\,u\,\}=\omega^2[\,m\,]\{u\,\}. \tag{4}$$

Equation (4) has nontrivial solutions if and only if:

$$\det\left([\,k\,]-\omega^2[\,m\,]\right)=0. \tag{5}$$

If n is the dimension of the system, Equation (5) represents a polynomial:

$$\left(\omega^2\right)^n-I_1\left(\omega^2\right)^{n-1}+I_2\left(\omega^2\right)^{n-2}-\ldots.+(-1)^nI_n=0, \tag{6}$$

with the real and positive solutions: $\omega_1^2,\omega_2^2,\ldots\quad\ldots,\omega_n^2$. Equation (6) possesses, in general, n distinct roots, referred to as eigenvalues. For every ω_i^2, the linear homogeneous system (4) is written in the form:

$$\left([\,k\,]-\omega^2[\,m\,]\right)\{u\,\}=0, \tag{7}$$

we obtain the solution $\{u_i\,\}, i=1,2,\ldots\quad,n$. These vectors are the eigenmodes of vibration.

The case of multiple roots is not considered herein. The square roots of the eigenvalues are the system of natural frequencies,ω_r, usually presented in order of increasing magnitude (see Table 1).

Table 1. Eigenfrequencies (in Hz).

Nr.	25,000 Tons	35,000 Tons	54,000 Tons
1	2.27	1.90	1.53
2	2.43	2.03	1.64
3	2.49	2.08	1.68
4	3.38	2.82	2.28
5	3.40	2.84	2.29
6	3.45	2.89	2.33
7	3.60	3.01	2.42
8	3.66	3.06	2.47
9	3.89	3.26	2.62
10	3.92	3.28	2.64
11	4.00	3.34	2.69
12	4.04	3.38	2.73
13	4.14	3.46	2.78
14	4.38	3.66	2.95
15	4.42	3.69	2.96

Table 1. *Cont.*

Nr.	25,000 Tons	35,000 Tons	54,000 Tons
16	4.43	3.71	2.98
17	4.52	3.78	3.04
18	4.57	3.82	3.07
19	4.60	3.85	3.10
20	4.64	3.88	3.12
21	4.72	3.95	3.19
22	4.79	4.00	3.23
23	4.97	4.16	3.36
24	5.04	4.21	3.4
25	5.06	4.23	3.41
26	5.18	4.33	3.49
27	5.33	4.46	3.6
28	5.42	4.54	3.66
29	5.57	4.66	3.76
30	5.64	4.72	3.8
31	5.67	4.74	3.83
32	5.84	4.88	3.94
33	6.05	5.06	4.08
34	6.3	5.27	4.25
35	6.38	5.33	4.31
36	6.62	5.53	4.47
37	6.69	5.59	4.51
38	6.90	5.77	4.66
39	7.10	5.93	4.79
40	7.13	5.96	4.81

In the paper, a calculation was made of the eigenfrequencies (resonance frequencies) and of the eigenmodes of vibration of the concrete slab. In Figure 13, Figure 14, Figure 15, Figure 16, Figure 17, Figure 18, Figure 19, Figure 20, Figure 21, and Figure 22, the first 10 modes of vibration are represented. Modes 10–20 are represented in Appendix C (Figures A5–A15).

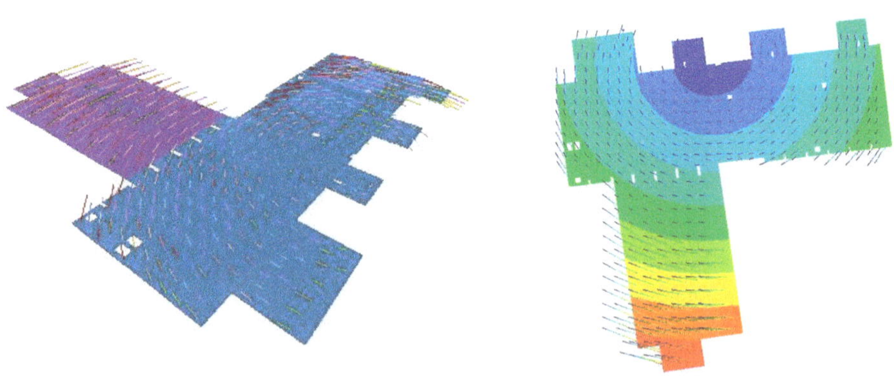

(**a**) perspective (**b**) top view

Figure 13. Eigenmode 1. $\omega_1 = 1.90\ Hz$.

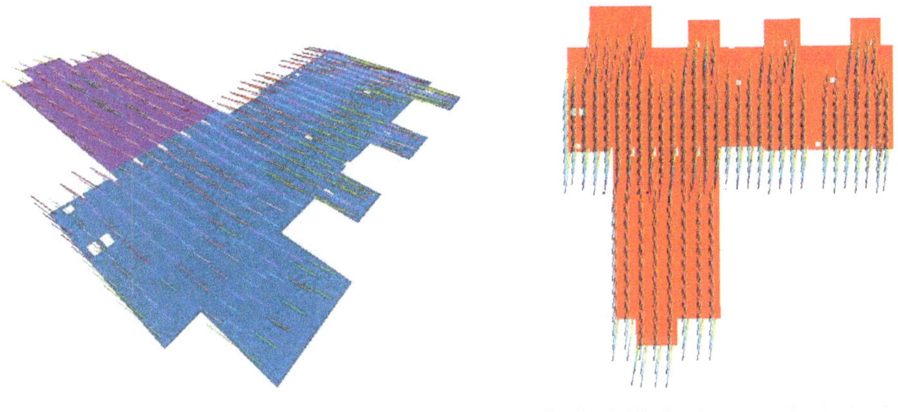

(a) perspective (b) top view

Figure 14. Eigenmode 2. $\omega_2 = 2.03\ Hz$.

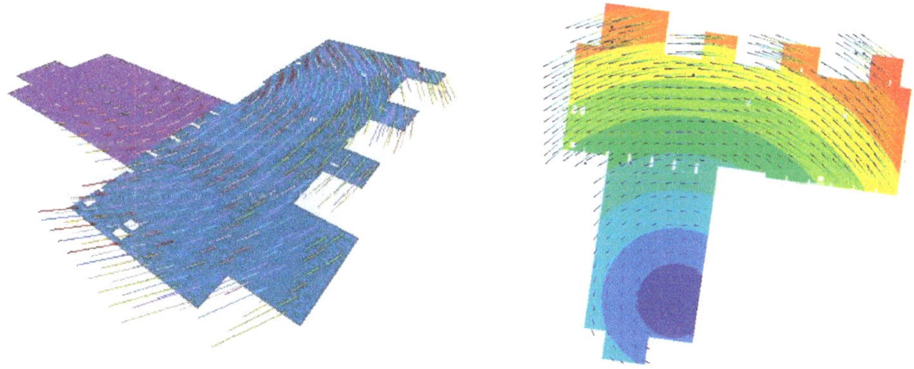

(a) perspective (b) top view

Figure 15. Eigenmode 3. $\omega_3 = 2.08\ Hz$.

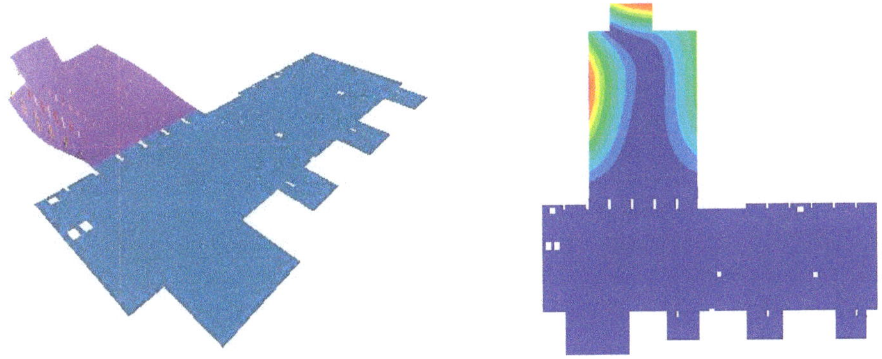

(a) perspective (b) top view

Figure 16. Eigenmode 4. $\omega_4 = 2.82\ Hz$.

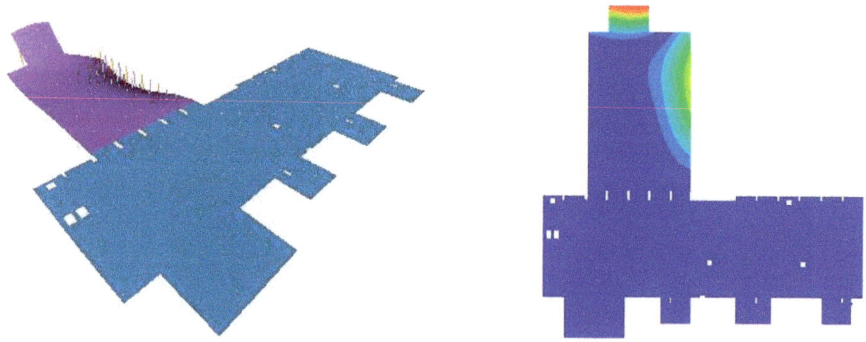

(**a**) perspective (**b**) top view

Figure 17. Eigenmode 5. $\omega_5 = 2.86\ Hz$.

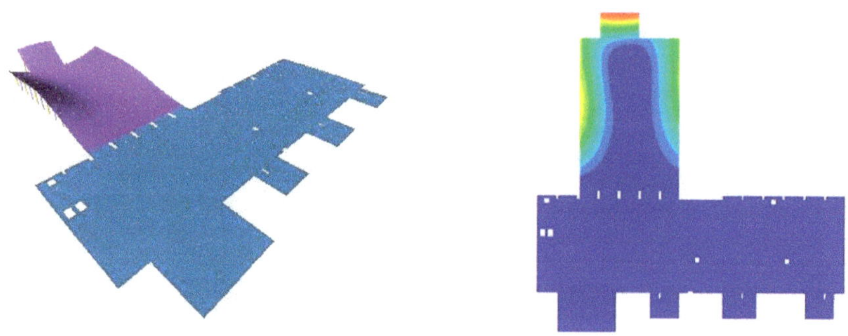

(**a**) perspective (**b**) top view

Figure 18. Eigenmode 6. $\omega_6 = 2.89\ Hz$.

(**a**) perspective (**b**) top view

Figure 19. Eigenmode 7. $\omega_7 = 3.01\ Hz$.

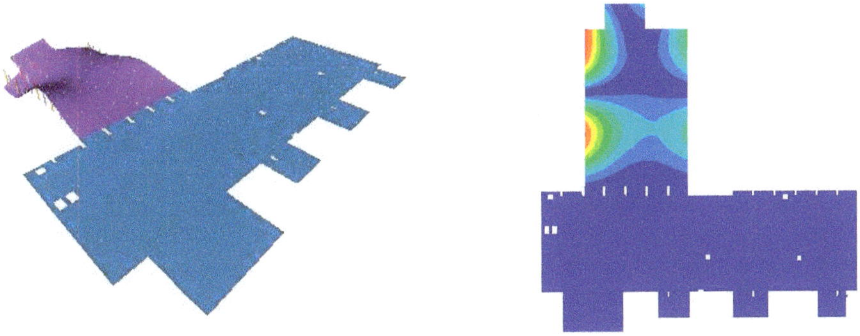

(a) perspective (b) top view

Figure 20. Eigenmode 8. $\omega_8 = 3.06\ Hz$.

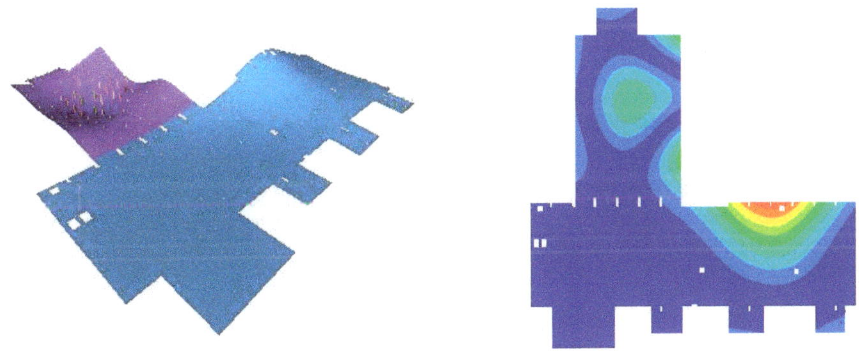

(a) perspective (b) top view

Figure 21. Eigenmode 9. $\omega_9 = 3.26\ Hz$.

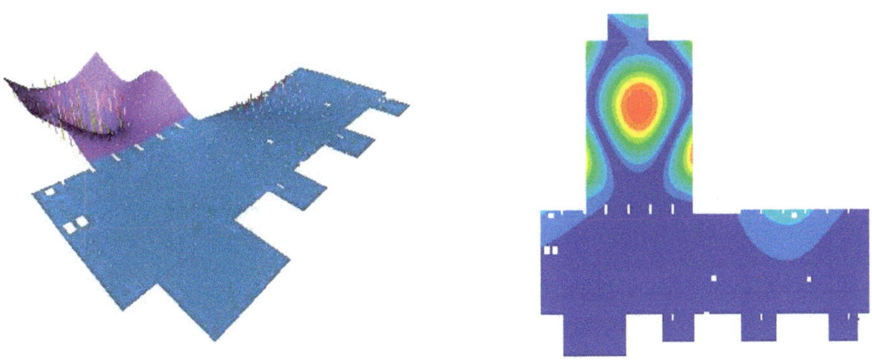

(a) perspective (b) top view

Figure 22. Eigenmode 10. $\omega_{10} = 3.28\ Hz$.

5. Discussion

The paper studies the response of the inertial platform supported on sets of spring batteries on ground base to identify if excessive vibrations exist and how are the different ground vibration transmitted to the platform. On the platform, a lot of devices and instruments are distributed implying in the scientific researches and the necessary annexes. Usually, vibrations coming from the vehicles and other anthropic activities and transmitted through the ground are nonstationary vibrations. To reduce the influence of these vibrations on the inertial platform, this is suspended on a set of spring batteries and dampers that can be activated or not. This suspension constitutes an excellent tool to reduce the transfer of the unwanted vibrations to the studied platform.

The deformability of the concrete platform under the action of its own weight and under the action of the weight of the walls is also studied. It is found that the relative deformations and stresses that occur in the platform material are relatively small, so there are no problems related to the strength and cracking of the platform, even if it is charged at high loads.

The platform consists of two plates, one main, on which are found most of the instruments and devices and where the massive partition walls are raised, which is about 1600 mm, and the second platform, which I called secondary, smaller and thinner, is only 600 mm, and weighs about half the weight of the unloaded main board. If the main board is loaded with all the equipment and partitions, the platform weighs about 1/6 of the total weight of the platform. This platform is found to have significant amplitudes in its eigenmodes, so in the case of external excitations with frequencies close to the natural frequencies of the entire platform, we can expect an amplification of these excitations at the secondary platform. The problem can be solved by activating additional springs from the spring batteries that support the secondary platform.

Additionally, another problem that can be mentioned is the fact that the eigenfrequencies of the platform are found in the range where the earthquakes that take place in the area have the excitation frequencies.

6. Conclusions and Future Works

The results based on the obtained model allow the formulation of some conclusions regarding the achievement of the objectives proposed within the project. The static analysis of the platform indicates that under the action of its own weight, supported on the elastic springs, the platform deforms the most by 26 mm, which corresponds to a strain of about 0.1%. In addition, under the action of additional weights that could appear on the platform (mainly these weights are represented by different dividing concrete walls), the deformations that can occur are not important. The calculation made with masses of 1000 tons (much exaggerated compared to the real situation—the hardest wall we identified has less than 250 tons) placed in different parts of the platform indicated to us, in the most unfavorable case, a deformation of about 250 mm corresponding to a strain of 1%. This situation shows us that any constructions on the platform or changes in the architecture will not influence its flatness and, therefore, the operation of the measuring instruments that impose certain strict conditions.

The dynamic analysis of the platform indicates a large number of natural frequencies below 5 Hz, which is a negative aspect from the point of view of the platform's behavior in earthquakes. Anthropic activities will not influence the behavior of the system, taking into account the frequencies generated by these activities. Instead, the effects of a catastrophic event (earthquake) could amplify the possibilities of vibration in certain directions of the platform, which could lead to destruction and damage to the installation. For this reason, a careful calculation is required to the structure's response in case of an earthquake and its permanent monitoring for the analysis of the transmissibility of external excitations. The ELI-NP project has a decoupling system for earthquakes, the need for an additional isolation system of the entire mass should be studied. Such a study, the necessity of which has been demanded by the researches made, is to be carried out.

Author Contributions: Conceptualization, S.V., P.B., C.I., P.N.B. and D.L.; methodology, S.V. and C.I.; software, C.I.; validation, C.I., S.V., P.B. and D.L.; formal analysis, P.N.B., S.V., P.B. and D.L.; investigation, C.I. and S.V.; resources, S.V.; data curation, C.I. and S.V.; writing—original draft preparation, C.I., S.V., P.B., D.L. and P.N.B.; writing—review and editing, C.I., S.V., P.B., P.N.B. and D.L.; visualization, S.V. and P.B.; supervision, S.V.; project administration, S.V.; funding acquisition, P.B. All authors have read and agreed to the published version of the manuscript.

Funding: This research was funded by IFA-MG, grant number 16/2016. The APC was funded by the Transylvania University of Brasov.

Acknowledgments: We want to thank the reviewers who have read the manuscript carefully and have proposed pertinent corrections that have led to the improvement of our manuscript.

Conflicts of Interest: The authors declare no conflict of interest. The funders had no role in the design of the study; in the collection, analyses, or interpretation of data; in the writing of the manuscript; or in the decision to publish the results.

Appendix A

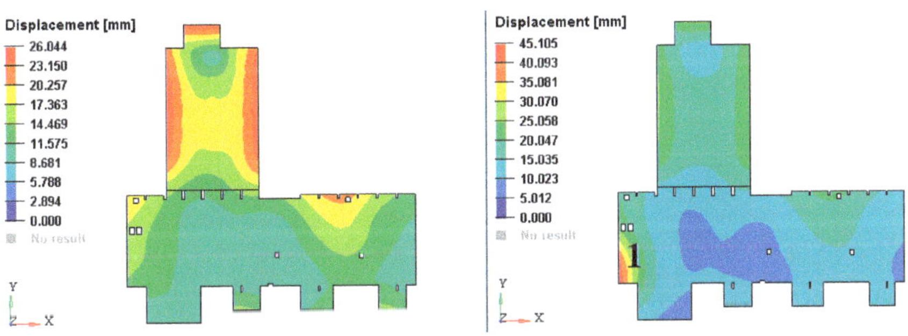

Figure A1. Deformation of the concrete platform placed on the spring batteries: (**a**) under its own weight and (**b**) loaded with 1000 tons in point 1.

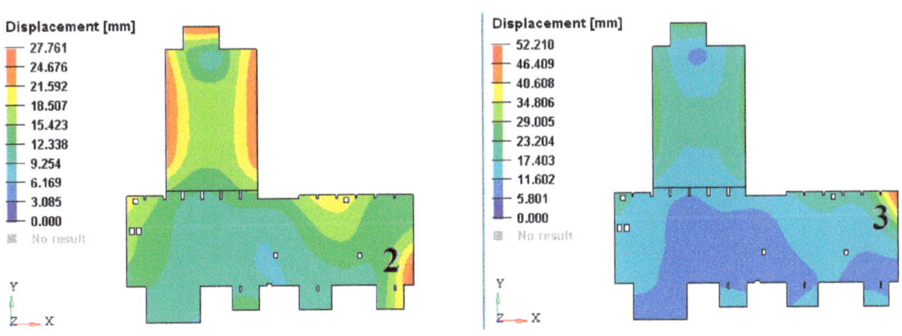

(**a**) (**b**)

Figure A2. Deformation of the concrete platform (**a**) loaded with 1000 tons in points 2 and (**b**) loaded with 1000 tons in point 3.

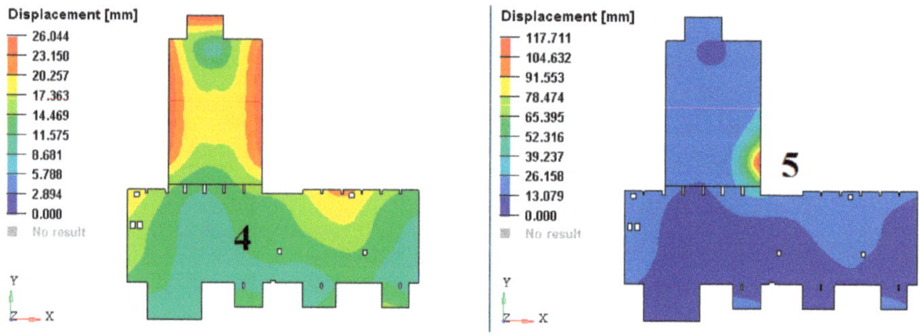

(a) (b)

Figure A3. Deformation of the concrete platform (**a**) loaded with 1000 tons in points 4 and (**b**) loaded with 1000 tons in point 5.

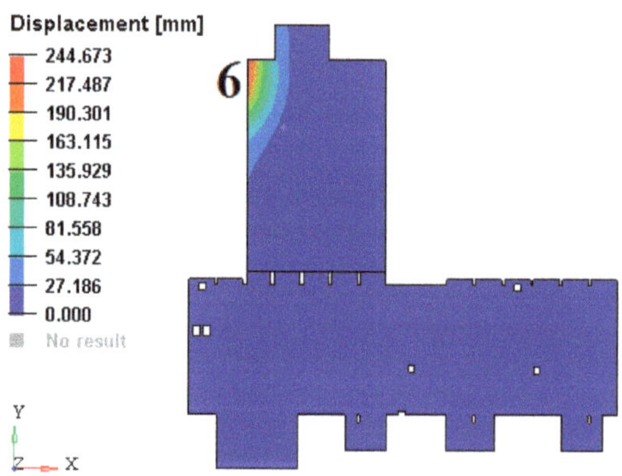

Figure A4. Deformation of the concrete platform loaded with 1000 tons in points 6.

Comment: The load of 1000 tons represents only a hypothetical load, to see how the platform deforms in extreme situations. In practical cases, it is hard to believe that a task will be added, even 10 times less, in point 6 studied. Usually, a large load can occur when a wall is realized or removed. A real situation was encountered when a 250 tons wall was removed. The deformation in this case was much smaller than the one calculated for the hypothetical case. A simulation with a heavier wall was made in Figure 12. It has been found that in this case, the difference between deformations was much smaller. At the same time, the dimensions of the structure are of the order of tens of meters and 250 mm compared to 25 m representing a strain of 1%.

Appendix B. Stress Field in the Platform (Expressed in MPa)

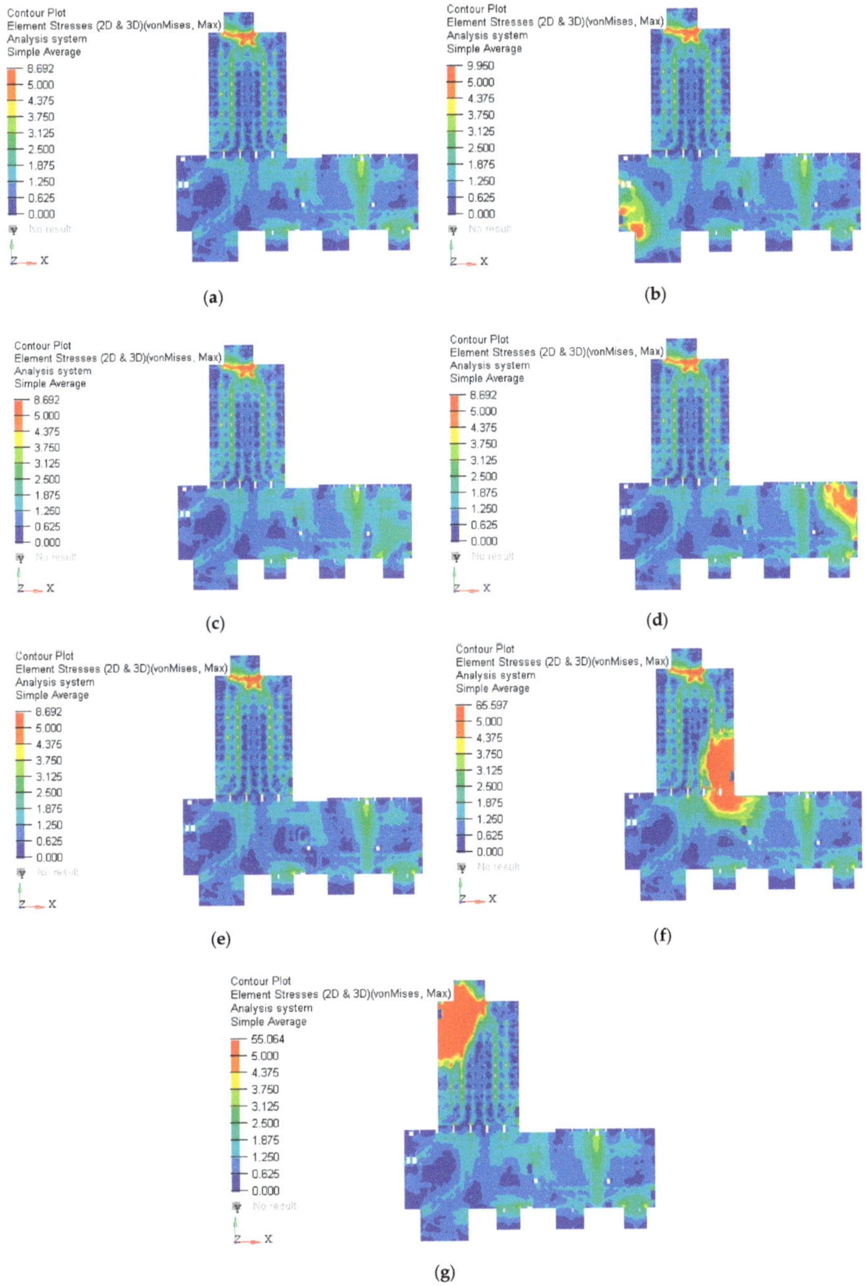

Figure A5. Stresses in the concrete platform expressed in MPa. (**a**) under its own weight; (**b**) loaded with 1000 tons in point 1; (**c**) loaded with 1000 tons in points 2 and (**d**) loaded with 1000 tons in point 3; (**e**) loaded with 1000 tons in points 4 and (**f**) loaded with 1000 tons in point 5; (**g**) loaded with 1000 tons in points 6.

Appendix C

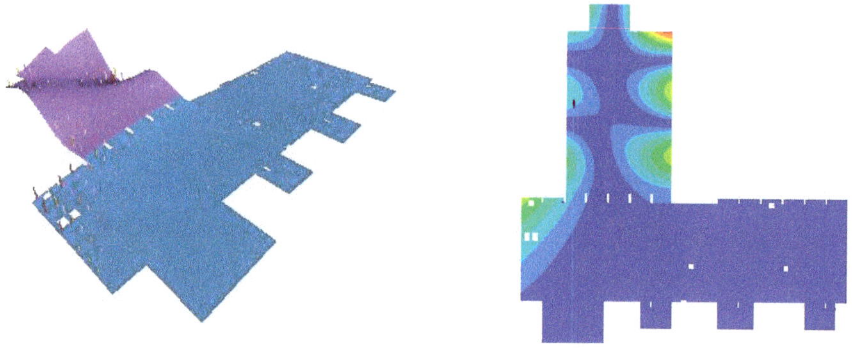

(**a**) perspective (**b**) top view

Figure A6. Eigenmode 11.

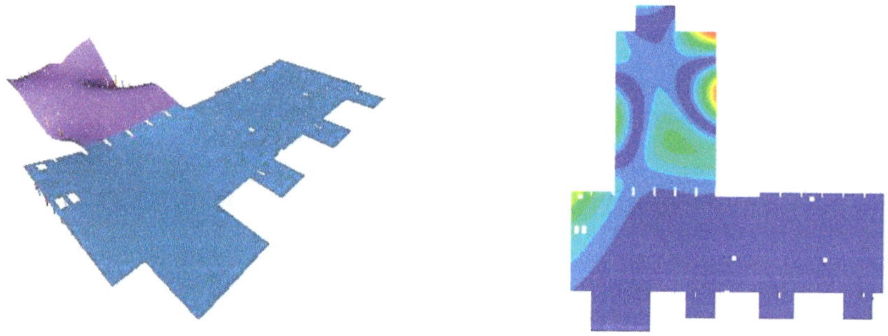

(**a**) perspective (**b**) top view

Figure A7. Eigenmode 12.

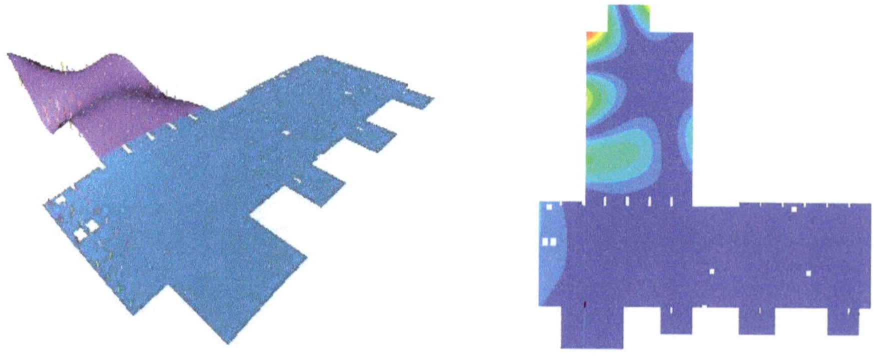

(**a**) perspective (**b**) top view

Figure A8. Eigenmode 13.

(**a**) perspective (**b**) top view

Figure A9. Eigenmode 14.

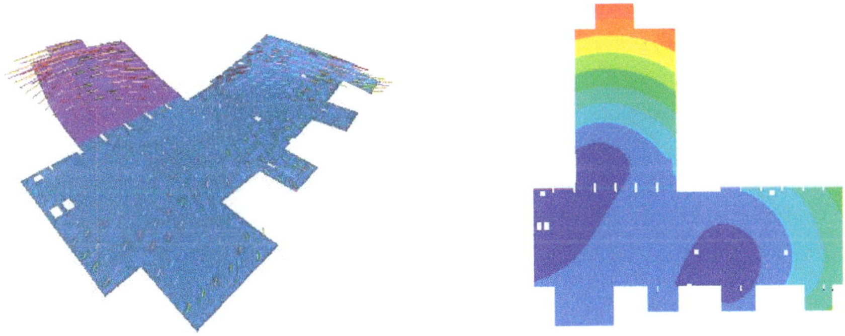

(**a**) perspective (**b**) top view

Figure A10. Eigenmode 15. (**a**) perspective and (**b**) top view.

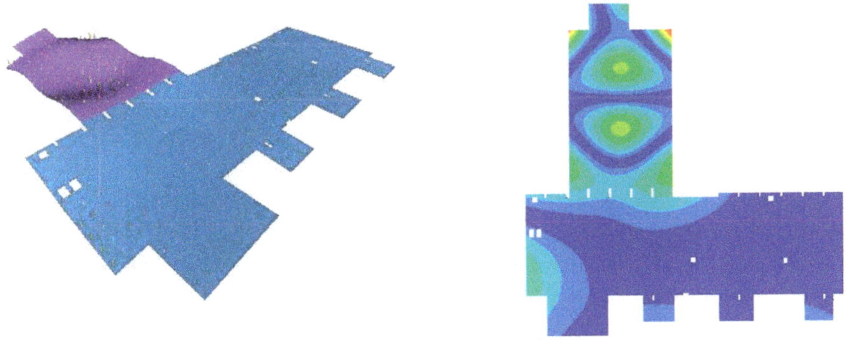

(**a**) perspective (**b**) top view

Figure A11. Eigenmode 16.

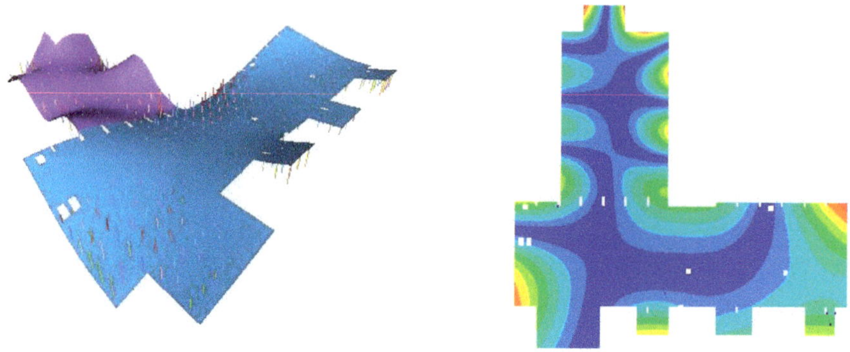

(a) perspective (b) top view

Figure A12. Eigenmode 17.

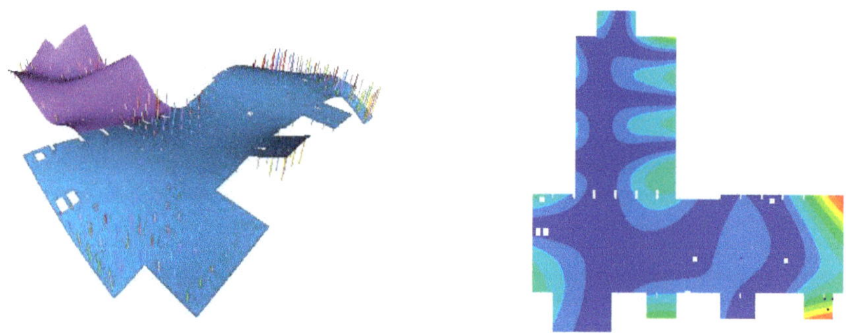

(a) perspective (b) top view

Figure A13. Eigenmode 18. (a) perspective and (b) top view.

(a) perspective (b) top view

Figure A14. Eigenmode 19. (a) perspective and (b) top view.

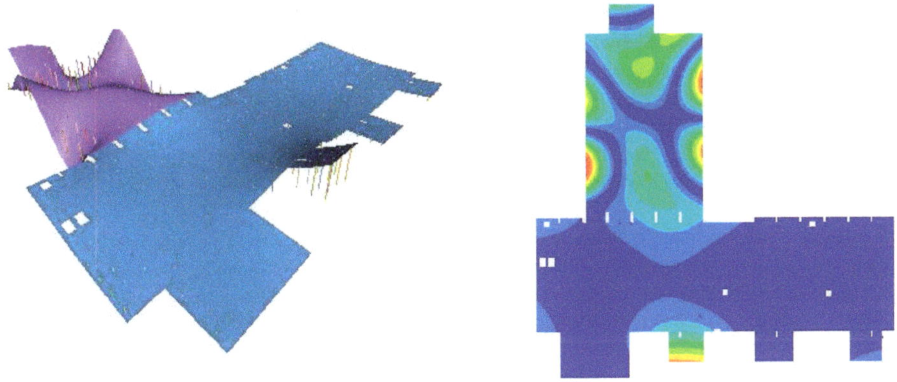

(a) perspective (b) top view

Figure A15. Eigenmode 20.

References

1. The White Book of ELI Nuclear Physics Bucharest-Magurele, Romania. Available online: http://www.eli-np.ro/whitebook.php (accessed on 31 October 2020).
2. Serafini, L. EuroGammaS proposal for the ELI-NP Gamma beam System. Technical Design Report, July. 2014. Technical Design Report. Available online: https://arxiv.org/ftp/arxiv/papers/1407/1407.3669.pdf (accessed on 31 October 2020).
3. Habs, D.; Tajima, T.; Zamfir, V. Extreme Light Infrastructure–Nuclear Physics (ELI–NP): New Horizons for Photon Physics in Europe. *Nucl. Phys. News* **2011**, *21*, 23–29. [CrossRef]
4. J. Welch Ground Vibration and Siting of the Cryogenics Facility for Cornell ERL Prototype. Technical Design Report. Available online: https://www.classe.cornell.edu/public/ERL/2002/ERL02-3/ERL02-3.pdf (accessed on 25 November 2020).
5. Arnold, C. Seismic Design & Devices for Detaching Building from the Ground. *Archit. AIA J.* **1987**, *76*, 64–67.
6. Tarics, A.G. Base-Isolation—A New Strategy for Earthquake Protection of Buildings. *J. Archit. Plan. Res.* **1987**, *4*, 64–76.
7. Teramoto, T. Recent Current of Nase Isolation and Seismic Control-System for Building Structure. *Tetsu Hagane J. Iron Steel Inst. Jpn.* **1987**, *73*, S348.
8. Jiang, M.; Rui, X.; Zhu, W.; Yang, F.; Zhang, Y. Optimal design of 6-DOF vibration isolation platform based on transfer matrix method for multibody systems. *Acta Mech. Sin.* **2020**, 1–11. [CrossRef]
9. Han, P.; Wang, T.; Wang, D.H. Modeling and control of a Stewart platform based six-axis hybrid vibration isolation system. In Proceedings of the 2008 7th World Congress on Intelligent Control. and Automation, Chongqing, China, 25–27 June 2008; Volume 1–23, pp. 1613–1618. [CrossRef]
10. Geng, Z.; Haynes, L.S. Six-degree-of-freedom active vibration isolation using a stewart platform mechanism. *J. Robot. Syst.* **1993**, *10*, 725–744. [CrossRef]
11. Liu, L.; Wang, B. Development of stewart platforms for active vibration isolation and precision pointing. In Proceedings of the International Conference on Smart Materials and Nanotechnology in Engineering, Harbin, China, 1–4 July 2007.
12. Ma, R.; Bi, K.; Hao, H. Heave motion mitigation of semi-submersible platform using inerter-based vibration isolation system (IVIS). *Eng. Struct.* **2020**, *219*, 110833. [CrossRef]
13. Xie, X.; Diao, J.; Xu, Y.; Zhang, Z. Performance of a low-frequency hybrid vibration isolation platform for vibration-sensitive devices. *J. Low Freq. Noise, Vib. Act. Control.* **2018**, *37*, 1164–1175. [CrossRef]
14. Xu, Z.-D.; Xu, F.-H.; Chen, X. Vibration suppression on a platform by using vibration isolation and mitigation devices. *Nonlinear Dyn.* **2015**, *83*, 1341–1353. [CrossRef]

15. Assmann, R.; Jeanneret, B.; Verdier, A.; Vos, L.; Wildner, E.; Zimmermann, F.; Brinkmann, R.; Montag, C.; Reyzl, I.; Walker, N.; et al. Stability considerations for final focus systems of future linear collinder. In Proceedings of the 7th European Conference, EPAC 2000, Vienna, Austria, 26–30 June 2000; p. 447.
16. Bolzon, B. Etude des Vibrations et de la Stabilisation à L'échelle Sous-Nanométrique des Doublets Finaux d'un Collisionneur Linéaire. Ph.D. Thesis, Université Chambery Annecy de Savoie, Chambéry, France, 2007.
17. A Multi-Tev Linear Collider Based on Clic Technology, Conceptual Design Report. KEK Report 2012-1, Geneva. 2012. Available online: http://cds.cern.ch/record/1601966/files/ILCTDR-VOLUME_1.pdf. (accessed on 31 October 2020).
18. Artoos, K. Compatibility and Integration of a CLIC Quadrupole Nanometre-Stabilization and Positioning system in a Large Accelerator Environment. In Proceedings of the 1st International Particle Accelerator Conference (IPAC'10), Kyoto, Japan, 23–28 May 2010; Volume 16, pp. 1274–1276.
19. Vlase, S.; Marin, M.; Öchsner, A.; Scutaru, M.L. Motion equation for a flexible one-dimensional element used in the dynamical analysis of a multibody system. *Contin. Mech. Thermodyn.* **2019**, *31*, 715–724. [CrossRef]
20. Marin, M. Some basic theorems in elastostatics of micropolar materials with voids. *J. Comput. Appl. Math.* **1996**, *70*, 115–126. [CrossRef]
21. Itu, C.; Öchsner, A.; Vlase, S.; Marin, M. Improved rigidity of composite circular plates through radial ribs. *Proc. Inst. Mech. Eng. Part. L J. Mater. Des. Appl.* **2019**, *233*, 1585–1593. [CrossRef]
22. Scutaru, M.L.; Vlase, S.; Marin, M. New analytical method based on dynamic response of planar mechanical elastic systems. *Bound. Value Probl.* **2020**, *104*. [CrossRef]
23. Nazreen, M.S.; Mohamed, R.N.; Ab Kadir, M.A.; Azillah, N.; Shukri, N.A.; Mansor, S.; Zamri, F. Characterization of lightweight concrete made of palm oil clinker aggregates. *MATEC Web Conf.* **2018**, *250*, 03002. [CrossRef]
24. Gul, M.; Bashir, A.; Naqash, J.A. Study of Modulus of Elasticity of Steel Fiber Reinforced Concrete. *Int. J. Eng. Adv. Technol.* **2014**, *3*. ISSN: 2249-8958. Available online: https://www.ijeat.org/wp-content/uploads/papers/v3i4/D2995043414.pdf (accessed on 31 October 2020).
25. Yusof, M.A.; Nor, N.M.; Fauzi, M.; Zain, M.; Ismail, A.; Sohaimi, A.; Zaidi, A.M. Mechanical Properties of Hybrid Steel Fibre Reinforced Concrete with Different Aspect Ratio. *Aust. J. Basic Appl. Sci.* **2011**, *5*, 159–166.
26. Eurocode Applied. Available online: https://eurocodeapplied.com/design/en1992/concrete-design-properties (accessed on 25 November 2020).
27. Vlase, S. Elimination of lagrangian multipliers. *Mech. Res. Commun.* **1987**, *14*, 17–22. [CrossRef]
28. Rades, M. *Mechanical Vibration, II, Structural Dynamics Modeling*; Printech Publishing House: Bangalore, Karnataka, India, 2010.
29. Vlase, S.; Negrean, I.; Marin, M.; Scutaru, M.L. Energy of Accelerations Used to Obtain the Motion Equations of a Three- Dimensional Finite Element. *Symmetry* **2020**, *12*, 321. [CrossRef]
30. Hu, R.; Iwamoto, S.; Feng, L.; Ju, S.; Hu, S.; Ohnishi, M.; Nagai, N.; Hirakawa, K.; Shiomi, J. Machine-Learning-Optimized Aperiodic Superlattice Minimizes Coherent Phonon Heat Conduction. *Phys. Rev. X* **2020**, *10*, 021050. [CrossRef]
31. Imani, M.; Ghoreishi, S.F. Bayesian Optimization Objective-Based Experimental Design. In Proceedings of the 2020 American Control Conference (ACC), Denver, CO, USA, 1–3 July 2020.

symmetry

MDPI

Article

Multibody System with Elastic Connections for Dynamic Modeling of Compactor Vibratory Rollers

Polidor Bratu [1,2]

1 Research Institute for Construction Equipment and Technology—ICECON SA, 021652 Bucharest, Romania; icecon@icecon.ro; Tel.: +40-21-202-55-00

2 Faculty of Engineering and Agronomy, "Dunărea de Jos" University of Galaţi, 810017 Brăila, Romania

Received: 21 August 2020; Accepted: 25 September 2020; Published: 29 September 2020

Abstract: The dynamic model of the system of bodies with elastic connections substantiates the conceptual basis for evaluating the technological vibrations of the compactor roller as well as of the parameters of the vibrations transmitted from the vibration source to the remainder of the equipment components. In essence, the multi-body model with linear elastic connections consists of a body in vertical translational motion for vibrating roller with mass m_1, a body with composed motion of vertical translation and rotation around the transverse axis passing through its weight center for the chassis of the car with mass m and the moment of mass inertia J and a body of mass m' representing the traction tire-wheel system located on the opposite side of the vibrating roller. The study analyzes the stationary motion of the system of bodies that are in vibrational regime as a result of the harmonic excitation of the m mass body, with the force $F(t) = m_0 r\omega^2 \sin \omega t$, generated by the inertial vibrator located inside the vibrating roller. The vibrator is characterized by the total unbalanced m_0 mass in rotational motion at distance r from the axis of rotation and the angular velocity or circular frequency ω.

Keywords: multibody; elastic bonds; vibrations; initial matrix; stiffness matrix

1. Introduction

The real-time assessment of the degree of compaction of the foundation soil both with stabilized natural soil as well as mixed with stone mineral aggregates or in the case of compaction of asphalt concrete layers, requires precision and high sensitivity of the dynamic response in amplitude of the compactor roller to the changes of soil rigidity as a result of the compaction process.

After each passage on the same compacted layer, the final rigidity of the soil has a new value, higher than the initial rigidity. In this case, after each passage, there can be estimated, through an appropriate instrumental system, the modified amplitude of vibration in correlation with the new state of compaction of the soil corresponding to modified rigidity.

Currently, there are several companies manufacturing vibration compactor machines that use instrumental and computer systems for capturing, treating, and processing the specific signal to the vibration of the vibrating roller. Usually, the dynamic calculation model used is reduced to that of the vibrating roller system with a single degree of freedom, without taking into account the effect of the other vibrating moving masses of the machine.

Frequently, for vibration regime at frequencies in the range of 40–50 Hz, the system ensures the degree of compaction in real time based on the change in rigidity with each passing on the same layer of land. In this case, the first two resonant frequencies are neglected, although they may be important in the work process.

At frequencies between 15 and 30 Hz, the automatic analysis of technological vibration systems produce errors 30% larger, which leads to major inconveniences. For these reasons, the current dynamic study highlights the influence of the masses of the body assembly at various dynamic regimes for functional frequencies from 15 Hz to 80 Hz. According to the category of the compaction technology, that is, the change in final rigidity after each passage of the compacted layer, there are many scientific and technical approaches with case studies on technologically defined sites that require a more complete dynamic approach, highlighting the influences of the body system on the dynamic response and of the degree of compaction [1,2].

The numerical data used for the case study represent parametric values established on an experimental basis both in the laboratory and "in situ". [2,3]

2. Multibody System Model

The dynamic multibody model of the vibrating roller is presented in Figure 1 [4–6], where the following notations are used:

I_1—elastic connection point of the vibrating roller with vertical translational movement;

I_2—connection point of the elastic system to the front side of the car chassis;

I_3—connection point between the rear of the car chassis to the traction unit consisting of tire-wheels;

m'—mass of the vibrating roller;

m—mass of the car chassis;

$J = J_z$—moment of mass inertia in relation to the transverse axis z passing through the center of mass C of the car chassis;

m_1—mass of the traction group;

k_1—rigidity of the compacted material;

k_2—rigidity of the elastic connection system and dynamic insulation between the vibrating roller and the front chassis;

k_3—combined rigidity of the traction wheel tires in contact with the compacted material;

a, b—distances of the C mass center in relation to the I_2 and I_3 ends of a chassis, so that $a + b = l$, where $l = I_2I_3$ is the equivalent length of the chassis;

x, $\varphi = \varphi_z$—instantaneous displacements of the chassis; and

x_1, x_2, x_3—absolute instantaneous displacements relative to a fixed reference system.

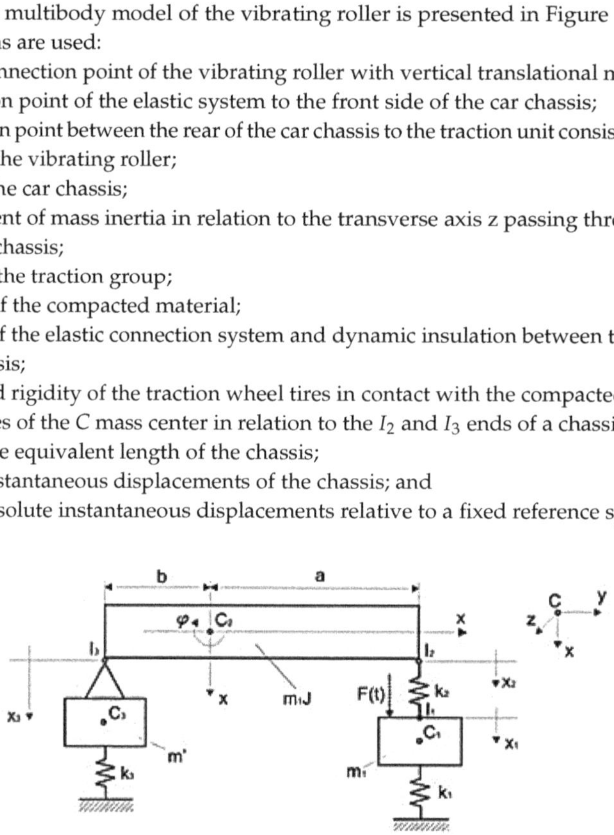

Figure 1. Dynamic multibody model with linear elastic connections.

Instantaneous displacements of points $i = 1, 2, 3$, can be determined with the following matrix relation [7,8]:

$$u_{I_i} = \begin{Bmatrix} x_i \\ y_i \\ z_i \end{Bmatrix} + \begin{bmatrix} 0 & -\varphi_z & \varphi_y \\ \varphi_z & 0 & -\varphi_z \\ -\varphi_y & \varphi_x & 0 \end{bmatrix} \begin{Bmatrix} x_i \\ y_i \\ z_i \end{Bmatrix} \quad (1)$$

where x, y, z are the tri-orthogonal instantaneous linear coordinates of the mass center belonging to each rigid body I_1 and C, respectively.

φ_x, φ_y, φ_z—the tri-orthogonal instantaneous angular coordinates relative to the competing x,y,z axes in the center of mass of each rigid body C_1 and C_2, respectively.

For the m_1 mass body with vertical translational motion and the null instantaneous angular coordinates, that is $\varphi_x = \varphi_y = \varphi_z = 0$, the displacement of the $I_1 \equiv C_1$ point is

$$u_{I_1} = \begin{Bmatrix} x_1 \\ 0 \\ 0 \end{Bmatrix} + \begin{bmatrix} 0 & 0 & 0 \\ 0 & 0 & 0 \\ 0 & 0 & 0 \end{bmatrix} \begin{Bmatrix} 0 \\ 0 \\ 0 \end{Bmatrix} = x_1 \tag{2}$$

For the mass body m and moment of inertia $J_z = J$, with the instantaneous angular coordinates $\varphi_x = \varphi_y = 0$ and $\varphi_z = \varphi$, it has a plane motion (x, φ), so that the displacements of points I_2 and I_3 can be determined as follows:

$$u_{I_2} = \begin{Bmatrix} x \\ 0 \\ 0 \end{Bmatrix} + \begin{bmatrix} 0 & -\varphi & 0 \\ \varphi & 0 & 0 \\ 0 & 0 & 0 \end{bmatrix} \begin{Bmatrix} 0 \\ a \\ 0 \end{Bmatrix} = x - a\varphi \tag{3}$$

$$u_{I_3} = \begin{Bmatrix} x \\ 0 \\ 0 \end{Bmatrix} + \begin{bmatrix} 0 & -\varphi & 0 \\ \varphi & 0 & 0 \\ 0 & 0 & 0 \end{bmatrix} \begin{Bmatrix} 0 \\ -b \\ 0 \end{Bmatrix} = x + b\varphi \tag{4}$$

2.1. Kinetic Energy of the Multibody System

Taking into account the motion of body C_1 of mass m_1 with translation coordinate x_1 and of the assembled body C_2C_3 with mass $m + m'$, moment of mass inertia $J + m'b^2$, with coordinates x, φ (vertical translation and rotation), the kinetic energy of the assembly of bodies is [9,10]

$$2E = \left\langle \dot{q},\, M\dot{q} \right\rangle = \dot{q}^T M\dot{q}. \tag{5}$$

where $\dot{q}$ is the column vector of the generalized velocity with $\dot{q}^T = \begin{bmatrix} \dot{x}_1 & \dot{x} & \dot{\varphi} \end{bmatrix}$;

M—symmetric and positively defined inertia matrix; and

$\left\langle \dot{q},\, M\dot{q} \right\rangle$—scalar product between vectors $\dot{q}$ and $M\dot{q}$.

Matrix M of the entire system of instantaneous moving bodies with generalized coordinates x_1, x, and φ, consists of inertial elements of zero order m_1, $m + m'$, one order $m'b$ and two order $J + m'b^2$, placed on the main diagonal and symmetrically in relation to it, highlighting an inertial coupling due to a C_3 body eccentrically assembled on body C_2. In this case, matrix M can be written as follows:

$$M = \begin{bmatrix} m_1 & 0 & 0 \\ 0 & m + m' & m'b \\ 0 & m' & J + m'b^2 \end{bmatrix} \tag{6}$$

The analytical expression of the kinetic energy, based on relations (5) and (6), can be developed in the form of

$$2E = \widetilde{m}_1\dot{x}_1^2 + \widetilde{m}_2\dot{x}^2 + \widetilde{m}_3\dot{\varphi}^2 + 2\widetilde{m}_{23}\dot{x}\dot{\varphi} \tag{7}$$

where the following notations were used for the inertia coefficients m_2, m_3, and m_{23}, so $\widetilde{m}_2 = m + m'$; $\widetilde{m}_3 = J + m'b^2$; $\widetilde{m}_{23} = m'b$.

2.2. Elastic Potential Energy

For the elastic elements, modeled as linear springs with rigidities k_1, k_2, k_3, the vector of the elastic deformations v, with $v^T = \begin{bmatrix} v_1 & v_2 & v_3 \end{bmatrix}$ has the following components [7,11]:

$$v_1 = x_1$$

$$v_2 = u_{I_2} - x_1 = x - a\varphi - x_1 \tag{8}$$

$$v_3 = x_3 = x + b\varphi$$

Thus, vector $\boldsymbol{v}$ can be written as

$$\boldsymbol{v} = \left\{ \begin{array}{c} x_1 \\ x - a\varphi - x_1 \\ x + b\varphi \end{array} \right\} \tag{9}$$

The transition from the elastic deformations vector $\underline{\underline{v}}$ to the vector of instantaneous displacements $\boldsymbol{q}$ with $q^T = \left[\begin{array}{ccc} x_1 & x & \varphi \end{array} \right]$ can be done by the linear transformation of

$$\boldsymbol{v} = \boldsymbol{A}\boldsymbol{q} \tag{10}$$

where $\boldsymbol{A}$ is the matrix of the linear transformation as an operator of influence of the displacements on deformations.

Taking into account relations (9) and (10), matrix $\boldsymbol{A}$ can be formulated as follows:

$$\boldsymbol{A} = \left[\begin{array}{ccc} 1 & 0 & 0 \\ -1 & 1 & -a \\ 0 & 1 & b \end{array} \right] \tag{11}$$

The potential elastic energy $2V$ can be formulated based on the use of the scalar product between vectors $\boldsymbol{v}$ and $\boldsymbol{K_0}\,\boldsymbol{v}$, where $K_0 = diag\left[\begin{array}{ccc} k_1 & k_2 & k_3 \end{array} \right]$, as follows:

$$2V = \langle \boldsymbol{v},\ \boldsymbol{K_0 v} \rangle \tag{12}$$

Using the linear transformation (10) where A has the property of a self-adjoint operator inside the scalar product, relation (12) becomes

$$2V = \langle Aq, K_0 Aq \rangle$$

or

$$2V = \left\langle q, A^T K_0 A q \right\rangle = \langle q, Kq \rangle \tag{13}$$

where K is the rigidity matrix of the multibody elastic system.

In this case, matrix $K = A^T K_0 A q$ can be written as

$$K = A^T K_0 A q = \left[\begin{array}{ccc} k_1 + k_2 & -k_2 & ak_2 \\ -k_2 & k_2 + k_3 & -ak_2 + bk_3 \\ ak_2 & -ak_2 + bk_3 & a^2 k_2 + b^2 k_3 \end{array} \right] \tag{14}$$

It is found that matrix K is symmetrical and positively defined with elastic coupling elements symmetrically placed in relation to the main diagonal. In general form, matrix K can be written as follows:

$$K = \left[\begin{array}{ccc} k_{11} & k_{12} & k_{13} \\ k_{21} & k_{22} & k_{23} \\ k_{31} & k_{32} & k_{33} \end{array} \right] \tag{15}$$

where elements k_{ij} are those in formulation (15), that is:

$$\begin{gathered} k_{11} = k_1 + k_2;\ k_{12} = -k_2;\ k_{13} = ak_2 \\ k_{21} = -k_2\ ;\ k_{22} = k_2 + k_3;\ k_{23} = ak_2 + bk_3 \\ k_{31} = ak_2;\ k_{32} = -ak_2 + bk_3;\ k_{33} = a^2 k_2 + b^2 k_3 \end{gathered}$$

The potential elastic energy in analytical form, in this case, can be formulated in the form of $2V = \Phi$, as follows

$$2V = (k_1 + k_2)x_1^2 + (k_2 + k_3)x^2 + \left(a^2k_2 + b^2k_3\right)\varphi^2 - 2k_2x_1x + 2(-ak_2 + bk_3)x\varphi + 2ak_2x_1\varphi = \Phi \quad (16)$$

Elastic force Q_j, which corresponds to the generalized coordinate q_j can be written as follows:

$$Q_j^V = -\frac{\partial V}{\partial q_j} \quad (17)$$

In this case, deriving the relation (16) in the form of $2V = \Phi$ in relation to coordinate q_j, that is $\frac{\partial(2V)}{\partial q_j} = \frac{\partial \Phi}{\partial q_j}$ leads to $\frac{\partial V}{\partial q_j} = \frac{1}{2}\frac{\partial \Phi}{\partial q_j}$, and thus we obtain

$$\begin{cases} Q_1^V = -\frac{\partial V}{\partial q_1} = -\frac{1}{2}\frac{\partial \Phi}{\partial q_1} \\ Q_2^V = -\frac{\partial V}{\partial q_2} = -\frac{1}{2}\frac{\partial \Phi}{\partial q_2} \\ Q_3^V = -\frac{\partial V}{\partial q_3} = -\frac{1}{2}\frac{\partial \Phi}{\partial q_3} \end{cases} \quad (18)$$

Taking into account function Φ in relation (16) and the fact that $q_1 = x_1$, $q_2 = x$ and $q_3 = \varphi$, applying relations (18), we obtain

$$\begin{cases} Q_1^V = -\frac{1}{2}\frac{\partial \Phi}{\partial q_1} = -(k_1 + k_2)x_1 + k_2x - ak_2\varphi \\ Q_2^V = -\frac{1}{2}\frac{\partial \Phi}{\partial q_2} = -(k_2 + k_3)x + k_2x_1 - (-ak_2 + bk_3)\varphi \\ Q_3^V = -\frac{1}{2}\frac{\partial \Phi}{\partial q_3} = -\left(a^2k_2 + b^2k_3\right)\varphi - (-ak_2 + bk_3)x - ak_2x_1 \end{cases} \quad (19)$$

2.3. Disruptive Force

The harmonic excitation is given by the disruptive force $F(t) = F_0 \sin \omega t$, where the amplitude of the force is $F_0 = m_0 r\omega^2$. This is applied on body C_1 in order to generate forced vibrations in the vertical direction so that the mass body m_1 and coordinate x_1 only have vertical translational movement.

In this case, the vector of disruptive forces is

$$f^T = \begin{bmatrix} F_0 \sin \omega t & 0 & 0 \end{bmatrix}$$

The generalized force corresponding to the disruptive force after the generalized coordinated q_j can be determined as follows:

$$Q_j^F = \frac{\delta L_j}{\delta q_j} \quad (20)$$

where δL_j is the virtual mechanical work of force F;

δq_j—virtual variation of coordinate q_j,

In this case, forces Q_1^F, Q_2^F, Q_3^F emerge as

$$Q_1^F = \frac{F\delta x_1}{\delta x_1} = F = F_0 \sin \omega t \quad (21)$$

and

$$Q_2^F = Q_3^F = 0 \text{ because } \delta L_2 = \delta L_3 = 0.$$

3. Analysis of Forced Vibrations

The response of the multibody system with elastic connections is given by the excitation given by the harmonic force $F(t) = F_0 \sin \omega t$. $F_0 = m_0 r\omega^2$ defines the inertial force of mass m_0 in the rotational motion at distance r with the circular frequency ω in relation to the axis of rotation of the vibrating device placed symmetrically inside the vibrating roller [1,2,8].

For the multibody system, the Lagrange equations of the second kind can be applied as follows [5,11]:

$$\frac{d}{dt}\left(\frac{\partial E}{\partial \dot{q}_j}\right)-\frac{\partial E}{\partial q_j}=Q_j^V+Q_j^F,\ \mathrm{j}\ =\ 1,\ 2,\ 3 \tag{22}$$

where E is the kinetic energy expressed by relation (7), and the generalized forces Q_j^V and Q_j^F are given by the relations (19) and (21), respectively.

Taking into account relations (7), (19), and (21), respectively, the Lagrange equations of the second kind given by relation (22), for each degree of freedom, can be written in the form

$$\begin{cases} m_1\ddot{x}_1+(k_1+k_2)x_1-k_2x+ak_2\varphi\ =F_0\sin\omega t \\ \widetilde{m}_2\ddot{x}+\widetilde{m}_{23}\ddot{\varphi}+(k_2+k_3)x-k_2x_1+(-ak_2+bk_3)\varphi=0 \\ \widetilde{m}_{23}\ddot{x}+\widetilde{m}_3\ddot{\varphi}+\left(a^2k_2+b^2k_3\right)\varphi+(-ak_2+bk_3)x+ak_2x_1=0 \end{cases} \tag{23}$$

In stationary forced mode, the dynamic response is given by the solutions of the system of linear differential Equation (23), as follows:

$$\begin{cases} x_1=A_1\sin\omega t \\ x=A_x\sin\omega t \\ \varphi=A_\varphi\sin\omega t \end{cases} \tag{24}$$

which introduced together with $\ddot{x}_1$, $\ddot{x}$ and $\ddot{\varphi}$ in system (23) results in an algebraic system having as unknown amplitudes A_1, A_x, and A_φ, as

$$\begin{cases} a_{11}A_1+a_{12}A_x+a_{13}A_\varphi=F_0 \\ a_{21}A_1+a_{22}A_x+a_{23}A_\varphi=0 \\ a_{31}A_1+a_{32}A_x+a_{33}A_\varphi=0 \end{cases} \tag{25}$$

Coefficients a_{ij} $i,j = 1, 2, 3$ have the following expressions thus determined:

$$\begin{cases} a_{11}=k_1+k_2-m_1\omega^2 \\ a_{22}=k_2+k_3-\widetilde{m}_2\omega^2 \\ a_{33}=a^2k_2+b^2k_3-\widetilde{m}_3\omega^2 \\ a_{12}=a_{21}=-k_2 \\ a_{13}=a_{31}=-ak_2 \\ a_{23}=a_{32}=-ak_2+bk_3-\widetilde{m}_{23}\omega^2 \end{cases} \tag{26}$$

The determinant of the unknown coefficients based on relation (25) emerges as follows:

$$D=a_{11}a_{22}a_{33}+2a_{12}a_{13}a_{23}-a_{11}{a_{23}}^2-a_{22}{a_{13}}^2-a_{33}{a_{12}}^2 \tag{27}$$

Condition $D = 0$ generates the pulse equation, from where there emerges three real values of ω that coincide with the three own pulses ω_{nj}, $j = 1, 2, 3$.

Amplitudes A_1, A_x, and A_φ are obtained by solving out the algebraic equation system (25) applying Cramer's method, so that we have

$$A_1=\left(a_{22}a_{33}-{a_{23}}^2\right)\frac{m_0r\omega^2}{D} \tag{28}$$

$$A_x=A_2=(a_{13}a_{23}-a_{12}a_{33})\frac{m_0r\omega^2}{D} \tag{29}$$

$$A_\varphi=A_3=(a_{12}a_{23}-a_{13}a_{22})\frac{m_0r\omega^2}{D} \tag{30}$$

For a vibrating equipment modeled as a multibody system, the parametric values resulting from the numerical evaluation are given as follows: $m_1 = 2{\cdot}10^3$ kg; $\widetilde{m}_2 = 4.5{\cdot}10^3$ kg; $\widetilde{m}_3 = 32{\cdot}10^3$ kgm^2; $\widetilde{m}_{23} = 10^6$ kgm; $k_1 = (1;2;4;6){\cdot}10^7$ N/m; $k_2 = 10^6$ N/m; $k_1 = 1.25{\cdot}10^6$ N/m; $m_0 r = 2$ kgm; $a = 1$ m; $b = 2$ m.

For the variation of ω in the range of values (0 ÷ 400) rad/s, the response curves of amplitudes $A_1(\omega)$, $A_2(\omega)$, and $A_3(\omega)$ were drawn and represented in Figures 2–4 for four discrete values of rigidity k_1. Thus, three own pulses emerge of which the first two at the values $\omega_{n1} = 12.23$ rad/s, $\omega_{n2} = 22.24$ rad/s, are common and constant for the four values of rigidity $k_1 = (1;2;4;6){\cdot}10^7$ N/m; the last value of the own pulse ω_{n3} is different according to rigidity k_1. In this case, for k_j, $j = 1, 2, 3$, we have $k_1 = 10^7$ N/m, $\omega_{n3}^{(1)} = 74.73$ rad/s, $k_1 = 2{\cdot}10^7$ N/m, $\omega_{n3}^{(2)} = 102.1$ rad/s, $k_1 = 4{\cdot}10^7$ N/m, $\omega_{n3}^{(3)} = 142.6$ rad/s, and $k_1 = 6{\cdot}10^7$ N/m, $\omega_{n3}^{(4)} = 174.1$ rad/s. It can be found that in the post-resonance regime for $\omega > \omega_{nj}$, amplitude A_1 tends asymptomatically toward a constant value and stable motion at the value $A_1 = 1.245$ mm, and amplitudes A_2 and A_3 tend toward very small values, of the order 1.87×10^{-3} mm, respectively, 3×10^{-7} rad.

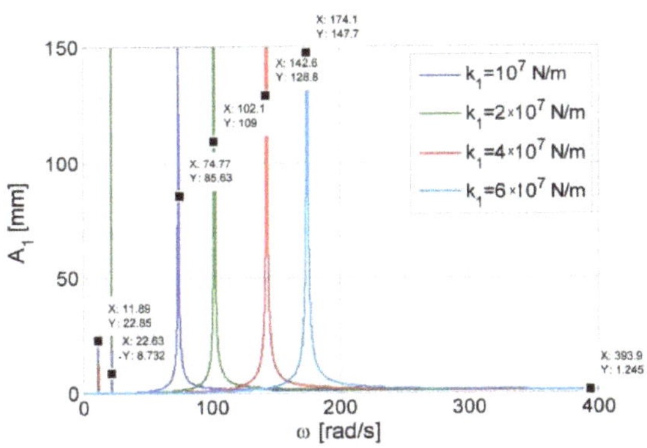

(a)

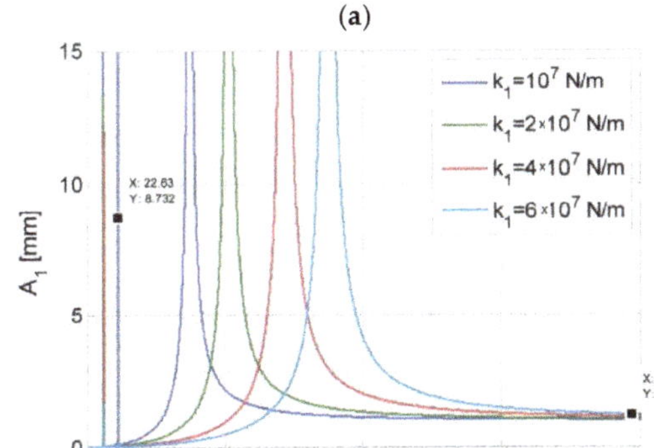

(b)

Figure 2. Family of curves for amplitude A_1. (**a**) Normal representation. (**b**) Enlarged scale representation.

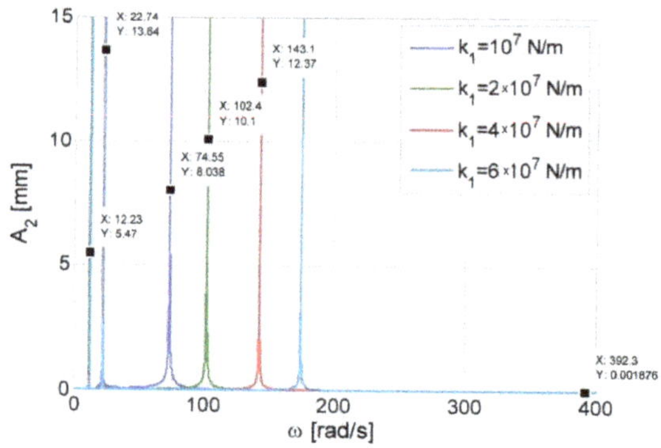

(**a**)

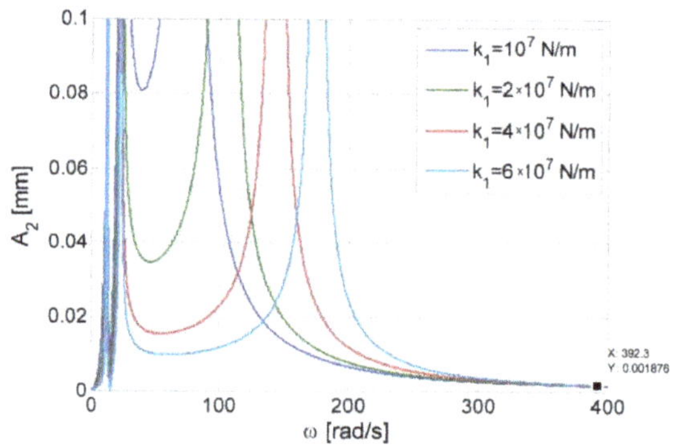

(**b**)

Figure 3. Family of curves for amplitude A_2. (**a**) Normal representation. (**b**) Enlarged scale representation.

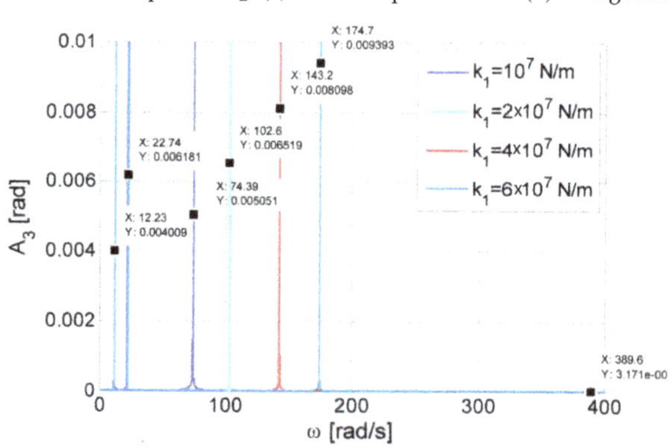

(**a**)

Figure 4. *Cont.*

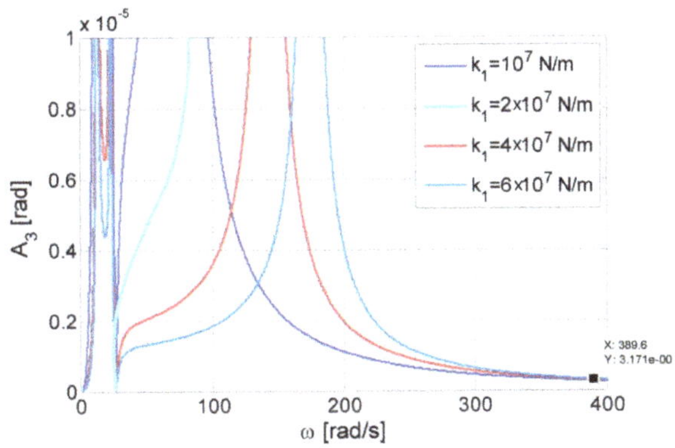

(**b**)

Figure 4. Family of curves for amplitude A_3. (**a**) Normal representation. (**b**) Enlarged scale representation.

In order to determine the resonance pulses to ensure a post-resonance regime, only the significant linear elastic case was considered, obviously with the neglect of the viscous effects.

The low numerical values of amplitudes A_2 and A_3 in post-resonance highlight the fact that the forced vibrations transmitted from body C_1 to body C_2 are negligible.

The amplitude variation curves in Figures 2–4 were numerically elevated for the previously specified parametric data for a towed vibrating roller, with a hydrostatic system for continuously changing the excitation pulsation (i.e., the angular velocity of the eccentric mass of the vibrator). Thus, the resonance pulses were measured for each case, with an accuracy of ±5 Hz compared to the numerically obtained value. A Bosch hydrostatic control system and a Bruel & Kjaer vibration measurement system were used.

4. Conclusions

The structural assembly of a vibrating roller can be modeled as a system of two rigid bodies with linear elastic connections so that two contradictory desiderata can be achieved simultaneously, namely: achieving technological vibrations for body C_1 (vibrating roller) and the significant reduction of the vibrations transmitted to body C_2 (machine chassis) in the control cabin was assembled with the working space for the operating mechanic and the drive unit.

In this context, the modeling of the multi-body system was conducted taking into account the inertial characteristics in direct correlation with the possible and significant movements of the two rigid bodies. Thus, the vertical translational motion of body C_1 of mass m_1 is characterized by a coordinate or a single dynamic degree of freedom that describes the vertical instantaneous displacement.

The motion of the C_2 body is characterized by two degrees of dynamic freedom defined by the x and φ coordinates. They describe the instantaneous vertical translational motion and respectively, the instantaneous angular rotational motion around the horizontal axis passing through the center of gravity of body C_2. In this case, the multibody system is characterized by three degrees of dynamic freedom noted with x_1, x, and φ.

As a result of the dynamic study developed in the paper, based on the numerical analysis and the experimental results obtained on five categories of equipment, the presented model faithfully describes the dynamic behavior of the tested equipment. In this context, the following conclusions can be drawn.

(a) The dynamic model of the multibody system with elastic connections is characterized by the inertia matrix M and by the rigidity matrix K, both symmetrical in relation to the main diagonal;

(b) The elements of inertial coupling $m_{23} = m'b$ and of elastic coupling $-k_2$, ak_2 and $-ak_2 + bk_3$ are found in the differential equations of motion (23) with significant effects on the equation of own pulses (27) and of amplitudes A_1, A_2, A_3 as a dynamic response to the harmonic excitation $F(t) = m_0 r\omega^2 \sin \omega t$.

(c) The numerical and experimental analysis on a vibrating roller equipment, with mass, elastic and excitation data, for the evaluated case study, provides the following conclusions:

- the first two own pulses with relatively low values $\omega_{n1} = 12.23$ rad/s and $\omega_{n2} = 22.24$ rad/s were influenced by the fact that the inertial effect is large enough and rigidity k_2 of the elastic connection system between bodies C_1 and C_2 is low enough for good post-resonance vibration isolation at $\omega > \omega_{n3}$;
- the last own pulse ω_{n3}, is mainly influenced by rigidity k_1 of the compaction soil. Thus, for four distinct values of k_1, which correspond to successive passages on the same layer of road structure, in the compaction process, there emerged four distinct values of the own pulses (resonance) ω_{n3}, in ascending order, as follows: 74.73 rad/s, 102.1 rad/s, 142.6 rad/s, 174.1 rad/s [12].

(d) The family of curves for amplitudes A_1, A_2, and A_3 represented in Figures 2–4 highlights the fact that in the post-resonance regime for $\omega > \omega_{n3}$, amplitude A_1 of the technological vibrations is constant and stable for $\omega \in (300\ldots400)$ rad/s, and amplitudes A_2 and A_3 tend toward small values, assuring the favorable effect of dynamic insulation for body C_2.

(e) The analytical relations (26), (27), and (28) can be used for the parametric optimization of the dynamic response, as follows:

- amplitude A_1 of the technological vibrations, which must be constant and stable at the excitation pulse ω, must meet the post-resonance operating condition $\omega > 1.5\ \omega_{n3}$. Practically, it is recommended that $\omega = 2\omega_{n3}$ to achieve the technological requirements of efficient compaction;
- amplitudes A_2 and A_3 of body C_2 must have low values so that the degree of isolation of the vibrations transmitted from the body C_1 to be $I_v \geq 95\%$; and
- the first two own pulses or resonance circular frequencies must be within the range $(10 \div 60)$ rad/s, so that the influence of the two resonance zones for $\omega = \omega_{n1}$ and $\omega = \omega_{n2}$ becomes negligible for stable optimal operation [13].

Given the above, the analytical approach of the dynamic behavior of multibody systems with effective applications for vibratory rollers for compacting road structures can be useful in the stage of establishing technical design solutions as well as in the parametric optimization stage.

This can be achieved by adjustments and tuning that can be made during the working process such as the discrete change in steps, the static moment $m_0\underline{r}$, and/or the continuous modification of the excitation pulse ω that can be achieved with the hydrostatic actuation of the vibrator [14].

Funding: This research received no external funding.

Conflicts of Interest: The author declare no conflict of interest.

References

1. Bratu, P.; Dobrescu, C.F. Dynamic Response of Zener-Modelled Linearly Viscoelastic Systems under Harmonic Excitation. *Symmetry* **2019**, *11*, 1050. [CrossRef]
2. Dobrescu, C.F. The Zener rheological viscoelastic modelling of the dynamic compaction of the ecologically stabilized soils. *Acta Tech. Napoc.* **2019**, *62*, 281–286.
3. Tonciu, O.; Dobrescu, C.F. Optimal dynamic regimes of vibration compaction equipment. *Acta Tech. Napoc.* **2020**, *63*, 101–106.
4. Shabana, A.A. *Dynamics of Multibody Systems*; Cambridge University Press: Cambridge, UK, 2014.
5. Bratu, P. Dynamique des compacteurs autopropulses avec deux rouleaux vibrateurs. *Rev. Roum. Sci. Tech. Mec. Appl.* **1988**, *33*, 349–366.
6. Pandrea, N.; Stănescu, N.-D. *Dynamics of the Rigid Solid with General Constrains by a Multibody Approach*; John Wiley & Sons: Hoboken, NJ, USA, 2016.
7. Bratu, P. Establishing the equations of motion for the rigid with elastic connections. Studies and researches of applied mechanics, SCMA, Romanian Academy, 1990. Volume 50, no. 6. Available online: https://books.google.ro/books?id=ejEeAQAAMAAJ&q (accessed on 28 September 2020).

8. Bratu, P. Evaluation of quality dynamic performances in case of vibrating rollers used for road works. In *Proceedings of the SISOM 2007 and Homagial Session of the Commission of Acoustics, Bucharest, Romania, 29–31 May 2007*, 1st ed.; Research Institute for Construction Equipment and Technology ICECON SA: Bucharest, Romania, 2008; pp. 91–104.
9. Schiehlen, W. Computational dynamics: Theory and applications of multibody systems. *Eur. J. Mech. A Solids* **2006**, *25*, 566–594. [CrossRef]
10. Huston, R.L. Multibody dynamics—Modeling and analysis methods. *Appl. Mech. Rev.* **1991**, *4*, 109–117. [CrossRef]
11. Wittenburg, J. *Dynamics of Multibody Systems*; Springer: Berlin/Heidelberg, Germany, 2007.
12. Dobrescu, C.F. Analysis of the dynamic regime of forced vibrations in the dynamic compacting process with vibrating roller. *Acta Tech. Napoc. Appl. Math. Mech. Eng.* **2019**, *62*, 1.
13. Dobrescu, C.F. Highlighting the Change of the Dynamic Response to Discrete Variation of Soil Stiffness in the Process of Dynamic Compaction with Roller Compactors Based on Linear Rheological Modelling. *Appl. Mech. Mater* **2015**, *801*, 242–248. [CrossRef]
14. Dobrescu, C.F.; Brăguţă, E. Optimization of Vibro-Compaction Technological Process Considering Rheological Properties. In *Proceedings of the Acoustics and Vibration of Mechanical Structures—AVMS-2017: 14th AVMS Conference, Timisoara, Romania, 25–26 May 2017*; Herisanu, N., Marinca, V., Eds.; Springer: Cham, Switzerland, 2018; pp. 287–293.

symmetry

MDPI

Article

Parameterized Design and Dynamic Analysis of a Reusable Launch Vehicle Landing System with Semi-Active Control

Chen Wang [1,*], Jinbao Chen [1], Shan Jia [1] and Heng Chen [2]

1 Key Laboratory of Exploration Mechanism of the Deep Space Planet Surface, Ministry of Industry and Information Technology, Nanjing University of Aeronautics and Astronautics, Nanjing 211100, China; chenjbao@nuaa.edu.cn (J.C.); jiashanazz@nuaa.edu.cn (S.J.)

2 Field Engineering College, Army Engineering University of PLA, Nanjing 210001, China; hengchen.123@nuaa.edu.cn

* Correspondence: nuaawangchen@nuaa.edu.cn

Received: 17 August 2020; Accepted: 17 September 2020; Published: 22 September 2020

Abstract: Reusable launch vehicles (RLVs) are a solution for effective and economic transportation in future aerospace exploration. However, RLVs are limited to being used under simple landing conditions (small landing velocity and angle) due to their poor adaptability and the high rocket acceleration of current landing systems. In this paper, an adaptive RLV landing system with semi-active control is proposed. The proposed landing system can adjust the damping forces of primary strut dampers through semi-actively controlled currents in accordance with practical landing conditions. A landing dynamic model of the proposed landing system is built. According to the dynamic model, an light and effective RLV landing system is parametrically designed based on the response surface methodology. Dynamic simulations validate the proposed landing system under landing conditions including the highest rocket acceleration and the greatest damper compressions. The simulation results show that the proposed landing system with semi-active control has better landing performance than current landing systems that use passive liquid or liquid–honeycomb dampers. Additionally, the flexibility and friction of the structure are discussed in the simulations. Compared to rigid models, flexible models decrease rocket acceleration by 51% and 54% at the touch down moments under these two landing conditions, respectively. The friction increases rocket acceleration by less than 1%. However, both flexibility and friction have little influence on the distance between the rocket and ground, or the compression strokes of the dampers.

Keywords: reusable launch vehicles; soft landing; magnetorheological fluid; numerical simulation; multibody systems with flexible elements

1. Introduction

As one of the most important technologies for aerospace exploration, advances in launch vehicles have greatly promoted aerospace developments [1]. Reusable launch vehicles (RLVs) can achieve fast and cheap launches by dividing the launch costs into several launch missions. Since the 1950s, many countries have focused on developing RLVs. From American X-series spacecraft to the Falcon-series rockets of SpaceX, RLVs have always been a hot topic in aerospace technology [2].

The landing system is a critical subsystem of RLVs, the malfunction of which can cause recycle failure [3]. The design of RLV landing systems is quite difficult, because it requires high reusability, effective impact absorption, reliability, and heat resistance [4]. The Delta Clipper proposed by the McDonnell-Douglas Corporation was the first RLV to use a vertical soft-landing system. The Delta Clipper was to be a single-stage-to-orbit vehicle that took off vertically and landed vertically [5].

However, the project was aborted due to lack of funding, and it has not been used in practical engineering. The New Shepard proposed by the Blue Origin Corporation completed a sub-orbital experiment and landed vertically via retractable landing legs. However, it was only just able to reach the Kármán Line (100 km a.s.l.) [6]. The Falcon-series rockets proposed by SpaceX were the first to realize the application of vertical landing in aerospace missions. The landing gears of the Falcon-series rockets use four sets of landing gears with liquid dampers that include primary struts, auxiliary struts, and locking mechanisms [7–9]. The Ariane Group, the French Space Agency, and the Deutsches Zentrum für Luft- und Raumfahrt have also proposed a new low-cost reusable rocket project called Callisto. Callisto uses four landing legs to absorb the impact energy, which is similar to the Falcon-series rocket [10,11]. However, it is still in the design stage. Yue [12–15] and Lei [16] proposed a vertical landing system with novel oleo–honeycomb dampers and conducted many landing experiments. The oleo–honeycomb dampers were able to improve the landing performance of the RLV under dangerous conditions. However, it was necessary to replace the aluminum honeycomb after every landing. In summary, the current landing systems employed by RLVs use passive liquid or liquid–honeycomb dampers to absorb impact energy. These two kinds of passive dampers have complex structures, as well as greater mass and rocket accelerations. Additionally, they are not able to adjust damping forces in order to meet practical landing conditions, which require a low landing velocity and angle, making the recycling of rockets more difficult. There is limited research on controllable landing systems in RLVs.

In this paper, an RLV landing system with semi-active control is proposed. Its dynamic landing model, control approach, and parameterized design are introduced. Furthermore, the proposed landing system is validated by means of multiple rigid bodies and multiple coupled flexible–rigid dynamic simulations. The influence of structural flexibility and friction during the RLV landings is also discussed in the dynamic simulations. Section 2 introduces the overall scheme, working principles, and control approach of the proposed RLV landing system. Section 3 introduces the landing dynamic model and parameterized design of the RLV landing system. Section 4 validates the proposed landing system and discusses the influence of structural flexibility and friction on the RLV landing performance with reference to the simulations.

2. Working Principles of the RLV Landing System

2.1. Overall Scheme of the RLV Landing System with MRF Dampers

Figure 1 shows the overall scheme of the RLV landing system with magnetorheological fluid (MRF) dampers, which consists of a rocket and four sets of landing gears. Each set of landing gear includes a primary strut (including a damper and deployment), an auxiliary strut, and a pad. The primary strut damper is full of MRF, whose profile is shown in Figure 2. The primary strut damper includes a master cylinder (the inner diameter is D), a piston (compose of a magnetic core and a group of coils, the length is L), a piston rod (the diameter is d), a nitrogen accumulator, and a gap between the piston and cylinder (the width is h). The magnetic properties of the primary strut materials are shown in Table 1. The MRF-132DG produced by the Lord Company is used as an example in this paper. Its main specifications are shown in Table 2.

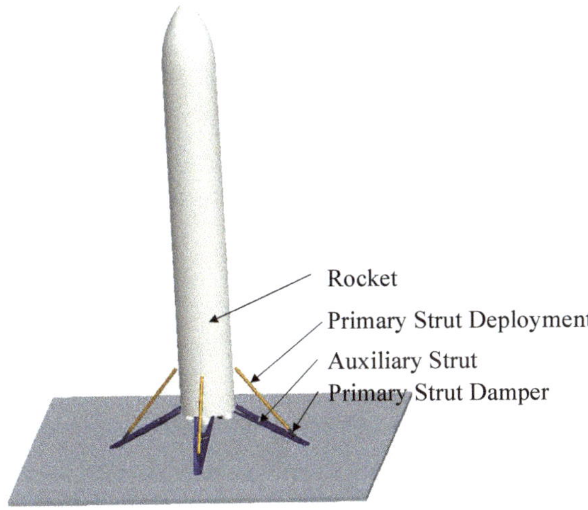

Figure 1. The overall scheme of the RLV landing system with MRF dampers.

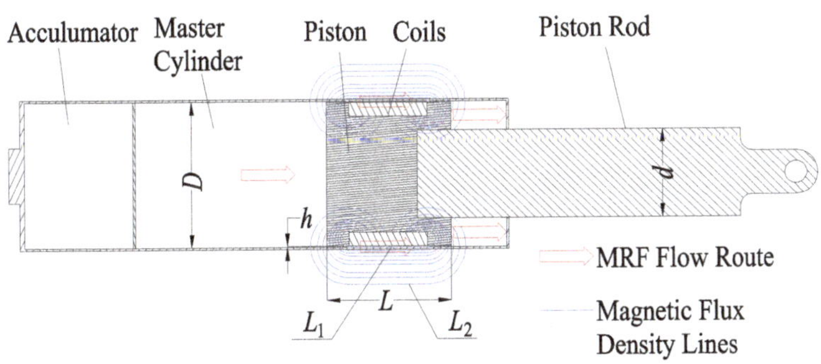

Figure 2. The cross-section diagram of the primary strut damper.

Table 1. The magnetic properties of primary strut materials.

Structure	Material	Relative Permeability	Conductivity (S/m)	Relative Permittivity
Cylinders	Aluminum	1	3.774×10^7	1.000022202
Magnetic core	AISI1010	500	6.452×10^6	1
Coils	Copper	1	5.998×10^7	0.9999935542
Gas in the accumulator	Nitrogen	1	0	1

Table 2. Main specifications of MRF-132DG.

Parameters	Value
Viscosity	0.112 Pa s
Density	2.95 g/cm^3
Solid content by Weight	20.98%
Maximum Yield Stress	48 kPa
Operating Temperature	−40~+130 °C

There are N turns coils on the magnetic core of the piston. The length of the magnetic flux density lines along the coils is L_1, the length of the magnetic flux density lines in the gap is h, and the length of the arched magnetic flux density lines out the cylinder is L_2. Based on the Maxwell equation and Ampere circuit rule,

$$\int_{L_1} B_1 \cdot dl + \int_{2h} B_2 \cdot dl + \int_{L_2} B_3 \cdot dl = \mu_m NI \quad (1)$$

where B_1 is the magnetic flux density in the magnetic core area, B_2 is the magnetic flux density in the gap, B_3 is the magnetic flux density out the cylinder, and μ_m is the magnetic constant. The magnetic flux density B is in T, the magnetic field intensity H is in A/m, the electric current I is in A. Due to the magnetic field intensity in the magnetic core area and gap is much larger than that out the cylinder, the magnetic field intensity in the gap is

$$H_2 = \frac{B_2}{\mu_m} = \frac{NI - H_1 L_1}{2h} \quad (2)$$

The relationship between yield stress and magnetic field intensity of MRF-132DG [17,18] is

$$\tau = 2.717 \times 10^5 \times 0.32^{1.5239} tanh(6.33 \times 10^{-3} \cdot H_2) \quad (3)$$

During the landing, the viscosity and plasticity of the MRF change quickly under the magnetic fields produced by energized coils. When the piston moves and pushes the MRF to flow through the gap between the cylinder and piston, the coils energize and produce magnetic fields. The MRF becomes semi-solid from the liquid in milliseconds, and its yield stress is controlled by different magnetic fields to absorb the impact energy. After landing, MRF will return to the liquid state without the magnetic fields [19].

2.2. Working Principles of the RLV Landing System

When the rocket approaches the recycle-platform, four sets of landing gears deploy simultaneously and prepare for the landing impact. After the sensors of pads touch the recycle-platform and the RLV entries landing state, four primary strut dampers absorb the impact energy by their compressions and extensions. Their damping forces of primary strut dampers consist of the controllable parts and uncontrollable parts. The uncontrollable parts are determined by the viscosity and velocity of the MRF, and the air-spring forces of accumulators. The controllable parts are related to the yield stress of MRF, which are controlled according to the acceleration, jerk, pitch angle, and roll angle of the rocket [20].

2.3. Control Approach of the RLV Landing System

Due to the landing process is quite short, which requires a fast and robust control approach. Fuzzy control is suitable for complex systems and can decrease the response time significantly. Furthermore, the nonlinear characteristics of fuzzy control can increase the system robustness [21–24].

During the landing, the acceleration, jerk, pitch angle, and roll angle of the rocket are set as inputs. These four inputs are from rocket sensors to the control system of damping forces. Meanwhile, the yield stresses of four primary struts controlled by currents are set as outputs. The currents can control the damping forces of every primary strut, respectively. These four outputs are from the control system of damping forces and act on four primary strut dampers.

The highest acceleration of the RLV should be smaller than 2 g to protect the precise electronic instruments. Its jerk is set as $[-2a_{max}, 2a_{max}]$. The pitch angle and roll angle are set as [–3, 3], due to landing angles of the current RLVs are from $-3°$ to $3°$ [2]. The output yield stresses of MRF in four primary strut dampers are set as [0, 100% Maximum] to adapt to different landing conditions. Considering the control accuracy and efficiency, the acceleration a is divided into four equal fuzzy sets (Z, S, M, B). The jerk da is divided into two equal fuzzy sets (Z, B). Both the pitch angle $alpha$ and roll angle $beta$ are divided into three equal fuzzy sets (N, Z, P). The output yield stresses are divided into seven equal fuzzy sets (Z, S, SM, M, SB, MB, B). The membership affiliations between physical parameters and fuzzy sets for the inputs are shown in Figure 3. The membership affiliations between physical parameters and fuzzy sets for the outputs are shown in Figure 4.

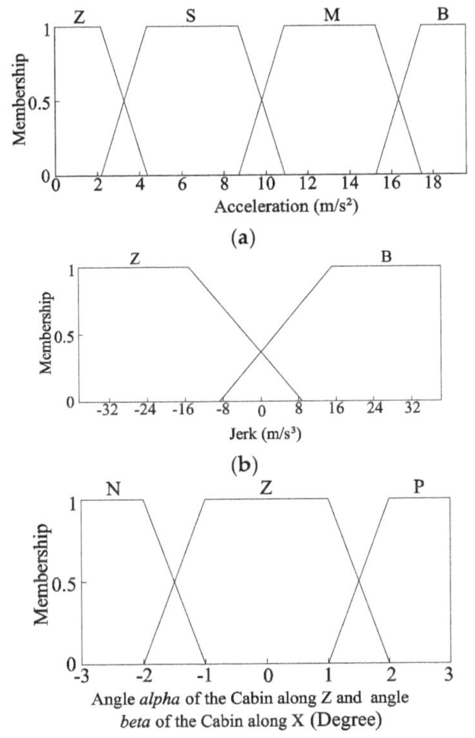

Figure 3. Membership Affiliations of inputs: (**a**) Membership affiliation of *a*; (**b**) Membership affiliation of *da*; (**c**) Membership affiliations of *alpha* and *beta*.

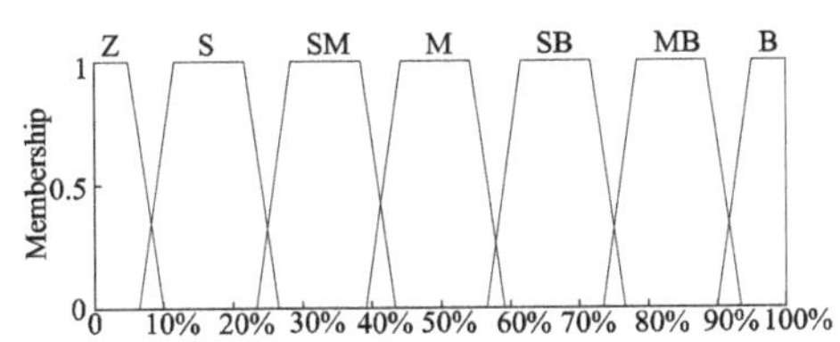

Figure 4. Membership Affiliations of outputs.

The control principles are (a) While the acceleration is increasing and less than the setting value, controllable damping forces are small. (b) While the acceleration is increasing and larger than the setting value, controllable damping forces are zero. (c) While the acceleration is decreasing and larger than the setting value, controllable damping forces are zero. (d) While the acceleration is decreasing and less than the setting value, controllable damping forces are big. Moreover, the output damping forces are also determined by the pitch angle and roll angle of the rocket. Detailed fuzzy control rules for inputs and outputs are shown in the Appendix A.

3. Landing Dynamic Analysis and Parameterized Design of RLV Landing System

3.1. Landing Dynamic Analysis of the RLV Landing System

Based on the working principles and control approach of the proposed RLV landing system, its landing dynamic model is required to design the detailed structures. The RLV is composed of an elastic part and four non-elastic parts, as shown in Figure 5. The elastic part includes the rocket, four primary strut deployments, and cylinders of four primary strut dampers. The non-elastic parts include piston rods of four primary strut dampers, four auxiliary struts, and four pads [25]. Due to the RLV being symmetric, a quarter landing dynamic model of the RLV is built, as shown in Figure 6.

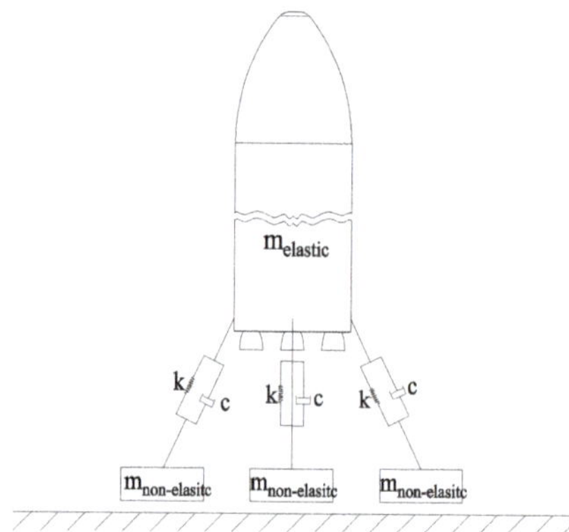

Figure 5. Elastic and non-elastic parts of the RLV.

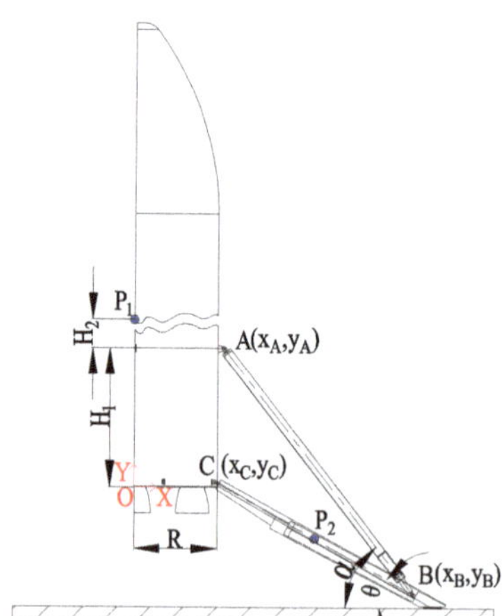

Figure 6. Quarter landing dynamic model of the RLV landing system.

The coordinate system is at the center of the bottom surface of the rocket. The revolute joint between the primary strut and rocket is A (x_A, y_A). The sphere joint between the primary strut and auxiliary strut is B (x_B, y_B). The projection of the revolute joint between the auxiliary strut and rocket is C (x_C, y_C). The horizontal distance x_C between the origin and C is R. The angle between the primary strut and ground is α. The angle between the auxiliary strut and ground is θ. The vertical distance y_A between the origin and A is H_1. The mass center of the elastic part is P_1 (0, $H_1 + H_2$). Due to auxiliary struts occupy most mass of non-elastic parts, whose center can be simplified as the mass center of non-elastic parts. It is 0.5($x_B + x_C, y_B + y_C$), which is shown as

$$\mathrm{P2}\left(\frac{1}{2}\frac{H_2}{\tan\alpha-\tan\theta}+R,\frac{1}{2}\left(3H_1+H_2+R^2\tan\alpha+\frac{H_2R\tan\alpha}{\tan\alpha-\tan\theta}\right)\right) \tag{4}$$

The landing dynamic models of elastic parts are

$$\begin{cases} m_{\mathrm{P}_1}\ddot{x}_{\mathrm{P}_1} = -F_{\mathrm{p}}\cos\alpha - F_{\mathrm{a}}\cos\theta \\ m_{\mathrm{P}_1}\ddot{y}_{\mathrm{P}_1} = F_{\mathrm{p}}\sin\alpha + F_{\mathrm{a}}\sin\theta - m_{\mathrm{P}_1}g \end{cases} \tag{5}$$

where F_{p} is the damping force of the primary strut, F_{a} is the damping force of the auxiliary strut. The landing dynamic models of non-elastic parts are

$$\begin{cases} m_{\mathrm{P}_2}\ddot{x}_{\mathrm{P}_2} = F_{\mathrm{p}}\cos\alpha + F_{\mathrm{a}}\cos\theta - \mu F_{\mathrm{n}} \\ m_{\mathrm{P}_2}\ddot{y}_{\mathrm{P}_2} = F_{\mathrm{n}} - F_{\mathrm{p}}\sin\alpha - F_{\mathrm{a}}\sin\theta - m_{\mathrm{P}_2}g \end{cases} \tag{6}$$

where μ is the friction coefficient. F_{n} is the contact force between the pad and ground, shown as follows [26]

$$F_{\mathrm{n}} = \begin{cases} 0, \ q > q_0 \\ k(q_0-q)^e - c\dot{q}step(q, q_0-d, 1, q_0, 0), \ q \le q_0 \end{cases} \tag{7}$$

where q is the distance criterion of the impact function, q_0 is the trigger distance of the impact function. k is the stiffness, e is the contact force exponent, c is the contact damping, and d is the penetration depth.

According to the cross-section diagram of the primary strut damper in Figure 2, the damping force of the primary strut damper F_{p} is

$$\begin{aligned} \mathrm{F_p} &= F_{\mathrm{c}} + F_{\mathrm{u}} + F_{\mathrm{Ni}} \\ &= \tfrac{3\pi D^2 L\tau}{4h} + \tfrac{3\eta L[\pi(D^2-d^2)]^2}{4\pi D h^3}v + 0.5\rho K_{\mathrm{entry}}\tfrac{{A_{\mathrm{p}}}^4}{{A_{\mathrm{gap}}}^2}v^2 \\ &+0.5\rho K_{\mathrm{exit}}\tfrac{{A_{\mathrm{p}}}^4}{{A_{\mathrm{gap}}}^2}v^2 + A_{\mathrm{n}}P_0\left(\tfrac{V_0}{V}\right)^{1.1} \end{aligned} \tag{8}$$

where F_{u} is the uncontrollable damping force of the MRF damper. F_{c} is the controllable damping force of the MRF damper. F_{Ni} is the air-spring force caused by the accumulator [27–29]. F_{entry} is local resistance caused by the abrupt enlargement, and F_{exit} is local resistance caused by the abrupt contraction. ρ is the density of MRF. K_{entry} is the local resistance coefficient of the entry, and K_{exit} is the local resistance coefficient of the exit. v is the piston velocity. A_{p} is the piston area. A_{gap} is the gap area between the master cylinder and piston, and A_{n} is the cross-section area of the master cylinder. P_0 is the initial pressure of the accumulator. V_0 is the initial volume of the accumulator, and V is the volume of the accumulator during the landing.

3.2. Parameterized Design of the RLV Landing System

According to the proposed landing dynamic model, H_1, α, and θ determine the buffer effects of F_{a} and F_{p}, and the efficiency and performance of landing systems [30]. Hence, a parameterized design of the RLV landing system is proposed according to these three parameters to get an effective landing system. The lower and upper limits of these three parameters are given in Table 3. The rocket acceleration, compression strokes of dampers, and the distance between the rocket and ground are the most important indexes for the design of a landing system [31]. A large rocket acceleration will damage structures and instruments [32]. Large compressions of primary strut dampers will cause the rocket to incline or tip over. The distance between the rocket and the ground should be large enough for a safe landing [33]. The mass is also an important index for spacecraft, a lighter landing system means a lower launch cost. Hence, these four design targets of the landing system are selected as responses, as shown in Table 4. The parameterized design principle based on the response surface methodology (RSM) is shown as follows

$$\begin{cases} \text{Minimize}\{D\} = \text{Min}\left\{\sqrt[4]{R_1R_2(R_{P_1} - R_3)R_4}\right\} \\ \quad s.t.\begin{cases} 24 \le \theta \le 30 \\ 42 \le \alpha \le 53 \\ 1200 \le H_1 \le 1800 \end{cases} \end{cases} \tag{9}$$

where R_{P1} is the initial distance between the mass center of the rocket and the ground.

Table 3. Lower and upper boundaries of parameters [34,35].

Codes	Design Parameters	Lower Limits	Upper Limits
A	The angle between the auxiliary strut and ground (θ)	24°	30°
B	Angle between the primary strut and ground (α)	42°	53°
C	The vertical distance between point A and C (H_1)	1200 mm	1800 mm

Table 4. Design target parameters.

Responses	Design Targets	Goal
R_1	Highest rocket acceleration (m/s^2)	Minimize
R_2	Greatest compression stroke (mm)	Minimize
R_3	Distance between rocket and ground (mm)	Maximum
R_4	Mass of a set of landing gear (kg)	Minimize

$\{D\}$ is the desirability function, which shows the desirable ranges for each response R_i. The function combines these four responses in a non-dimensional way. Its design goal is the smallest rocket acceleration, compression stroke, the mass of a set of landing gear, and the largest distance between the rocket and ground.

The RSM builds an approximate model between the codes (design parameters) and responses (design targets) via function fitting. The RSM assumes every code is an n-dimensional vector $x \in E^n$, which is the independent variable of its response function y. Their relationship is $y = f(x)$. Based on lots of simulation data, an approximate function of the response $\widetilde{y}$ is obtained by the undetermined coefficient method. Considering the efficiency and accuracy, a quadratic function with cross terms is used, which is shown as follows

$$\widetilde{y} = a_0 + \sum_{j=1}^{n} a_j x_j + \sum_{i=1}^{n}\sum_{j=1}^{n} a_{ij} x_i x_j \tag{10}$$

where a_0 is the undetermined coefficient of the constant term, a_j is the undetermined coefficient of the one-degree term, and a_{ij} is the undetermined coefficient of the quadratic term. $\widetilde{y}$ is close to y by keeping their sum of error squares smallest via the least square principle [36].

According to the ranges of the three parameters (factors) in Table 3, landing dynamic simulations are carried to get corresponding four design targets (responses) under different parameter combinations. Their results are the RSM sampling, as shown in the appendix. Based on the RSM sampling and fit function in Equation (10), accurately fitted functions between codes and responses are obtained by the undetermined coefficient method, as shown in Table 5. These code coefficients of functions show the influences of codes on responses [37]. The influences of three codes and their extended codes on R_1 is C > B > A > BC > AB > A^2 > C^2 > AC > B^2. The influences of three codes and their extended codes on R_2 is B^2 > A^2 > A > AC > C^2 > BC > C > B > AB. The influences of three codes and their extended codes on R_3 is A > C > AC > C^2 > B^2 > A^2> AB > BC > B. The influences of three codes and their extended codes on R_4 is C > A > AC > A^2, B^2, C^2 > BC > AB > B.

Table 5. Fitted functions between codes and responses.

Response	Fitted Functions
R_1	$R_1 = +61.879 + 0.531\mathrm{A} + 2.978\mathrm{B} + 5.535\mathrm{C} - 4.194\mathrm{AB} - 13.779\mathrm{AC} - 0.248\mathrm{BC} - 8.078\mathrm{A}^2 - 25.670\mathrm{B}^2 - 10.901\mathrm{C}^2$
R_2	$R_2 = +21.472 + 7.281\mathrm{A} - 14.109\mathrm{B} - 6.472\mathrm{C} - 24.521\mathrm{AB} + 7.265\mathrm{AC} - 0.111\mathrm{BC} + 24.440\mathrm{A}^2 + 34.093\mathrm{B}^2 + 6.040\mathrm{C}^2$
R_3	$R_3 = +1121.000 + 317.666\mathrm{A} - 520.112\mathrm{B} + 263.320\mathrm{C} - 118.180\mathrm{AB} + 194.039\mathrm{AC} - 130.021\mathrm{BC} - 35.514\mathrm{A}^2 + 40.594\mathrm{B}^2 + 129.553\mathrm{C}^2$
R_4	$R_4 = +87.578 + 3.815\mathrm{A} - 11.215\mathrm{B} + 5.924\,\mathrm{C} - 3.3477\mathrm{AB} + 3.329\mathrm{AC} - 2.125\mathrm{BC}$

Based on these four functions, the predicted values versus actual values of four responses are shown in Figures 7–10. The points above or below the line indicate that they are over or under prediction. The data points of plots are randomly scattered along the 45° oblique line, which suggests that these four functions are accurate. These fitted functions can provide powerful support for the following parameterized design.

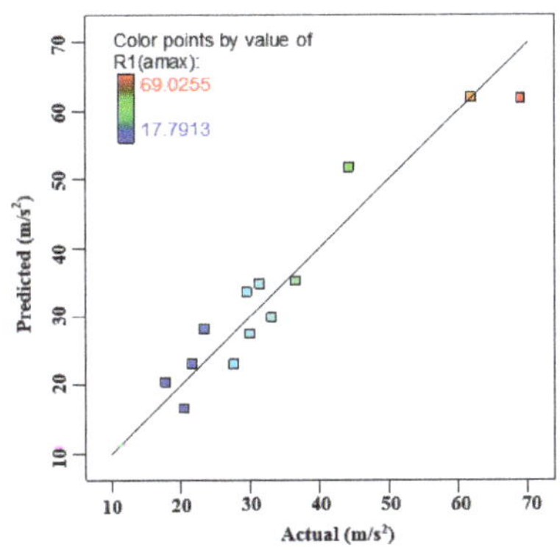

Figure 7. Predicted versus actual values of the highest rocket acceleration.

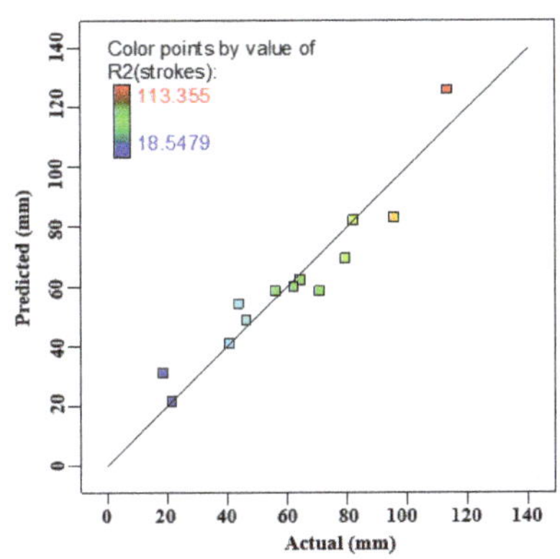

Figure 8. Predicted versus actual values of damper greatest compression stroke.

Combining the design principle in Equation (10) and the fitted functions in Table 5, the final design result based on RSM is shown in Table 6.

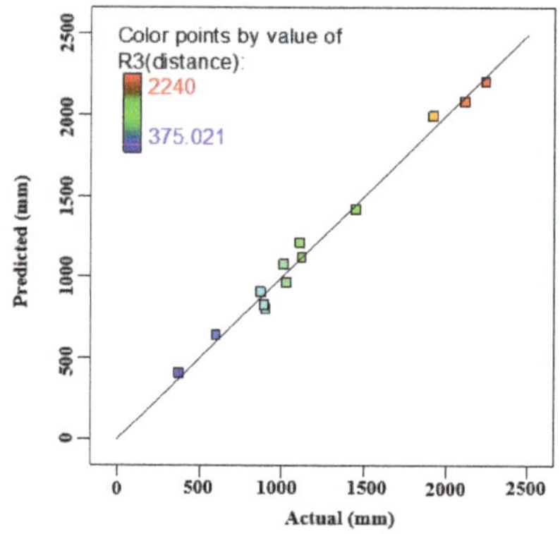

Figure 9. Predicted versus actual values of the distance between rocket and ground.

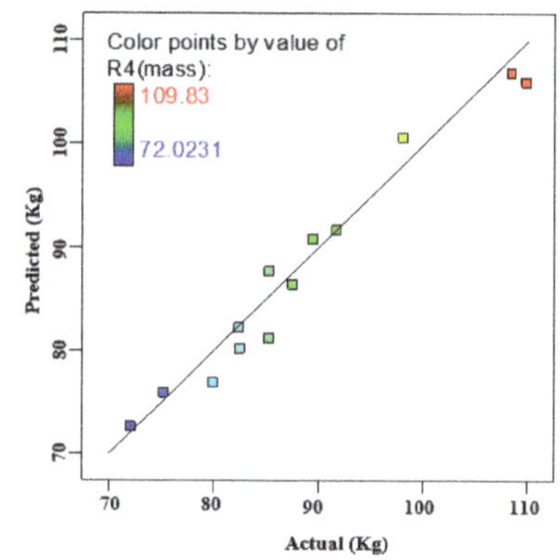

Figure 10. Predicted versus actual values of mass.

Table 6. Final design parameters and targets.

Parameters (Design Parameters)			Responses (Design Targets)			
A (θ)	B (α)	C (H_1)	Highest rocket acceleration	Greatest compression stroke	Distance between rocket and ground	Mass
29.58°	52.42°	1800 mm	17.79 m/s^2	47.31 mm	1286.82 mm	85.12 Kg

4. Landing Dynamic Simulations

Based on the design parameters in Section 3, a dynamic model of the RLV with semi-active control is built in MSC Adams to validate the proposed landing system, as shown in Figure 11. The rocket diameter is 2250 mm. The entire RLV weighs 5200 kg. Its center of mass is located at (0, 6017 mm, 0). The coordinate system is at the center of the bottom surface of the rocket, as shown in Figure 6. The highest rocket acceleration and greatest damper compression conditions are selected as examples because they are two of the most important design parameters of landing gears. These two conditions are shown in Figure 12. Their motion parameters are shown in Table 7.

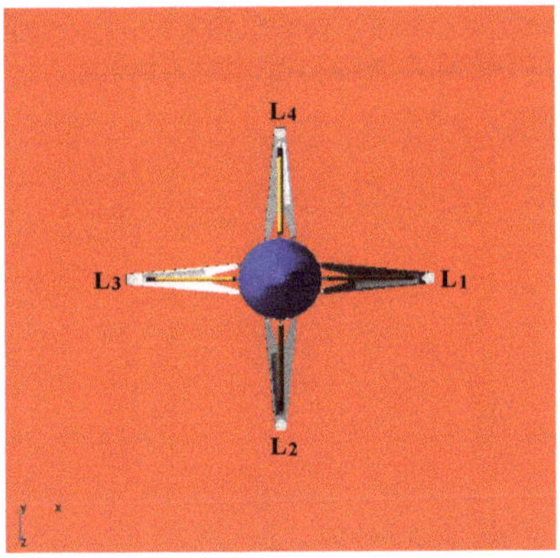

Figure 11. Top view of the RLV.

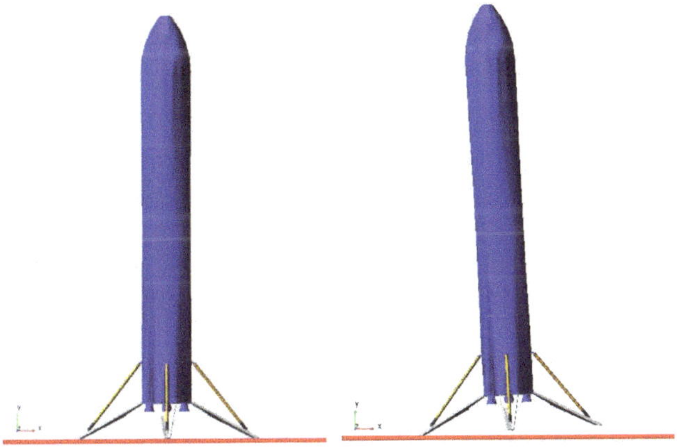

Figure 12. Two typical landing conditions of the RLV.

Table 7. Parameters of two typical landing conditions.

Landing Condition	Vertical Velocity	Horizontal Velocity	Pitch Angle
Highest acceleration	−2 m/s	1 m/s	0°
Greatest compression	−2 m/s	1 m/s	3°

Furthermore, the influences of structural flexibility and friction on landing performance are discussed in dynamic simulations. The end centers of the primary strut deployments and damper cylinders are fixed in their modal analysis. Their 20 order models are calculated in MSC Patran to obtain flexible primary struts. The flexible structures are imported into MSC Adams to conduct multiple coupled flexible–rigid dynamic simulations. Structural flexibility and friction will influence rocket acceleration, energy absorption, and compression strokes [38]. Different combinations of rigid structures, flexible structures, and frictions are simulated to analyze the proposed RLV landing system more accurately.

4.1. Highest Rocket Acceleration Condition

Under the highest rocket acceleration condition, four sets of landing gears touch the ground at the same time. The accelerations and the distances between the rocket and the ground are shown in Figures 13 and 14, respectively. L_1 is taken as an example, whose damping forces and damper compression strokes are shown in Figures 15 and 16, respectively.

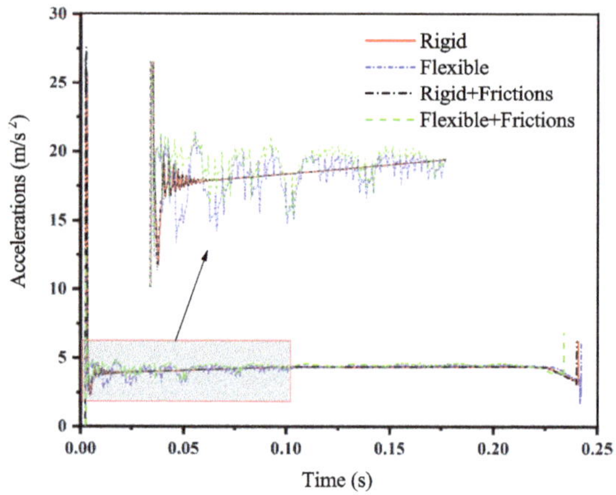

Figure 13. Rocket accelerations of the RLV under the highest acceleration landing condition.

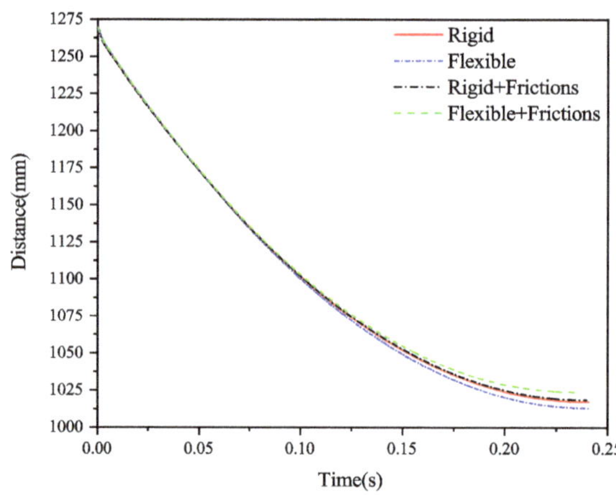

Figure 14. Distance between rocket and ground under the highest acceleration landing condition.

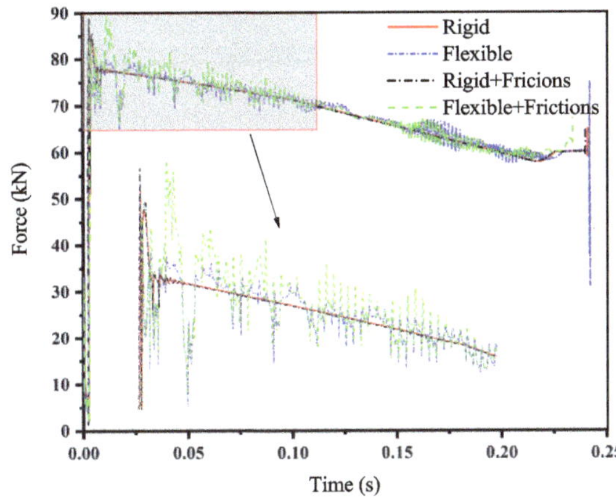

Figure 15. Damping forces of L_1 of the RLV under the highest acceleration landing condition.

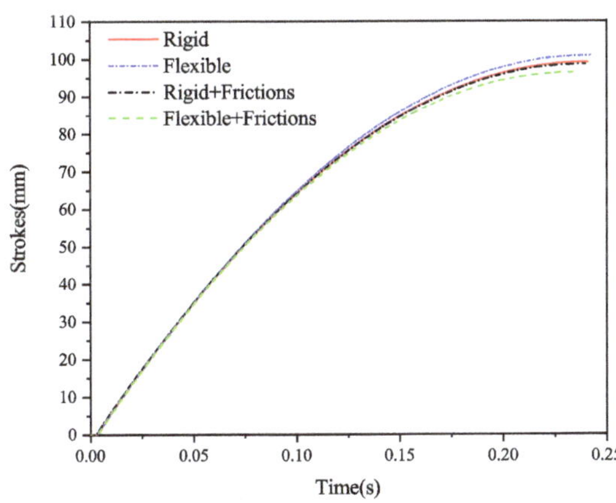

Figure 16. Damper compression strokes of L_1 of the RLV under the highest acceleration landing condition.

Figure 13 shows that all rocket accelerations of these four situations possess the same tendency. At about 0.003 s, four pads touch the ground, and peaks appear vertically. Subsequently, the rocket accelerations decrease vertically and remain at about 4 m/s^2. During the landing, the controllable damping forces F_c of four primary strut dampers belong to Z and S. Their uncontrollable damping forces F_u slowly decrease versus time due to compression velocity decrease. The highest rocket acceleration of the entire rigid model is 25.95 m/s^2. The highest rocket acceleration of the entire rigid model with friction is 27.57 m/s^2, which is the largest in these four situations. The highest rocket acceleration of the model with flexible primary struts is 12.73 m/s^2, which is the smallest in these four situations. Additionally, structural flexibility causes fluctuations in acceleration and damping force. Adding friction to the flexible model, the highest rocket acceleration increases to 13.13 m/s^2, and the fluctuations of the rocket acceleration and damping force also increase. The highest rocket accelerations for these two flexible situations decrease by about 51% at the touch down moment. At the same time, the damping force peaks of L_1 decrease by about 5%, because the flexible structures absorb parts of the impact energy. However, after the instantaneous contact, rocket accelerations and damping forces of these four situations are close to each other.

The highest rocket acceleration of current landing systems with passive liquid dampers is 37.2 m/s^2 under the highest acceleration landing condition [2,17]. Compared to this, the highest rocket acceleration of the proposed landing system with semi-active control decrease about 30.2%. By controlling the damping forces of the four primary strut dampers, the RLV has much lower rocket accelerations and impact forces, which can protect the structures and instruments better during rocket recycle. As shown in Figures 14 and 16, the distance between the rocket and the ground and the compression strokes of L_1 are close to each other in these four situations. In conclusion, friction has little influence on landing performance. However, structural flexibility has a strong influence on rocket acceleration and the damping forces of primary struts.

4.2. Greatest Damper Compressions Condition

Under the greatest damper compressions condition, L_3 touches the ground first. Second, L_2 and L_4 touch the ground together. Finally, L_1 touches the ground. In brief, it is a kind of 1–2–1 landing condition. The rocket accelerations and the distances between the rocket and the ground are shown in Figures 17 and 18. Because L_3 touches the ground first, L_3 is taken as an example. The damping forces and compression strokes of L_3 are shown in Figures 19 and 20.

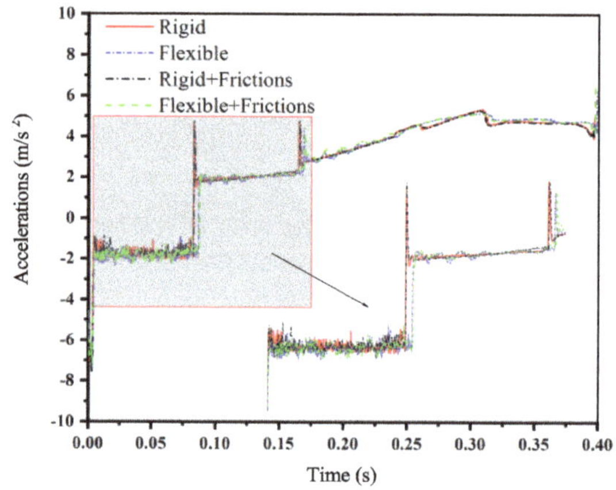

Figure 17. Accelerations of RLV under the greatest compressions landing condition.

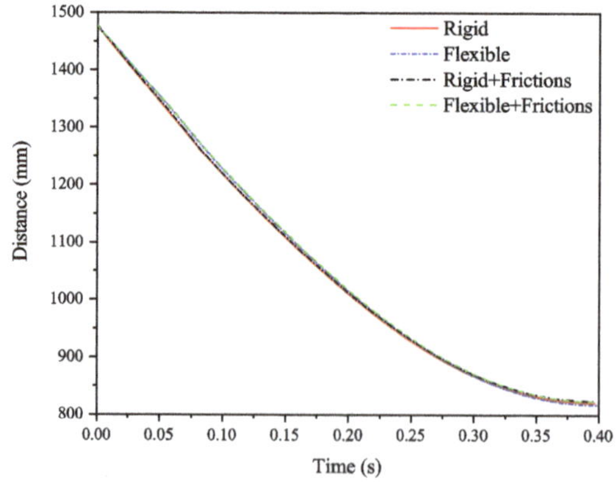

Figure 18. Distance between rocket and ground under the greatest compressions landing condition.

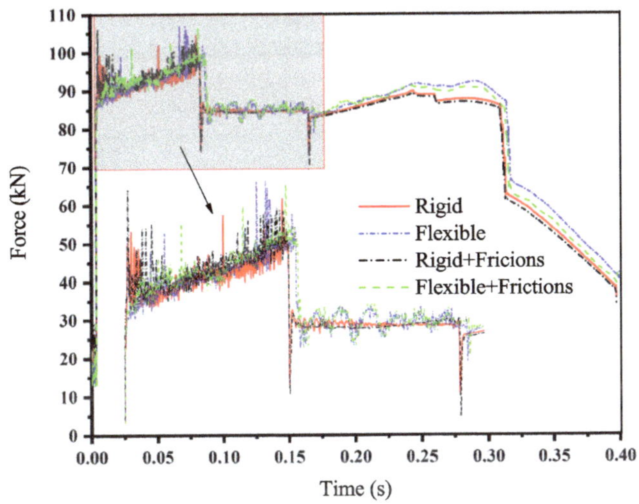

Figure 19. Force of primary strut 1 of RLV under the greatest compressions landing condition.

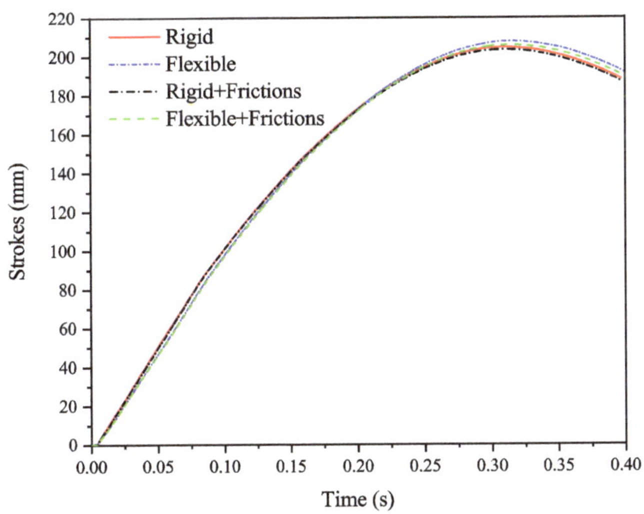

Figure 20. Strokes of primary strut 1 of RLV under the greatest compressions landing condition.

At about 0.005 s, L_3 touches the ground, and rocket accelerations and damping forces increase vertically. At about 0.085 s, L_2 and L_4 touch the ground at the same time. The rocket accelerations increase vertically again. At about 0.167 s, L_1 touches the ground, and the rocket accelerations increase vertically for a third time. From 0.167 s to 0.310 s, the controllable damping force F_c of L_1 belongs to Z, and the controllable damping forces F_c of L_2, L_3, and L_4 belong to S. Their uncontrollable damping forces F_u decrease versus time slowly due to the decrease of compression velocities. The air-spring forces F_{Ni} increase because damper compressions increase. Hence, their resultant forces remain basically stable. Additionally, the pitch angle gradually decreases to 0 due to the horizontal velocity and control of the damping forces. The rocket accelerations increase slightly during this time. After 0.310 s, four controllable damping forces F_c belong to Z together, and the rocket accelerations have a small vertical decrease.

Under the greatest damper compressions condition, the compression strokes of the proposed landing system with semi-active control are close to those of current landing systems. However, the highest rocket acceleration of current landing systems with passive liquid dampers is about 22.5 m/s^2 [2]. The highest rocket acceleration of the proposed landing system with a rigid model is 5.37 m/s^2, which is a decrease of about 76.1%. Additionally, the highest rocket accelerations also decrease by about 54% at the three touch down moments in these two flexible situations. At the same time, damping forces decrease by about 11% in these two flexible situations. Except for the three touch down moments, the rocket accelerations and damping forces of flexible situations are a little higher than those of rigid situations. Structural flexibility also causes an approximately 0.005 s delay, and fluctuations of rocket accelerations and damping forces. The compression strokes of primary strut dampers and distances between the rocket and ground are also close in these four situations.

In conclusion, these two typical landing conditions prove that the proposed landing system has better landing performance than current landing systems with passive liquid or liquid–honeycomb dampers. On one hand, structural flexibility decreases rocket acceleration and damping force. On the other hand, friction increases rocket acceleration and damping force a little. Both flexibility and friction have little influence on the compression strokes of the primary strut dampers and the distances between the rocket and the ground. Structural flexibility should be considered in the design of RLV landing systems.

5. Conclusions

A landing system for reusable launch vehicles with semi-active control was proposed in this paper. Its control approach and landing dynamic model were built. According to the dynamic model, an effective and light landing system was parametrically designed based on the response surface methodology. The parameterized design achieved the best-desired design targets under limited ranges of design parameters, guiding the design by fitted functions between design parameters and targets. The parameterized design provided a fast and high-efficiency approach to designing a landing system. Dynamic landing simulations validated the proposed landing system under landing conditions with the highest rocket acceleration and greatest damper compressions. The simulation results proved that the proposed landing system with semi-active control has better landing performance than currently available landing systems that use passive liquid or liquid–honeycomb dampers. Additionally, the simulation results show that structure flexibilities decrease rocket accelerations by about 50% at the touch down moments. At the same time, they also decrease the damping forces of the primary strut dampers by 5% and 11% at the touch down moments under two typical conditions. However, the friction has little influence on landing performance.

Author Contributions: Conceptualization, C.W. and J.C.; Methodology, C.W. and S.J.; Data curation, C.W. and H.C.; Software, C.W.; Validation, C.W.; Formal analysis, C.W.; Investigation, C.W. and J.C.; Resources, C.W. and J.C.; Writing—original draft preparation, C.W.; Writing—review and editing, H.C.; Supervision, J.C. All authors have read and agreed to the published version of the manuscript.

Funding: This research was funded by [National Natural Science Foundation of China] grant number [51675264], [National Natural Science Foundation of China] grant number [52075242], [National Natural Science Foundation of China] grant number]11902157], [Basic Research Program (Natural Science Foundation) of Jiangsu Province] grant number [BK20180417], and [Talents Startup Foundation of Nanjing University of Aeronautics and Astronautics] grant number [1011-YAH20114]. And The APC was funded by [National Natural Science Foundation of China] grant number [51675264].

Conflicts of Interest: The authors declare no conflict of interest.

Notation

The following symbols are used in this paper:

α	Angle between the primary strut and ground
θ	Angle between the auxiliary strut and ground
μ	Friction coefficient
τ	Maximum yield stress of MRF
ρ	Density of MRF
η	Viscosity of MRF
a_0	undetermined coefficient of the constant term
a_{j}	the undetermined coefficient of one-degree term
a_{ij}	the undetermined coefficient of the quadratic term.
c	Contact damping of the impact function
d	Penetration depth of the impact function
e	Contact force exponent of the impact function
k	Stiffness of the impact function
q	Distance function of the impact function
q_0	Trigger distance of the impact function
v	Piston velocity
A_{p}	Piston area
A_{gap}	The gap area between the master cylinder and piston
A_{n}	Cross-section area of the master cylinder
D	Diameter of piston
D	Diameter of piston rod
F_{a}	Force of auxiliary strut acting at point C
F_{c}	Controllable damping force of MRF damper
F_{p}	Force of primary strut acting at point A
F_{u}	Uncontrollable damping force of MRF damper
F_{Ni}	Air-spring force caused by the accumulator
K_{entry}	Local resistance coefficient of the entry
K_{entry}	Local resistance coefficient of the exit
H_1	Vertical distance between the origin and point A
H_2	Vertical distance between the mass center P1 of elastic parts and point A
L	Length of coils
R	Horizontal distance between the origin of rocket coordinate system and point C
R_1	Highest rocket acceleration
R_2	Greatest compression stroke
R_3	Distance between rocket and ground
R_4	Mass of a set of landing gear
R_{P1}	Initial distance between the mass center of the rocket and the ground
V	Volume of accumulator
V_0	Initial volume of accumulator

Appendix A

Table A1. Fuzzy control rules for inputs and outputs [39].

Inputs				Outputs			
Acceleration	**Jerk**	***alpha***	***beta***	τ_1	τ_2	τ_3	τ_4
B	All	All	All	Z	Z	Z	Z
Z	Z	Z	Z	B	B	B	B
Z	B	Z	Z	MB	MB	MB	MB
S	Z	Z	Z	SB	SB	SB	SB
S	B	Z	Z	M	M	M	M
M	Z	Z	Z	SM	SM	SM	SM
M	B	Z	Z	S	S	S	S
Z	Z	P	Z	S	SB	MB	SB
Z	B	P	Z	S	M	SB	M
Z	Z	N	Z	MB	SB	S	SB
Z	B	N	Z	SB	M	S	M
Z	Z	Z	P	SB	S	SB	MB
Z	B	Z	P	M	S	M	SB
Z	Z	Z	N	SB	MB	SB	S
Z	B	Z	N	M	SB	M	S
Z	Z	N	P	MB	SB	SB	MB
Z	B	N	P	SB	M	M	SB
Z	Z	P	N	SB	MB	MB	SB
Z	B	P	Z	M	SB	SB	M
Z	Z	N	N	SB	MB	MB	SB
Z	B	N	N	SB	SB	M	M
Z	Z	P	P	SB	SB	MB	MB
Z	B	P	P	M	M	SB	SB
S	Z	P	Z	Z	M	SB	M
S	B	P	Z	Z	SM	M	SM
S	Z	N	Z	SB	M	Z	M
S	B	N	Z	M	SM	Z	SM
S	Z	Z	P	M	Z	M	SB
S	B	Z	P	SM	Z	SM	M
S	Z	Z	N	M	SB	M	Z
S	B	Z	N	SM	M	SM	Z
S	Z	P	P	M	M	SB	SB
S	B	P	P	SM	SM	M	M
S	Z	N	N	SB	SB	M	M
S	B	N	N	M	M	SM	SM
S	Z	P	N	M	SB	SB	M
S	B	P	N	SM	M	M	SM
S	Z	N	P	SB	M	M	SB
S	B	N	P	M	SM	SM	M
M	Z	P	Z	Z	SM	M	SM
M	B	P	Z	Z	S	SM	S
M	Z	N	Z	SM	Z	M	SM
M	B	N	Z	S	Z	SM	S
M	Z	Z	N	SM	M	SM	Z
M	B	Z	N	S	SM	S	Z
M	Z	Z	P	SM	Z	SM	M
M	B	Z	N	S	SM	S	Z
M	Z	N	N	M	M	SM	SM
M	B	N	N	SM	SM	S	S
M	Z	N	P	M	SM	SM	M
M	B	N	P	SM	S	S	SM
M	Z	P	N	SM	M	M	SM
M	B	P	N	S	SM	SM	S
M	Z	P	P	SM	SM	M	M
M	B	P	P	S	S	SM	SM

Table A2. RSM sampling.

Run	Factor 1: A(θ/°)	Factor 2: B(α/°)	Factor 3: C(H_1/mm)	Response 1: R_1 (a_{max}/m/s^2)	Response 2: R_2 (Strokes/mm)	Response 3: R_3 (Distance/mm)	Response 4: R_4 (Mass/kg)
1	27	42	1800	23.3495	79.5787	2240	108.446
2	24	42	1500	17.7913	64.4743	1112.8	91.6878
3	30	42	1500	33.193	113.355	2113.5	109.83
4	27	42	1200	20.5145	82.2504	1448.6	89.412
5	30	53	1500	30.0239	46.4953	902.999	79.913
6	27	47.5	1500	61.8792	21.4722	1121	85.2949
7	27	47.5	1500	61.8792	21.4722	1121	85.2949
8	24	47.5	1800	69.0255	18.5479	1028.8	87.5495
9	24	47.5	1200	21.6645	56.0716	894.954	85.2949
10	27	47.5	1500	61.8792	21.4722	1121	85.2949
11	27	53	1800	29.5464	40.7394	873.652	82.5555
12	30	47.5	1200	44.3331	70.8267	1013.2	82.452
13	30	47.5	1800	36.5785	62.3615	1923.2	98.0229
14	27	53	1200	27.7053	43.8536	602.336	72.0231
15	24	53	1500	31.3973	95.6979	375.021	75.1617
16	27	47.5	1500	61.8792	21.4722	1121	85.2949
17	27	47.5	1500	61.8792	21.4722	1121	85.2949

References

1. Wang, Z.G.; Luo, S.B.; Wu, J.J. *Developments of Reusable Launch Vehicles*; National University of Defense Technology Press: Changsha, China, 2004. (In Chinese)
2. Lei, B. Design and Analysis of Landing Absorber System for Vertical Takeoff and Landing Launch Vehicle. Ph.D. Thesis, Nanjing University of Aeronautics and Astronautics, Nanjing, China, 2020.
3. Yue, S.; Nie, H.; Zhang, M.; Huang, M.; Zhu, H.; Xu, D. Dynamic analysis for vertical soft landing of reusable launch vehicle with landing strut flexibility. *Proc. Inst. Mech. Eng. Part G J. Aerosp. Eng.* **2018**, *233*, 1377–1396. [CrossRef]
4. Xu, D.; Zhang, Z.; Wu, K.; Li, H.; Lin, J.; Zhang, X.; Guo, X. Recent progress on development trend and key technologies of vertical take-off vertical landing reusable launch vehicle. *Chin. Sci. Bull.* **2016**, *61*, 3453–3463. [CrossRef]
5. Sgarlata, P.; Weegar, R. Operational lessons of the DC-X propulsion system operations. In Proceedings of the 31st Joint Propulsion Conference and Exhibit, San Diego, CA, USA, 10–12 July 1995. [CrossRef]
6. Wagner, E.; DeForest, C.E. *Opportunities for Suborbital Space and Atmospheric Research Facilities on Blue Origin's New Shepard Crew Capsule*; AGU Fall Meeting Abstracts: Washington, DA, USA, 2016.
7. Bjelde, B.; Capozzoli, P.; Shotwell, G. *The SpaceX Falcon 1 Launch Vehicle Flight 3 Results, Future Developments, and Falcon 9 Evolution*; Space Exploration Technologies: Hawthorne, LA, USA, 2008.
8. Dreyer, L. Latest developments on SpaceX's Falcon 1 and Falcon 9 launch vehicles and Dragon spacecraft. In Proceedings of the 2009 IEEE Aerospace Conference, IEEE, Big Sky, MT, USA, 7–14 March 2009. [CrossRef]
9. SpaceX. Falcon 9 Launch Vehicle Payload User's Guide. 2008. Available online: https://www.spaceflightnow.com/falcon9/001/f9guide.pdf (accessed on 17 September 2020).
10. Tatiossian, P.; Desmariaux, J.; Garcia, M. Callisto Project-Reusable First Stage Rocket Demonstrator. In Proceedings of the 7th European Conference for Aeronautics and Space Sciences (EUCASS), Paris, France, 3–6 July 2017. [CrossRef]
11. Dumont, E.; Ecker, T.; Witte, L.; Chavagnac, C.; Windelberg, J.; Klevanski, J.; Giagkozoglou, S. CALLISTO-Reusable VTVL launcher first stage demonstrator. In Proceedings of the Space Propulsion Conference, Sevilla, Spain, 14–18 May 2018; p. SP2018_00406.
12. Shuai, Y. Retraction and Landing Dynamic Analysis of Landing gear for Vertical Takeoff and Landing Reusable Launch Vehicle. Ph.D. Thesis, Nanjing University of Aeronautics and Astronautics, Nanjing, China, 2019.
13. Yue, S.; Nie, H.; Zhang, M.; Wei, X.; Gan, S. Design and landing dynamic analysis of reusable landing leg for a near-space manned capsule. *Acta Astronaut.* **2018**, *147*, 9–26. [CrossRef]
14. Yue, S.; Nie, H.; Zhang, M.; Huang, M.; Dafu, X. Optimization and Performance Analysis of Oleo-Honeycomb Damper Used in Vertical Landing Reusable Launch Vehicle. *J. Aerosp. Eng.* **2018**, *31*, 04018002. [CrossRef]

15. Yue, S.; Titurus, B.; Nie, H.; Zhang, M. Liquid spring damper for vertical landing Reusable Launch Vehicle under impact conditions. *Mech. Syst. Signal Process.* **2019**, *121*, 579–599. [CrossRef]
16. Bo, L.; Ming, Z.; Shuai, Y. Design and Optimization of a Crashworthy Damper Used for Reusable Launch Vehicles. *J. Astronaut.* **2019**. (In Chinese) [CrossRef]
17. LORD. Viscosity vs. Temp and Yield Stress vs. H [R]. 2010. Available online: https://mar-lord-mr.worldsecuresystems.com/lord-mr-products/mrf-132dg-magneto-rheological-fluid (accessed on 17 September 2020).
18. LORD. *Yield Strength versus H for any MR fluid [R]*; LORD: Cary, NC, USA, 2011.
19. Rabinow, J. The magnetic fluid clutch. *Electr. Eng.* **1948**, *67*, 1167. [CrossRef]
20. Zhao, X.; Chen, J.; Wang, C.; Wei, J. Research on Magnetorheological Fluid Damper for Lunar Lander and Its Design. *Mach. Build. Autom.* **2018**. (In Chinese) [CrossRef]
21. Heckenthaler, T.; Engell, S. Approximately time-optimal fuzzy control of a two-tank system. *IEEE Control. Syst.* **1994**, *14*, 24–30. [CrossRef]
22. Kickert, W.; Lemke, H.V.N. Application of a fuzzy controller in a warm water plant. *Automatica* **1976**, *12*, 301–308. [CrossRef]
23. King, P.; Mamdani, E. The application of fuzzy control systems to industrial processes. *Automatica* **1977**, *13*, 235–242. [CrossRef]
24. Dutta, S. Fuzzy logic applications: Technological and strategic issues. *IEEE Trans. Eng. Manag.* **1993**, *40*, 237–254. [CrossRef]
25. Qing, L. Investigation on Several Key Problems for Performance of Lunar Lander Soft Landing. Ph.D. Thesis, Nanjing University of Aeronautics and Astronautics, Nanjing, China, 2012.
26. Verheul, C. ADAMS Methodology-Contact Modeling", Benelux ADAMS User Meeting. 2012. Available online: https://www.insumma.nl/wp-content/uploads/SayField_Verheul_ADAMS_Contacts.pdf (accessed on 17 September 2020).
27. Yutang, T. Design and Fuzzy Control of Magnetorheological Shock Absorber. Ph.D. Thesis, Nanjing University of Science and Technology, Nanjing, China, 2006. [CrossRef]
28. Zhaochun, L. Performances and Control Method of Magnetorheological Shock Isolation Device. Ph.D. Thesis, Nanjing University of Aeronautics and Astronautics, Nanjing, China, 2012. [CrossRef]
29. Mikułowski Grzegorz, M. Adaptive Impact Absorbers Based on Magnetorheological Fluids. Ph.D. Thesis, Psnc Poznan Supercomputer & Networking Center, Warsaw, Poland, 2008. Available online: http://www.ippt.pan.pl/_download/doktoraty/2008mikulowski_g_doktorat.pdf (accessed on 17 September 2020).
30. Weixiong, R.G. Preliminary Design of Reusable Lunar Lander Landing System. Ph.D. Thesis, Luleå University of Technology, Luleå, Sweden, 2017.
31. Heng, C. Investigation on Energy-Absorption Nanomaterials for Lunar Lander and Analysis on Soft-Landing Performance. Ph.D. Thesis, Nanjing University of Aeronautics and Astronautics, Nanjing, China, 2017. CNKI:CDMD:1.1019.801155.
32. Saxena, D.; Rathore, H. Vibration control of MR damper landing gear. *Int. J. Adv. Res. Artif. Intell.* **2013**, *2*, 72–76. [CrossRef]
33. Taylor, A.; Benney, R.; Bagdonovich, B. Investigation of the Application of Airbag Technology to Provide a Softlanding Capability for Military Heavy Airdrop. In Proceedings of the 16th Aiaa Aerodynamic Decelerator Systems Technology Conference & Seminar, Boston, MA, USA, 21–24 May 2001. [CrossRef]
34. Zhu, G.N. Structure Design and Analysis on Landing Support of Vertical Takeoff and Vertical Landing Launch Vehicle. Ph.D. Thesis, Nanjing University of Aeronautics and Astronautics, Nanjing, China, 2019.
35. Mingyang, H. Control strategy of launch vehicle and lander with adaptive landing gear for sloped landing. *Acta Astronaut.* **2019**, *161*, 509–523. [CrossRef]
36. Yunkang, S.; Huiping, Y. *Improvement of Response Surface Methodology and Application to Engineering Optimization*; Science Press of China: Beijing, China, 2011. (In Chinese)
37. Anderson, M.J.; Whitcomb, P.J.; Shari, L.K. *Handbook for Experimenters*; Stat-Ease: Minneapolis, MN, USA, 2017; Available online: https://cdnm.statease.com/pubs/handbk_for_exp_sv.pdf (accessed on 27 November 2017).

38. Junlin, W.; Hong, N.; Lichun, L.; Jinbao, C.; Fuming, Z. Analysis of Landing Performance and Multi-factor Effects with Lunar Lander. *J. Nanjing Univ. Aeronaut. Astronaut.* **2010**, 288–293. [CrossRef]
39. Wang, C.; Nie, H.; Chen, J.; Lee, H.P. The design and dynamic analysis of a lunar lander with semi-active control. *Acta Astronaut.* **2019**, *157*, 145–156. [CrossRef]

symmetry

MDPI

Article

A Multibody Inertial Propulsion Drive with Symmetrically Placed Balls Rotating on Eccentric Trajectories

Attila Gerocs, Gilbert-Rainer Gillich, Dorian Nedelcu and Zoltan-Iosif Korka *

Doctoral School of Mechanical Engineering, "Eftimie Murgu" University of Resita, 320085 Resita, Romania; a.gerocs@uem.ro (A.G.); gr.gillich@uem.ro (G.-R.G.); d.nedelcu@uem.ro (D.N.)

* Correspondence: z.korka@uem.ro; Tel.: +40-745-911887

Received: 30 July 2020; Accepted: 25 August 2020; Published: 26 August 2020

Abstract: Starting with the last century, a lot of enthusiastic researchers have invested significant time and energy in proposing various drives capable to generate linear propulsion force. Regrettably, only a few of these devices passed the patent phase and have been practically materialized. The aim of this paper was to simulate the dynamic behavior of an inertial propulsion drive (IPD) developed by the authors, to demonstrate its functionality. The core of the IPD consists of two symmetric drivers that each performs rotation of eight steel balls on an eccentric path. We propose three solutions for the element which maintain the off-center trajectory of the balls. For the simulation, we used the multibody system approach and determine the evolution of the displacement, velocity, and power consumption. Further, we analyze the collisions between the elements of the system and the influence of this phenomenon on the dynamic behavior of the IPD. We found that collisions generate impact forces which affect the ball acceleration values achieved by simulation. We have concluded that the developed system is capable to generate linear movement. In addition, in terms of velocity and power consumption, the best constructive version of the retaining disk is that which has a cylindrical inner bore placed eccentric relative to the rotation center of the balls.

Keywords: multibody; propulsion drive; linear motion; eccentric trajectory

1. Introduction

Inertial propulsion drives belong to a promising field of research and have therefore gained the attention of scientists and engineers in recent decades. These mechanisms are multi-body systems with eccentric masses in motion, usually presenting symmetry on one or two axes to compensate undesired forces in the direction orthogonal to the displacement. Displacement is produced by a propulsion force developed as a reaction to the variable centrifugal forces which are acting on a number of masses rotating on an eccentric trajectory.

It is a huge controversy about the effectiveness of inertial force-based propulsion drives because they challenge Newton's laws of motion [1]. In a National Aeronautics and Space Administration (NASA) report [2] anti-gravitational systems are considered to be impossible, but no discussion on specific inertial propulsion drives (IPDs) is made. In opposition, Allan Jr. wonders rhetorically: "Why does classical mechanics forbid inertial propulsion devices when they evidently do exist?" Thus, in his book [3], he presents a series of working inertial propulsion devices [4–6].

It is also shown that the inertial system proposed by Couloumbe [7] is not linked to gravity and may work in space. While this IPD is not gravity-based, it does not fall under the "restrictions" formulated by Millis and Thomas [2]. Numerous other patents are known [8–25]. Sadly, only a few of them exchange the stage of patent to the practical materialization. Here, the Dean Drive [26], with applications in the construction of inertial pumps and vibration elevators [1], or the Thornson Apparatus [27], which was able to develop a velocity of 1.6 miles per hour installed on a canoe in a swimming pool, can be mentioned.

An example of a patent without practical materialization is the propulsion unit proposed by Booden [9]. It converts the energy of two identical electromagnetically spinning assemblies of weights, which are rotating in opposite directions, in a unidirectional thrust. The invention relies on rotating elements with variable gyration radius around a center axis in order to provide an imbalance of centrifugal force to generate propulsion in a given direction.

Similarly, Cuff [10] discloses a mechanism for varying the radius of rotation of a plurality of rotating weights and for selectively modifying the direction of the resultant unbalanced force generated by these rotating masses. More particularly, the device comprises of two commonly non- rotating circular cams, disposed eccentrically, which engage a pair of cam followers that are linked to connecting rods which are successively joined to the gyrating weights. The direction of the unbalanced resultant force can be adapted by simply rotating, in concordance, the two commonly non- rotating cams.

The propulsion apparatus suggested by Dobos [11] includes a platform and a shaft placed perpendicularly on it. The shaft supports a disk which carries on the circumferential edge portion a plurality of reservoirs filled with liquid and spaced completely around the circumference of the disk. Each reservoir contains a buoyantly positioned hollow piston with a road which extends outwardly from the reservoir. An adaptable movable cam with a cam track is mounted on the shaft assembly—this is placed eccentrically relative to the axis of the shaft assembly. The piston rod ends are placed in opposition to the cam track. Thus, the movement of the pistons generates a relative displacement of the liquid in the reservoirs in response to the contact of the piston rods with the cam track, creating an unbalanced centrifugal force, which moves the platform in a preselected linear direction.

Another IPD using fluids for converting rotary motion into linear displacement was more recently proposed by Deschamplain [21]. His motion imparting system contains a centrifuge with a chamber partially filled with a quantity of fluid. A motor imparts a rotary motion to the centrifuge, the water being moved outwardly under the action of centrifugal forces. During the rotation of the centrifuge, an object is at its exterior end positioned within the fluid and at its interior end positioned exterior of the fluid in the air. Parallel with the axis of the centrifuge, but laterally offset is placed a rod which is rotated by a driver with the same speed as the centrifuge. A coupling connects the rod with the object. The centrifugal forces which occur during the simultaneously rotation of the rod and centrifuge are converted to a linear force.

In addition, in a recent patent, Murray [23] comes up with a mechanical force generator for converting the energy of centrifugal force in propulsion force by rotating a cage assembly around its longitudinal axis. Thus, the cage rotates secondary shafts which turn sets of eccentrics for generating a net force in a direction which is transversal on the rotation axis of the cage assembly. Two pairs of eccentrics turn such that for each 90° rotation of the carrier cage, the pairs of eccentrics have their mass centers located rather between an unbalanced and a balanced condition, but, at each quarter of complete rotation, one pair of eccentrics is every time generating a power stroke.

The effect of friction between the IPD and the external supporting surface was studied in [28]. Like it was concluded in similar researches [29,30], it was found that friction is important for providing movement. Other authors [31,32] consider that IPD's are able to produce displacement on frictionless supports or in space because reaction forces are not necessary.

Provatidis [33] makes a critical review regarding the actual state-of-the-art of inertial devices capable to move the objects on which they are attached, concluding, same as Robertson and Webb [34] that "the replacement of rockets by other advanced means is an ongoing procedure" and "the death of rocket science is only a matter of time."

The aim of this paper is to present an IPD developed by the authors which is using two groups of rotating masses. Furthermore, the mathematical relations, deducted for the kinematic of these rotating masses are introduced, the analytical results being compared with the outcomes of simulation data obtained by involving a professional software. Three versions of trajectories for the rotating masses are considered, and the best solution in terms of displacement velocity and power consumption is proposed for the construction of the system. Finally, to validate the functionality of the system, the most efficient version is built and experimental proofs are carried out.

2. Description of the Proposed IPD

The propulsion of the device developed by the authors [35] is based on producing a resultant centrifugal force for driving the assembly. As shown in Figure 1, the multi-body system consists of two identical groups disposed symmetrical relative to the direction of displacement. Each group consists of 8 equal steel balls (1/1 ÷ 1/8) with radii of 5 mm, placed between two rotating plates (2/1 and 2/2), which are foreseen in radial direction with 8 hemispherical slots for guiding the steel balls. The trajectory of the balls is maintained by the inner bore of a retaining disk (3).

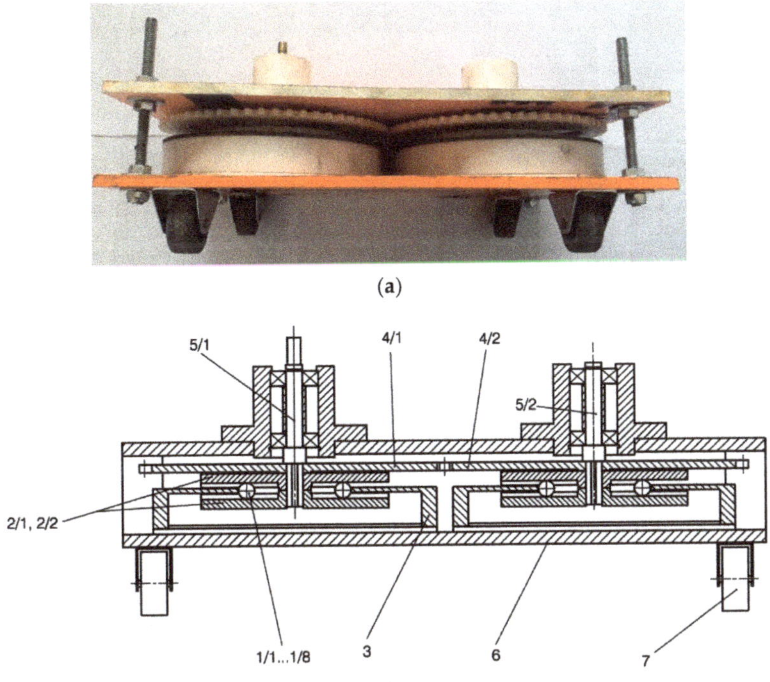

Figure 1. Construction of the inertial propulsion drive (IPD). (**a**) Lateral view; (**b**) cross-section.

The retaining disk is designed in three constructive versions, as shown in Figure A1 (see Appendix A):

- **Version 1**: Inner bore with a radius of R_1 = 41 mm and the center placed eccentric at a distance e = 20 mm, relative to the center of the slotted plates;
- **Version 2**: Inner bore consisting of a semicircle with a radius $R_{2\text{-}1}$ = 44.6 mm, two straight portions tangent to the semicircle and an arc of radius $R_{2\text{-}2}$ = 61 mm, concentric with the first radius semicircle;
- **Version 3:** Inner bore consisting of two semicircles with radii R_3 = 41 mm and centers placed at a distance of e = 20 mm, which are connected by two straight portions tangent to the semicircles.

For rotating the two constructive groups with the same angular speed, but in opposite directions, two spur gears (4/1 and 4/2) with the same number of teeth are used. The gears are supported by the shafts (5/1 and 5/2), which are also driving the slotted plates (2/1 and 2/2). The longer shaft (5/1) is connected with the driving motor which rotates the assembly with a constant speed n = 1200 rot/min. The retaining disk (3) is locked on the base plate (6) of the system, which lies on four rubber rollers of 40 mm diameter and fixed axis.

Rotating the plates 2/1 and 2/2 with a constant angular velocity ω, on each of the 8 steel balls acts a centrifugal force. As the balls are forced to follow a circular trajectory, but with variable radii $R_i(t)$, these centrifugal forces are variable in time and may be expressed in (1):

$$F_{ci}(t) = m_0 \cdot \omega^2 \cdot R_i(t), \tag{1}$$

where:

$Fc_i(t)$ [N]—centrifugal force acting on the balls;
m_0 [kg]—mass of the balls;
ω [rad/s]—angular velocity of the rotating plates;
$R_i(t)$—trajectory radius of ball i.

The resultant of the centrifugal forces acting on the 8 steel balls produces the linear propulsion of the assembly.

3. Analytical Investigation of the Three Proposed Constructive Alternatives

In this study, the physical quantities which describe the kinematics of the system are calculated in regard with the three constructive alternatives of the retaining disk. Furthermore, the deduction of the analytical expressions for the displacement, velocity and acceleration of the steel balls is presented.

3.1. Version 1 of Retaining Disk: Cylindrical Bore Placed Eccentric Relative to the Center of the Slotted Plates

For finding the analytical expressions of the elements which characterize the kinematics of one of the balls, having the center in C_i, a Cartesian system denoted with xO_1y was attached to the center O_1 of the slotted plates (2/1 and 2/2). Furthermore, as shown in Figure 2, the circle with radius R_1 and center in O_2, represents the inner bore of the retaining disk (3). As mentioned before, the retaining disk is placed eccentric, relative to the center of the slotted plates, at a distance *e*.

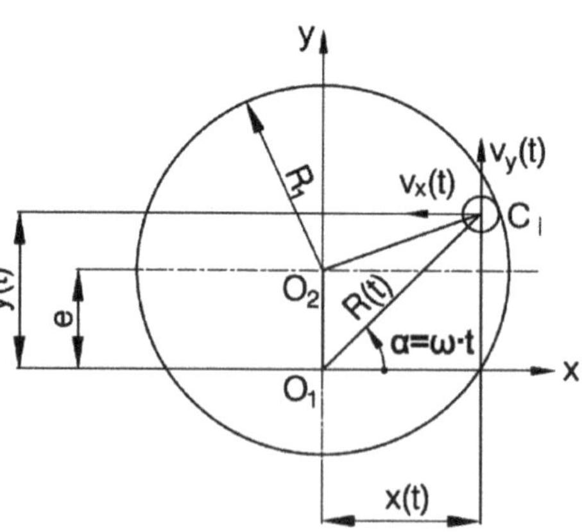

Figure 2. Kinematic of a ball at Version 1 of the retaining disk.

Following the procedure described in detail in [36,37] and the notations from Figure 2, the Cartesian coordinates and the trajectory radius of the center C_i may be written as in (2) and (3):

$$x(t) = \frac{e}{2}\sin 2\omega t + \cos\omega t\sqrt{(R_1 - r)^2 - e^2\cos^2\omega t}, \tag{2}$$

$$y(t) = e\sin^2\omega t + \sin\omega t\sqrt{(R_1 - r)^2 - e^2\cos^2\omega t}, \tag{3}$$

By deriving Equations (2) and (3), the components of the ball velocity are obtained as follows in (4) and (5):

$$v_x(t) = \dot{x}(t) = \omega e\cos 2\omega t - \omega\sin\omega t\sqrt{(R_1 - r)^2 - e^2\cos^2\omega t} + \frac{\omega e^2\cos\omega t \cdot \sin 2\omega t}{2\sqrt{(R_1 - r)^2 - e^2\cos^2\omega t}}, \tag{4}$$

$$v_y(t) = \dot{y}(t) = \omega e\sin 2\omega t + \omega\cos\omega t\sqrt{(R_1 - r)^2 - e^2\cos^2\omega t} + \frac{\omega e^2\sin\omega t \cdot \sin 2\omega t}{2\sqrt{(R_1 - r)^2 - e^2\cos^2\omega t}}. \tag{5}$$

Furthermore, deriving the expressions of the velocities, the components of the ball acceleration along the x and y axis are expressed as follows in (6) and (7):

$$a_x(t) = -2\omega^2 e\sin 2\omega t - \omega^2\cos\omega t\sqrt{(R_1 - r)^2 - e^2\cos^2\omega t} + \frac{\omega^2 e^2\cos 3\omega t}{\sqrt{(R_1-r)^2 - e^2\cos^2\omega t}} - \frac{\omega^2 e^4\cos\omega t \cdot \sin^2 2\omega t}{8\cdot\left[(R_1-r)^2 - e^2\cos^2\omega t\right]^{3/2}} \tag{6}$$

$$a_y(t) = 2\omega^2 e\cos 2\omega t - \omega^2\sin\omega t\sqrt{(R_1 - r)^2 - e^2\cos^2\omega t} + \frac{\omega^2 e^2\sin\omega t}{\sqrt{(R_1-r)^2 - e^2\cos^2\omega t}} - \frac{\omega^2 e^4\sin\omega t \cdot \sin^2 2\omega t}{8\cdot\left[(R_1-r)^2 - e^2\cos^2\omega t\right]^{3/2}} \tag{7}$$

3.2. Version 2 of Retaining Disk: Inner Bore Consisting of a Semicircle, a Circular Arc and Two Straight Portions

Similar to the previous case, a Cartesian system denoted with xOy was attached to the center O of the slotted plates (2/1 and 2/2). The inner bore of the retaining disk has at this constructive version a special design, consisting of a semicircle with center in O and radius $R_{2\text{-}1}$, a circular arc with radius $R_{2\text{-}2}$, concentric with the semicircle and two straight portions of length $R_{2\text{-}1}$, which are tangent to the semicircle (see Figure 3).

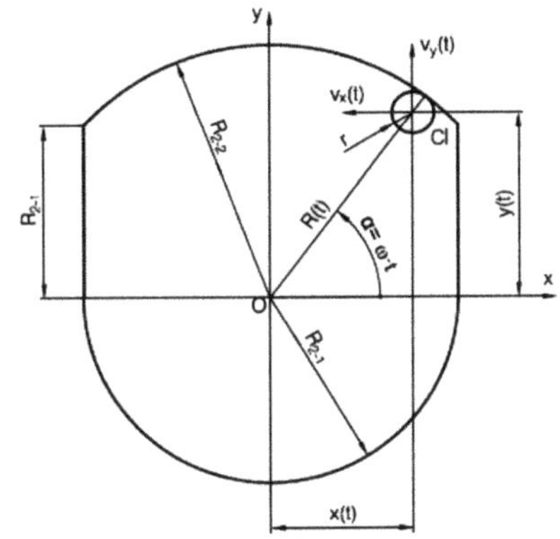

Figure 3. Kinematic of a ball at Version 2 of the retaining disk.

Using the notations from Figure 3, the Cartesian coordinates and the trajectory radius of the center Ci of a ball, may be written, depending on the position angle α, as follows in (8):

$$x(t)=\begin{cases}R_{2-1}-r & \alpha\in(0;45]\\(R_{2-2}-r)\cos(\omega t) & \alpha\in(45;135]\\r-R_{2-1} & \alpha\in(135;180]\\(R_{2-1}-r)\cos(\omega t) & \alpha\in(180;360]\end{cases},\quad y(t)=\begin{cases}(R_{2-1}-r)\tan(\omega t) & \alpha\in(0;45]\\(R_{2-2}-r)\sin(\omega t) & \alpha\in(45;135]\\(r-R_{2-1})\tan(\omega t) & \alpha\in(135;180]\\(R_{2-1}-r)\sin(\omega t) & \alpha\in(180;360]\end{cases},\quad (8)$$

The components of the ball velocity v_x and v_y are obtained in (9) by deriving (8):

$$v_x(t)=\dot{x}(t)=\begin{cases}0 & \alpha\in(0;45]\\\omega(r-R_{2-2})\sin(\omega t) & \alpha\in(45;135]\\0 & \alpha\in(135;180]\\\omega(r-R_{2-1})\sin(\omega t) & \alpha\in(180;360]\end{cases}\quad v_y(t)=\dot{y}(t)=\begin{cases}\frac{\omega(R_{2-1}-r)}{\cos^2(\omega t)} & \alpha\in(0;45]\\\omega(R_{2-2}-r)\cos(\omega t) & \alpha\in(45;135]\\\frac{\omega(r-R_{2-1})}{\cos^2(\omega t)} & \alpha\in(135;180]\\\omega(R_{2-1}-r)\cos(\omega t) & \alpha\in(180;360]\end{cases}\quad (9)$$

Correspondingly, the components of the ball acceleration a_x and a_y are obtained in (10) by deriving (9):

$$a_x(t)=\dot{v}_x(t)=\begin{cases}0 & \alpha\in(0;45]\\\omega^2(r-R_{2-2})\cos(\omega t) & \alpha\in(45;135]\\0 & \alpha\in(135;180]\\\omega^2(r-R_{2-1})\cos(\omega t) & \alpha\in(180;360]\end{cases}\quad a_y(t)=\dot{v}_y(t)=\begin{cases}\frac{2\omega^2(R_{2-1}-r)\sin(\omega t)}{\cos^3(\omega t)} & \alpha\in(0;45]\\\omega^2(r-R_{2-2})\sin(\omega t) & \alpha\in(45;135]\\\frac{2\omega^2(r-R_{2-1})\sin(\omega t)}{\cos^3(\omega t)} & \alpha\in(135;180]\\\omega^2(r-R_{2-1})\sin(\omega t) & \alpha\in(180;360]\end{cases}\quad (10)$$

3.3. Version 3 of Retaining Disk: Inner Bore Consisting of Two Identical Semicircles with the Centres Located at a Distance "e" and Two Straight Portions Tangent to the Semicircles

For the analytical approach of this constructive version, the Cartesian system denoted with xOy was attached to the center O of the slotted plates. This time, the inner bore of the retaining disc consists of two equal semicircles of radius R_3. One of the centers of the semicircles is located in O, while the other is placed at the distance e in vertical direction. The two semicircles are connected by two vertical segments tangent to the semicircles (see Figure 4).

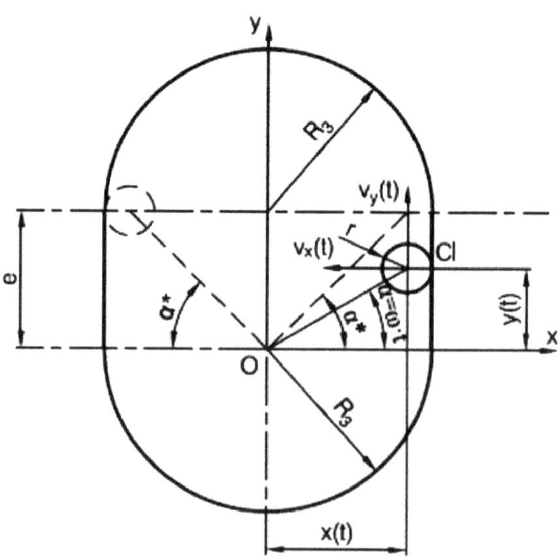

Figure 4. Kinematic of a ball at Version 3 of the retaining disk.

Depending on the position angle α and using the notations from Figure 4, the Cartesian coordinates of the center Ci of a ball, can be expressed in (11) and (12):

$$x(t)=\begin{cases} R_3-r & \alpha\in(0;\alpha^*] \\ \frac{e}{2}\sin 2\omega t+\cos\omega t\sqrt{(R_3-r)^2-e^2\cos^2\omega t} & \alpha\in(\alpha^*;180-\alpha^*] \\ r-R_3 & \alpha\in(180-\alpha^*;180] \\ (R_3-r)\cos(\omega t) & \alpha\in(180;360] \end{cases}, \tag{11}$$

$$y(t)=\begin{cases} (R_3-r)\tan(\omega t) & \alpha\in(0;\alpha^*] \\ e\sin^2\omega t+\sin\omega t\sqrt{(R_3-r)^2-e^2\cos^2\omega t} & \alpha\in(\alpha^*;180-\alpha^*] \\ (r-R_3)\tan(\omega t) & \alpha\in(180-\alpha^*;180] \\ (R_3-r)\sin(\omega t) & \alpha\in(180;360] \end{cases}, \tag{12}$$

where in (13):

$$\alpha^*=\arctan\frac{e}{R_3-r}. \tag{13}$$

α^* is the position angle of the inflection point, where the trajectory of the balls changes from circular to linear and vice versa.

In the same way, by deriving twice the expressions of the ball coordinates, the velocities and accelerations may be deducted as follows in (14)–(17):

$$v_x(t)=\begin{cases} 0 & \alpha\in(0;\alpha^*] \\ \omega e\cos 2\omega t-\omega\sin\omega t\sqrt{(R_3-r)^2-e^2\cos^2\omega t}+\frac{\omega e^2\cos\omega t\cdot\sin 2\omega t}{2\sqrt{(R_3-r)^2-e^2\cos^2\omega t}} & \alpha\in(\alpha^*;180-\alpha^*] \\ 0 & \alpha\in(180-\alpha^*;180] \\ \omega(r-R_3)\sin(\omega t) & \alpha\subset(180;360] \end{cases} \tag{14}$$

$$v_y(t)=\begin{cases} \frac{\omega(R_3-r)}{\cos^2(\omega t)} & \alpha\in(0;\alpha^*] \\ \omega e\sin 2\omega t+\omega\cos\omega t\sqrt{(R_3-r)^2-e^2\cos^2\omega t}+\frac{\omega e^2\sin\omega t\cdot\sin 2\omega t\cdot}{2\sqrt{(R_3-r)^2-e^2\cos^2\omega t}} & \alpha\in(\alpha^*;180-\alpha^*] \\ \frac{\omega(r-R_3)}{\cos^2(\omega t)} & \alpha\in(180-\alpha^*;180] \\ \omega(R_3-r)\cos(\omega t) & \alpha\in(180;360] \end{cases} \tag{15}$$

$$a_x(t)=\begin{cases} 0 & \alpha\in(0;\alpha^*] \\ -2\omega^2 e\sin 2\omega t-\omega^2\cos\omega t\sqrt{(R_3-r)^2-e^2\cos^2\omega t}+\frac{\omega^2e^2\cos 3\omega t}{\sqrt{(R_3-r)^2-e^2\cos^2\omega t}}- \\ \quad -\frac{\omega^2e^4\cos\omega t\cdot\sin^2 2\omega t}{8\cdot[(R_3-r)^2-e^2\cos^2\omega t]^{3/2}} & \alpha\in(\alpha^*;180-\alpha^*] \\ 0 & \alpha\in(180-\alpha^*;180] \\ \omega^2(r-R_3)\cos(\omega t) & \alpha\in(180;360] \end{cases} \tag{16}$$

$$a_y(t)=\begin{cases} \frac{2\omega^2(R_3-r)\sin(\omega t)}{\cos^3(\omega t)} & \alpha\in(0;\alpha^*] \\ 2\omega^2 e\cos 2\omega t-\omega^2\sin\omega t\sqrt{(R_3-r)^2-e^2\cos^2\omega t}+\frac{\omega^2e^2\sin\omega t}{\sqrt{(R_3-r)^2-e^2\cos^2\omega t}}- \\ \quad -\frac{\omega^2e^4\sin\omega t\cdot\sin^2 2\omega t}{8\cdot[(R_3-r)^2-e^2\cos^2\omega t]^{3/2}} & \alpha\in(\alpha^*;180-\alpha^*] \\ \frac{2\omega^2(r-R_3)\sin(\omega t)}{\cos^3(\omega t)} & \alpha\in(180-\alpha^*;180] \\ \omega^2(r-R_3)\sin(\omega t) & \alpha\in(180;360] \end{cases} \tag{17}$$

4. Motion Simulation of the IPD

Like in our previous researches [38,39], the resources of the Motion module [40] from SolidWorks (SW) program were used. SW Motion is a module powered by ADAMS technology, being a virtual prototyping tool for technicians involved in assessing the performance of their design

There are two equations which are governing the three dimensional (3D) motion of a rigid body in the Motion module. The first one is Newton's second law of motion, which affirms that the sum of externally applied forces on a rigid body is equal to the rate of change of linear momentum P, as written in (18):

$$\sum \mathrm{F} = \frac{\mathrm{dP}}{\mathrm{dt}}, \tag{18}$$

This equation for bodies with constant masses (m), simplifies to the more commonly known form (19):

$$\sum \mathrm{F} = \mathrm{ma}, \tag{19}$$

with a for the acceleration of the body.

The second equation starts from the premise that the sum of the moments about the center of mass of a rigid body, due to external forces and couples, is equal to the rate of change of the angular momentum H of this body, which may be expressed in (20):

$$\sum \mathrm{M} = \frac{\mathrm{dH}}{\mathrm{dt}}. \tag{20}$$

The program applies the modified Newton–Raphson iteration method in several time steps. Setting very small time steps, the program is able to predict the position of parts at the next time step, starting from the initial conditions or the precedent time step.

Following steps were employed in the motion study:

- Design of the components shown in Figure 1b;
- Generation of the assembly;
- Specification of the speed for the rotary motor (1200 min^{-1});
- Specification of gravity;
- Specification of solid body contacts;
- Specification of the mates.

To specify the rotary motor parameters, the motor icon was chosen, while the inner cylindrical face of the slotted plates (2/1 and 2/2) was selected, together with constant speed from the motor type list.

Furthermore, selecting the gravity icon, axis Z as direction of action and the value of 9806.65 mm/s^2, the gravitational forces acting on the mechanism were simulated. Finally, the mates between parts were applied in the motion study between the components of the assembly.

A Solid Body Contact 1 was imposed between the first group, defined as the left situated balls (1/1 ÷ 1/8) and the second group, defined as the left situated retaining disk (3) and the base plate (6) generated as a single part together with the left rotating plates (2/1 and 2/2).

Next, a Solid Body Contact 2 was imposed between the first group, defined as the right situated balls (1/1 ÷ 1/8) and the second group, defined as the right situated retaining disk (3) and the base plate (6) generated as a single part together with the left rotating plates (2/1 and 2/2).

For both of Solid Body Contacts one type of material applicable to touching faces during contact were imposed. As the retaining disks were manufactured by 3D printing, from polylactide (PLA) filament, the selected material combination was acrylic with steel. For every material combination the elastic properties of the materials involved in contact phenomenon are selected automatically by SolidWorks Motion.

The simulation was performed without considering any friction, because we intended to compare the simulation outcomes with the analytical results, where, as simplifying hypothesis for constructing the mathematical model, the friction was neglected. It's obvious that the friction is influencing the results, but we have to mention that in this case we are dealing with rolling friction, which may be neglected, in accordance with the most similar mathematical models.

The analysis time of the study was imposed to 0.2 s. Within this time, at a speed of the slotted plates $n = 1200$ rot/min, they rotating 4 times, the duration of a complete rotation being 0.05 s. Furthermore, for the motion analysis were set 1440 frames per second, namely 72 frames for one rotation, meaning that the results were obtained at every 5° indexing of the slotted plates. Moreover, during the simulation was used a "precise contact" with an accuracy of 0.0001, while the contact resolution was set to medium.

5. Results and Discussion

In the first stage of the simulations, the coordinates, velocities and accelerations of a ball where computed for a complete rotation of the slotted plates. Furthermore, the simulation outcomes were examined in contrast with the analytical calculations accomplished by using Equations (2)–(17). The obtained results regarding the physical quantities which describe the kinematics of the ball (x, y, v_x, v_y, a_x and a_y) as functions of the rotation angle α are presented, for the three constructive versions of the retaining disk, in Figures A2–A4 (see Appendix A).

As it can be observed from Figures A2 and A3, in the case of the displacements and velocities, the outcomes obtained by analytical calculation and SW simulation are almost identical, for all three constructive variants of the retaining disk. This confirms first that the analytical relations are correct and, second that the virtual model constructed in SW corresponds with the reality.

A deviation between the analytical and simulation approach can be observed at the retaining disk having the inner bore consisting of a semicircle, a circular arc and two straight portions (Version 2), the highest differences being observed around the points where the ball is changing the trajectory from circular to linear and vice versa.

Regarding the differences between the accelerations that are observed in Figure A4, these may be justified by the fact that in the simulation the effect of the collision between the ball and the inner bore of the retaining disk is taken into consideration, phenomenon which we neglect at the analytical calculation. Each collision generates an impact force which affects the acceleration values achieved by simulation.

Furthermore, the displacement and velocity of the whole system was simulated. Figure 5 depicts the displacements of the system for the three investigated versions of the retaining disk. As it can be noticed, immediately after starting the turning of the slotted plates, because of the high inertia, the systems get an impulse, but after the first turn (T > 0.05 s), they become steady and the displacements grow linearly.

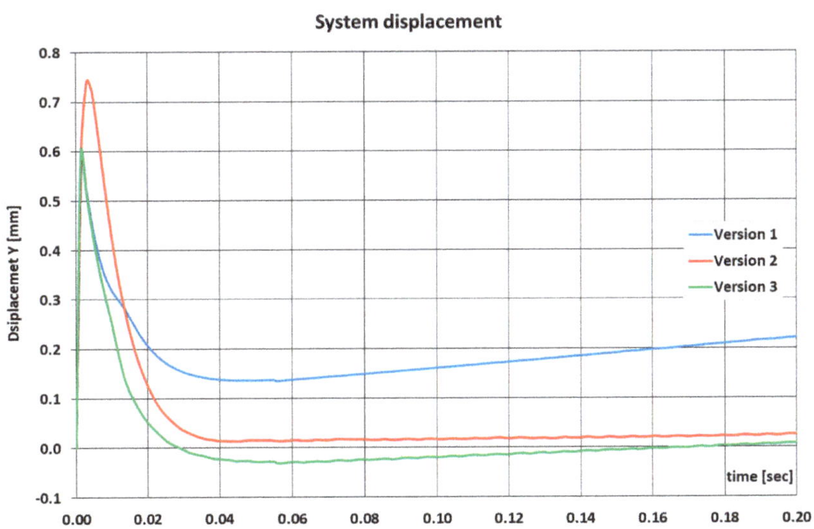

Figure 5. Displacement of the system.

Within the simulation time, the highest displacement (y = 0.22 mm) was shown by Version 1 of the retaining disk, the other two variants getting similar results. Note that the system modeled by Version 3 has, except for the start phase, a negative displacement up to the fourth rotation (t = 0.17 s).

In terms of velocities, the comparison between the three constructive alternatives of the retaining disk is depicted in Figure 6. Figure 6a shows the variation of the velocities during the whole simulation time (0.2 s), while Figure 6b illustrates the average velocities within the last complete rotation of the slotted plates. As it can be observed, the highest average velocity of the system is obtained at Version 1 (0.586 mm/s), while for Versions 2 and 3 the simulation highlighted speeds of 0.127 mm/s and 0.270 mm/s respectively.

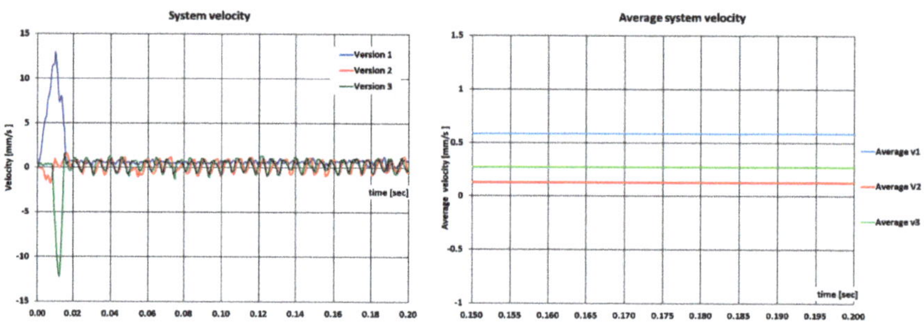

(**a**) (**b**)

Figure 6. Velocity of the system. (**a**) Entire simulation time; (**b**) average values at last rotation of the system.

Another benefit of the Motion module from SolidWorks is that it furnishes very easy information regarding the power consumption for driving the system, data which would be harder to attain by an analytical approach. Relevant information on this issue is shown in Figure 7, during the last rotation of the system and for the three investigated alternatives of the retaining disk.

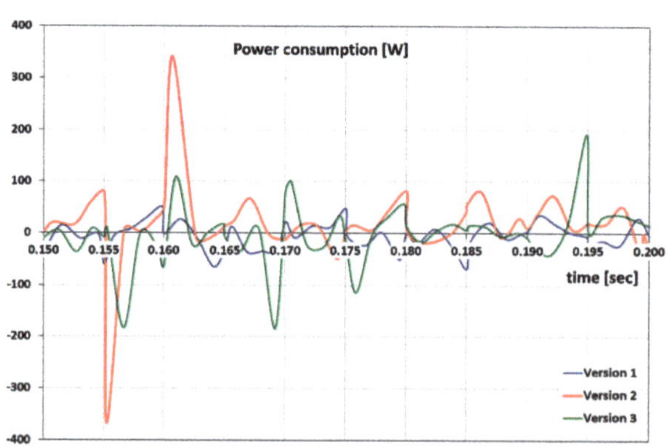

Figure 7. Power consumption of the system.

The theoretical background on which the computation of the power consumption in the Motion module is based starts from the well-known relation of the output power of a motor, which is the product of the torque (T) that the motor generates and the angular velocity (ω) of its output shaft. Moreover, applying Newton's Second Law for rotating bodies, the torque generated by the motor is equal with the product between the mass moments of inertia (I) of the body which is rotated about the axis of the motor and the angular acceleration (α) of this body.

As it can be observed, after the system became stable, the power consumption graph shows a few peaks, the most relevant one being experienced at Version 2. Version 3 shows also some power consumption peaks and Version 1 is the most stable in terms of power variation, requesting in average the smallest power for driving the system.

Summarizing, the most advantageous version of the IPD, in terms of velocities, displacements and power consumption, is Version 1 of the retaining ring.

6. Experimental Proof of the Concept

The obtained results encouraged the authors to build a prototype of the IPD [41]. For this, the additive manufacturing, using 3D printing was involved for producing the slotted disks (2/1 and 2/2), the gears (4/1 and 4/2), the retaining disks (3) and the two bearing covers which are sustaining the driving shafts. Figure 8 shows some details of these parts and the printer.

For driving the system, an angle grinder with a power of 1010 W and five steps of speed (3900, 7000, 9000, 10,500 and 12,000 rpm) was employed.

The tests aimed to determine experimentally the displacement speed of the IPD at the upper mentioned driving speeds. Therefore, the device covered a distance of 200 mm, the required traveling time being measured. For each driving speed step, three timings were made, the average displacement speed being calculated as the ratio between the distance traveled and the average travel time. Figure 9 shows a picture of the experimental setup, while Table 1 presents the measurement results and the calculated average displacement speed.

Figure 8. Additive manufacturing of some parts of the IPD.

Figure 9. Experimental setup for establishing the average displacement speed.

Table 1. Measurement results.

Distance [mm]	Driving Speed n [rpm]	Measured Time [s]			Average Time t_m[s]	Average Speed v_m [mm/s]
		t_1	t_2	t_3		
200	3900	15.0	15.1	15.0	15.03	13.31
	7000	12.0	12.1	12.1	12.07	16.57
	9000	9.1	9.1	9.0	9.07	22.05
	10,500	6.0	6.0	6.0	6.00	33.33
	12,000	4.0	4.0	4.1	4.03	49.63

It is obvious that the displacement speed grows with the increase of the driving speed. Thus, by amplifying the driving speed 3.08 times (from 3900 to 12,000 rpm), the average speed increases 3.73 times (from 13.31 to 49.63 mm/s). To have a better image regarding the influence of the driving speed on the average displacement speed of the IPD, Figure 10 provides a graphical representation of this interdependence.

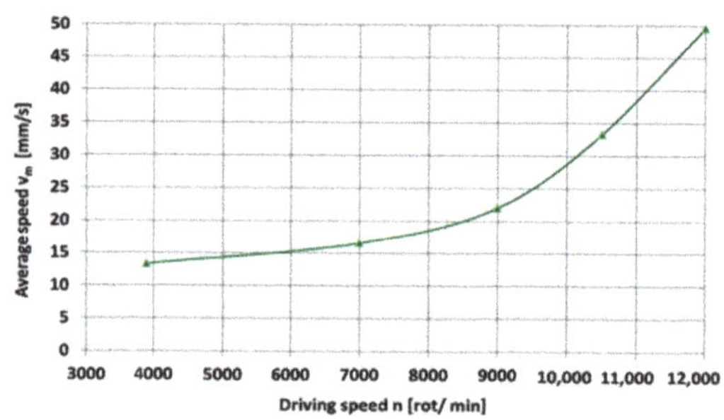

Figure 10. Influence of driving speed on the IPD's displacement speed.

Since the optimization of the IPD design was performed by numerical simulation at a speed lower than the speed at which the experimental measurements were done, the numerical simulation using Motion module from SW was redone at a driving speed of 3900 rpm. In order to maintain reasonable computation duration, the analysis time for the motion study was set to 4 s. The simulation outcomes regarding the displacement of the system are shown graphically in Figure 11.

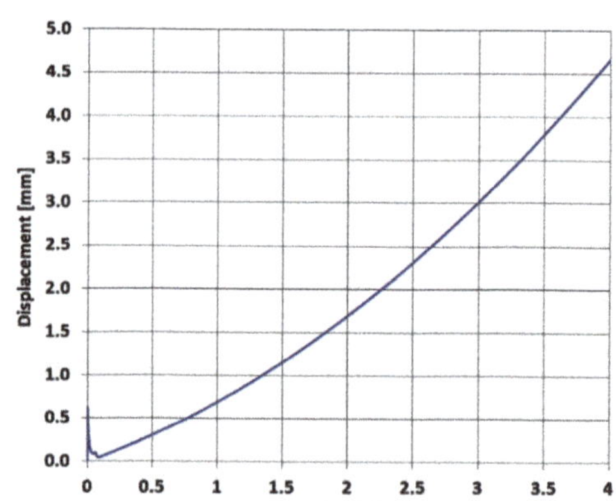

Figure 11. Results of movement simulation at n = 3900 rpm and T = 4 s.

As one can observe, the behavior of the IPD is identical to that shown during the simulation presented in Chapter 5. This means that, immediately after the system starts to be driven, its movement increases sharply, and then decreases, just as suddenly, after which, the distance traveled by the device increases smoothly. At the end of the analysis time, the IPD has covered a distance of 4.67 mm. The distances traveled by the IPD after intermediate durations of 2, 2.5, 3 and 3.5 s are shown in Table 2.

Table 2. Displacements resulted from SolidWorks (SW) simulation after different times.

Time t [s]	2	2.5	3	3.5	4
Distance y [mm]	1.69	2.32	3.02	3.80	4.67

For the experimental tests, the device displacements after durations of 2, 3 and 4 s, respectively, were measured with a dial gauge. For each of the above mentioned durations, three distance measurements were made, the average distance of the three measurements being compared with the simulation outcomes. Table 3 presents the measurement results, in opposition to the simulation outcomes and the percentage differences between the two approaches.

Table 3. Comparison of displacements obtained by experimental measurement and simulation.

Time t [s]	Measured Distance [mm]			Average Distance y_{med} [mm]	Simulation Distance y_{sim} [mm]	Deviation Δy [%]
	y_1	y_2	y_3			
2	1.55	1.49	1.53	1.52	1.69	10.06
3	2.85	2.83	2.86	2.85	3.02	5.74
4	4.25	4.15	4.23	4.21	4.67	9.85

As it can be observed, there is a good correlation between the simulation and the experimental results, with maximum differences of 10%. This confirms, on the one hand, that the experiments were set correctly and, on the other hand, that the 3D model and numerical simulation were performed correctly.

As the ability of the IPD concept to develop unidirectional movement has been also proven by experimental test, the future researches are aiming to study the efficiency of the drive and the influence of external factors, such as friction between wheels and ground, in order to improve the construction and to optimize the kinematic and dynamic behavior.

7. Conclusions

In this paper, a multibody inertial propulsion drive with symmetrically placed balls rotating on eccentric trajectories was presented. In this context, three versions of trajectories for the rotating masses of the IPD were investigated. Starting from the coordinate equations of the balls, their velocities and accelerations were analytically deducted, for each of the constructive version. Furthermore, using the Motion module from SolidWorks (SW), a kinematic and dynamic simulation was completed, the outcomes being examined in contrast with the analytical results. It was observed that in case of the displacements and velocities, the results obtained by the two approaches are almost identical, for all the three constructive variants of the retaining disk. The bigger differences in matter of accelerations between simulation and the analytic procedure were explained by the fact that in the analytical approach the collisions between the balls and the inner bore of the retaining disk were neglected. Furthermore, the 5° step with which the motion analysis was performed, in order to have a reasonable duration of the simulation, could be another reason for these differences.

Moreover, the kinematic and dynamic performances of the three constructive versions were compared. In terms of displacements, Version 1 ensures a 5, respective 10 times longer displacement within the simulation time, compared to Versions 2 and 3. Regarding velocities, the IPD equipped with Version 1 of the retaining disk is in average with 4.61, respective 2.17 faster than the IPD's foreseen with retaining disks constructed in Version 2 and 3 respectively. Moreover, Version 1 requires by far the smallest power consumption for driving the system. Therefore, it was concluded that the variant of retaining disk with cylindrical bore placed eccentric relative to the center of the slotted plates (Version 1) is the most advantageous version of the IPD regarding displacement, velocity and power consumption.

For proofing the ability of the device to generate linear movement, experimental tests were also performed. Thus, operating the system with five different driving speeds, it was obvious that the IPD's average displacement speed grows with the increase of the driver speed.

Finally, the present study confirms that the IPD developed by the authors is functional and capable to generate unidirectional linear movement, being especially suitable for spaces where the gravity is missing.

Author Contributions: Conceptualization, Z.-I.K. and D.N.; methodology, Z.-I.K.; software, D.N.; validation, Z.-I.K., G.-R.G. and D.N.; writing—original draft preparation, A.G.; writing—review and editing, Z.-I.K.; visualization, G.-R.G. and D.N.; supervision, Z.-I.K.; project administration, A.G. All authors have read and agreed to the published version of the manuscript.

Funding: This research received no external funding.

Conflicts of Interest: The authors declare no conflict of interest.

Abbreviations

This table lists the acronyms used in this manuscript:

Acronym	Meaning
3D	Three Dimensional
IPD	Inertial Propulsion Drive
NASA	National Aeronautics and Space Administration
PLA	Polylactide
SW	SolidWorks

Appendix A

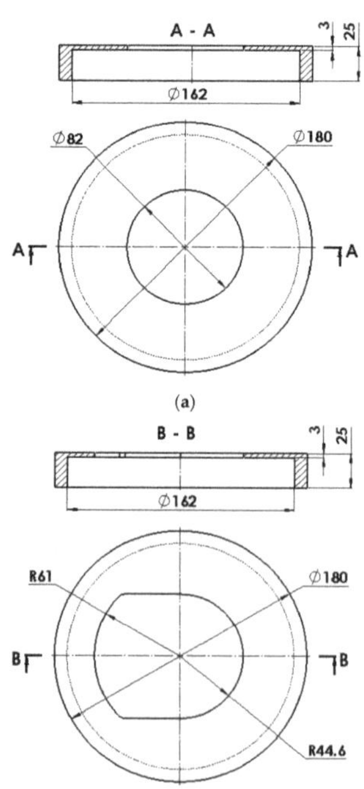

(a)

(b)

Figure A1. *Cont.*

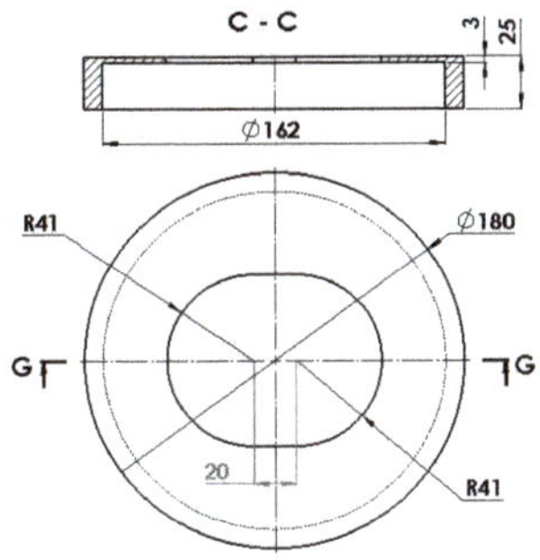

(c)

Figure A1. Constructive alternatives of the retaining disk. (**a**) Version 1; (**b**) Version 2; (**c**) Version 3.

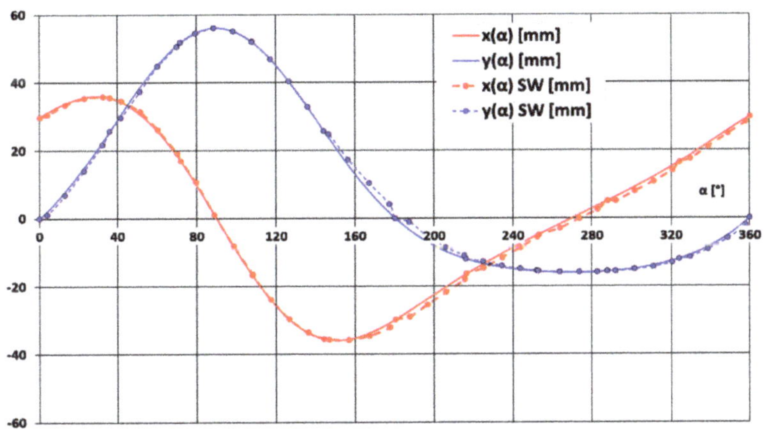

(**a**)

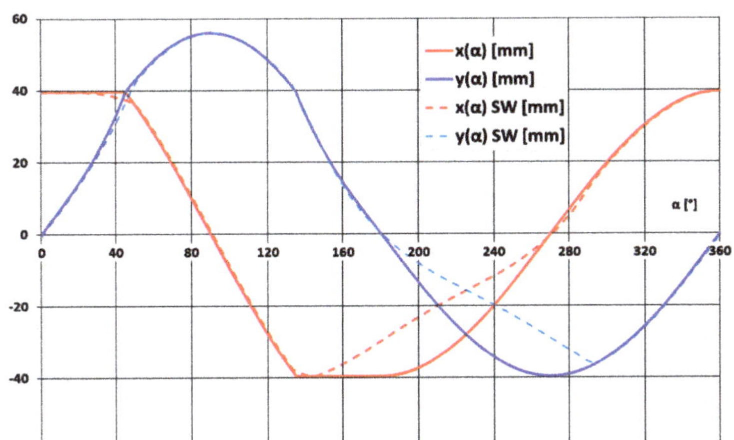

(**b**)

Figure A2. *Cont.*

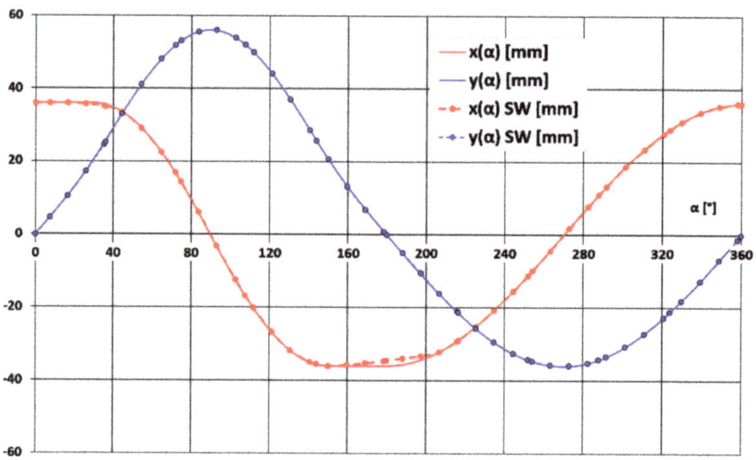

(c)

Figure A2. Comparison of simulation outcomes and analytical calculations for the coordinates of a ball. (**a**) Version 1; (**b**) Version 2; (**c**) Version 3.

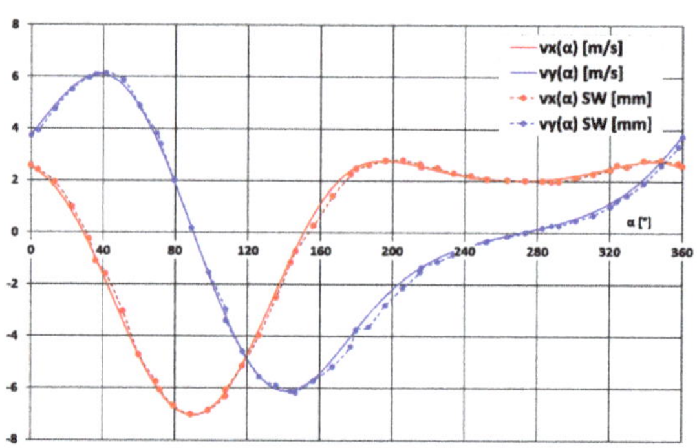

(a)

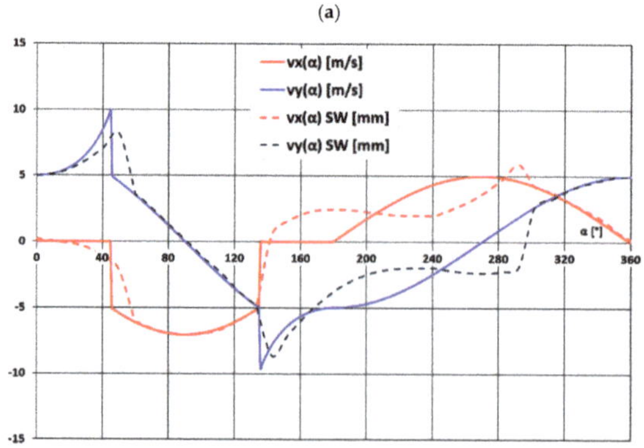

(b)

Figure A3. *Cont.*

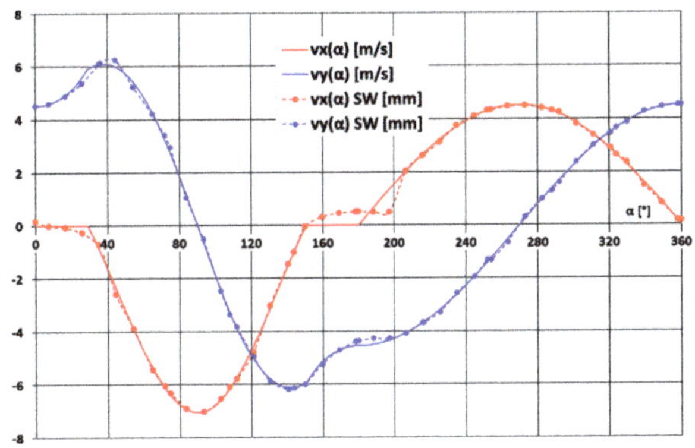

(c)

Figure A3. Comparison of simulation outcomes and analytical calculations for the velocities of a ball. (**a**) Version 1; (**b**) Version 2; (**c**) Version 3.

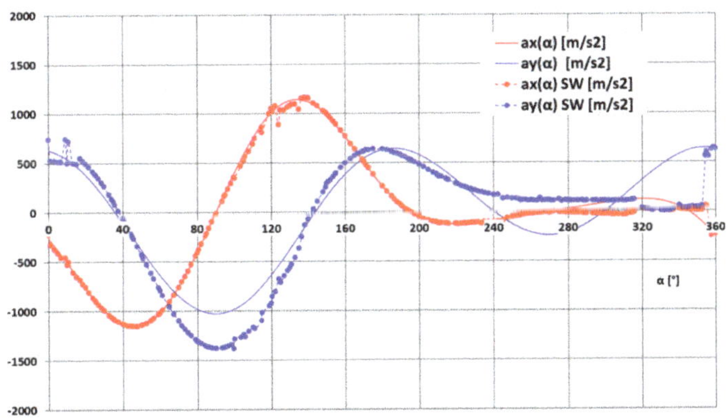

(**a**)

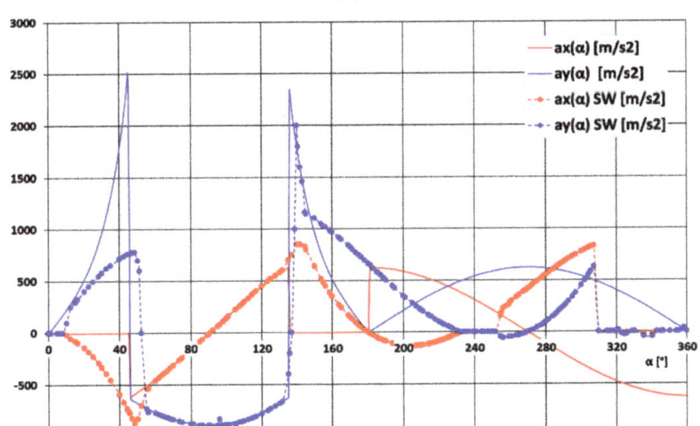

(**b**)

Figure A4. *Cont.*

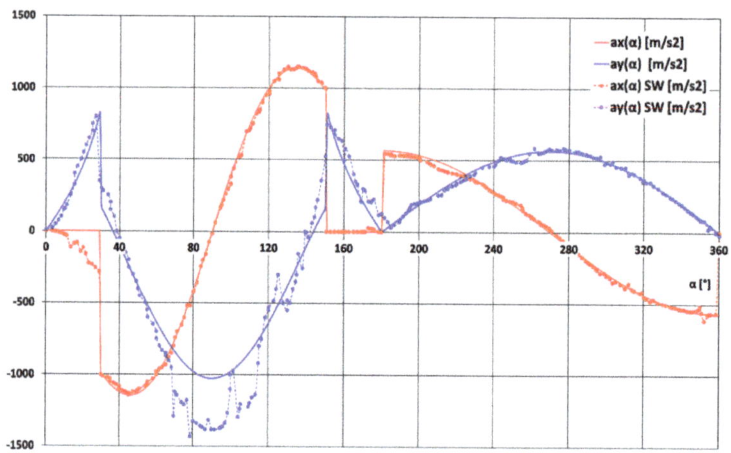

(c)

Figure A4. Comparison of simulation outcomes and analytical calculations for the accelerations of a ball. (**a**) Version 1; (**b**) Version 2; (**c**) Version 3.

References

1. Loukanov, I. Application of Inertial Forces for Generating Unidirectional Motion. *Sci. Rep. Univ. Ruse* **2014**, *53*, 9–19.
2. Millis, M.G.; Thomas, N.E. Responding to Mechanical Antigravity, NASA/TM-2006-214390, AIAA-2006-4913. In Proceedings of the 42nd AIAA/ASME/SAE/ASEE Joint Propulsion Conference & Exhibit, Sacramento, CA, USA, 12 July 2006.
3. Allen, D.P., Jr. *Why Does Classical Mechanics Forbid Inertial Propulsion Devices When They Evidently Do Exist? Is Newtonian Mechanics a Done Deal?* Createspace Independent Pub: Scotts Valley, CA, USA, 2018.
4. Gutsche, G.J. *Inertial Propulsion: The Quest for Thrust from Within;* Createspace Independent Pub: Scotts Valley, CA, USA, 2014.
5. Fiala, H.E. *An Inertial Propulsion Patient & Working Model. Presentation, Albuquerque, NM, USA, 29 July 2012;* Tesla Tech Inc.: San Carlos, CA, USA, 2012.
6. Gutsche, G.J. *Inertial Propulsion; and You Thought It's Impossible;* Createspace Independent Pub: Scotts Valley, CA, USA, 2018.
7. Coulombe, M. Differential Displacement Device under Simultaneous and Repetitive Electromagnetic Repulsive Forces. US Patent No. 7909669, 10 May 2010.
8. Benjamin, P.M. Centrifugal Thrust Motor. US Patent No. 3750484, 7 August 1973.
9. Booden, J.D. Electromagnetically Actuated Thrust Generator. US Patent No. 5782134, 21 July 1998.
10. Cuff, C.I. Device for Converting Rotary Motion into Unidirectional Motio. US Patent No. 4095460, 20 June 1978.
11. Dobos, E.M. Propulsion Apparatus. US Patent No. 4579011, 1 April 1986.
12. Farrall, A.W. Inertial Propulsion Device. US Patent No. 3266233, 16 August 1966.
13. Fulop, C. Flywheel. US Patent No. 4788882, 6 December 1988.
14. Haller, P. Propulsion Apparatus. US Patent No. 3177660, 13 April 1965.
15. Kellogg, H.D. Gyroscopic Inertial Space Drive. US Patent No. 3203644, 31 August 1965.
16. Mendez Llamozas, J.D. Direct Push Propulsion Unit. US Patent No. 2636340, 28 April 1953.
17. North, H. Apparatus for Producing a Force. US Patent No. 4712439, 15 December 1987.
18. Oades, R.A. Apparatus for Generating a Propulsion Force. US Patent No. 5890400, 6 April 1999.
19. Shimshi, E. Apparatus for Energy Transformation and Conservation. US Patent No. 5673872, 7 October 1997.
20. Schnur, N.J. Method and Apparatus for Propelling an Object by an Unbalanced Centrifugal Force with Continuous Motion. US Patent No. 3979961, 14 September 1976.

21. Deschamplain, D. Motion Imparting System. US Patent No. 6259177B1, 10 July 2001.
22. Marsh, R.O. Centrifugal Drive Machine. US Patent No. 5388470, 14 February 1995.
23. Murray, L.D. Mechanical Force Generator. US Patent No. 6290622B1, 18 September 2001.
24. Kunz, W.T. Centrifugal Propulsion System. US Patent No. 5937698, 17 August 1999.
25. Woltermg, H.M. Rotating Eccentric Weights Vibrator System. US Patent No. 5388469, 14 February 1995.
26. Dean, N.L. System for Converting Rotary Motion into Unidirectional Motion. US Patent No. 2886976, 19 May 1959.
27. Thornson, B.R. Apparatus for Developing Propulsion Force. US Patent No. 4631971, 30 December 1986.
28. Wang, Q.; Ju, J. Analysis of the effect of external factors to the propulsion by inertial centrifugal force of rotating mechanism. *Appl. Mech. Mat.* **2014**, *487*, 33–47. [CrossRef]
29. Malin, C.T.; Nedelcu, D.; Gillich, G.R.; Petrica, A.; Padurean, I. Comparison of the performance of friction pendulums with uniform and variable radii. *Vibroeng. Procedia* **2019**, *23*, 81–86.
30. Gillich, G.R.; Nedelcu, D.; Malin, T.C.; Iancu, V.; Hamat, C.O.; Gillich, N. The effect of the friction coefficient and the pendulum radius on the behavior of structures isolated with simple friction pendulums. *Rom. J. Acoust. Vibr.* **2018**, *15*, 130–135.
31. Anand, G.; Jobin, J.; Vijayan, K. Optimization of inertial propulsion system for future space application. *Am. Int. J. Res. Sci. Techn. Eng. Math.* **2014**, *7*, 95–100.
32. Oproescu, G.; Cautes, G. The model of terrestrial vehicle with inertial propulsion. In *The Annals of "Dunarea de Jos" University of Galati, 2005 Fascicle XIV Mechanical Engineering*; Galati University Press: Galati, Romania, 2005; pp. 33–37.
33. Provatidis, C.G. A device that can produce net impulse using rotating masses. *Engineering* **2010**, *2*, 648–657. [CrossRef]
34. Robertson, G.A.; Webb, D.W. The Death of Rocket Science in the 21st Century. *Phys. Procedia* **2011**, *20*, 319–330. [CrossRef]
35. Geröcs, A.; Korka, Z.I. Inertial Drive System. Patent application no. RO133571-A2, 30 August 2019.
36. Geröcs, A.; Korka, Z.I. Analytical investigations on the influence of the geometry of an inertial drive on the propulsion force. In *Annals of "Eftimie Murgu" University of Reşiţa*; Reşiţa University Press: Reşiţa, Romania, 2019; Volume 26, pp. 76–85.
37. Geröcs, A.; Korka, Z.I.; Biró, I.; Cojocaru, V. Analytical investigation of an inertial propulsion system using rotating masses. *J. Phys. Conf. Ser.* **2020**, *1426*, 012031. [CrossRef]
38. Nedelcu, D.; Gillich, G.R.; Bloju, A.; Padurean, I. The Kinematic and Kinetostatic Study of the Shaker Mechanism with SolidWorks Motion. *J. Phys. Conf. Ser.* **2020**, *142*, 012025. [CrossRef]
39. Nedelcu, D.; Nedeloni, M.D.; Daia, D. The Kinematic and Dynamic Analysis of the Crank Mechanism with SolidWorks Motion. In Proceedings of the 11th WSEAS International Conference on Signal Processing, Computational Geometry and Artificial Vision 2011, Florence, Italy, 23–25 August 2011; pp. 245–250.
40. Dassault Systems. *SolidWorks 2010 Motion*; SolidWorks Crop.: Concord, MA, USA, 2010.
41. Gerocs, A.; Korka, Z.I. Gerocs & Korka's Inertial Propulsion Drive. Available online: https://www.researchgate.net/publication/342601147_Gerocs_Korka\T1\textquoterights_Inertial_Propul-sion_Drive (accessed on 1 July 2020).

symmetry

MDPI

Article

Experimental Assessments on the Evaluation of Wire Rope Characteristics as Helical Symmetrical Multi-body Ensembles

Gina Diana Musca (Anghelache) [1,*], Carmen Debeleac [1] and Sorin Vlase [2]

1 Engineering and Agronomy Faculty in Braila, Research Center for Mechanics of Machines and Technological Equipments, "Dunarea de Jos" University of Galati, 810017 Braila, Romania; carmen.debeleac@ugal.ro

2 Department of Mechanical Engineering, Faculty of Mechanical Engineering, Transilvania University of Brașov, B-dul Eroilor 29, 500036 Brașov, Romania; svlase@unitbv.ro

* Correspondence: danghelache@ugal.ro; Tel.: +40-374-652-572

Received: 4 June 2020; Accepted: 24 July 2020; Published: 26 July 2020

Abstract: The existing literature provides various computational models related to the dynamic behavior of strand wire ropes. It starts from the simple longitudinally oscillating beam, to the complex nonlinear multi-body configuration based on helical structural symmetry. The challenge is the prior availability of characteristic parameters for material behavior, structural configuration, and functional capability. Experimental investigation is the main source for evaluation of these characteristics. However, tests have specifically been performed according to each case, minimizing the generalization aspect. This is the main frame of this study. Hereby, the authors propose an ensemble of spectral investigations, applied to a reduced set of experimental tests regarding wire rope dynamics. The research goal consists of wire rope characterization in terms of the flexible and adaptive groups of parameters, related to the conservative and dissipative behaviors. An experimental setup is considered here according to the rope exploitation conditions in order to enable an extension of the method application from the experimental mode to the operational mode. Experiments are conducted based on classical vibration measurement procedures. The analysis is performed using a spectral method ensemble, including discrete Fourier transform, time-frequency joint analysis, and the Prony method. The result show that the proposed assessments can provide suitable information related to a large group of wire rope models.

Keywords: strands wire rope; experimental transitory vibrating regime; stiffness; damping; joint time-frequency analysis; Prony method; matrix pencil method

1. Introduction

Strand wire ropes take part in most existing engineering elements. Wire ropes have a very large area of utilization, ranging from various types of anchorages, traction systems, and industrial applications, to cranes and hoisting systems. Being widely used and providing a very important structural/functional role, strand wire ropes have enabled many actual studies, specifically in terms of being able to accurately evaluate behavioral characteristics under operational conditions. It is also important to predict the extreme charging conditions and the risks in the exploitation of these flexible elements.

The authors of [1] present a co-simulation method based on both multi-body dynamics and finite element analysis. They have developed a dynamic model in the ADAMS/Cable software package related to preliminary evaluation of the time history response of dynamic force in wire rope. Secondly, they have implemented a finite element-based model within ABAQUS FEA software, followed by static and dynamic computational investigations. The computational results were compared with experimental data, and the results proved accuracy and reliability. Concluding, the authors present the stress distribution and spectrum of wire rope under dynamic loading during the hoisting process and show the most dangerous regions.

The work of [2] contains a model for the dynamic analysis of a load lifting system. The novelty of this analysis is sustained by the method used for solving the model, which is a high-precision direct integration method. Hereby, the author obtained more precise results in the same time with less computing time and high accuracy.

Within [3], Haniszewski proposed a test setup for experimental investigations regarding dynamic phenomena accompanying the lifting process of a load, which allows simultaneously monitoring of a large number of parameters in different regions of the crane system. Based on this test bench, the author developed both experimental and analytical investigations. The same author treated the problem of negative vibrations, which can appear during the use of hoisting systems [4]. Taking into account the various available models for rope and the major importance of rope within the entire hoisting system, he proposed both modified Bouc–Wen and Voigt–Kelvin models. The results were analyzed in the terms of a dynamic factor.

In his paper [5], Argatov presents a refined variant of a discrete mathematical model that is useful in the analysis of a simple helical wire rope. The model details the transverse contraction of a strand due to both the Poisson's ratio and the local contact deformations. The last aspect was approached as a frictionless unilateral plain strain problem. The author obtained an asymptotic model with constitutive equations in a closed form, being able to accurately predict the deformations of a wire rope strand subjected to both tensile and torsional loads.

A new theoretical method for the analysis of wire ropes simultaneously charged by both tensile and torsional loads was proposed in [6]. The study used the beam assumption and thin rod theory, applied to each wire within the strand structure. A kinematic model was developed on the basis of the hypothesis that is no relative sliding between adjacent wires.

Experiments based on monitoring the longitudinal vibration of a wire rope were developed by Hamilton and presented as part of a study [7]. The author used the principle of a spring-mass oscillator in order to evaluate the parameters of the rope in terms of the elastic and damping properties. The study investigated the transitory vibrating regime of the experimental system containing a mass suspended by a wire rope. The results revealed an acceptable accuracy for wire rope lengths less than 2 m and with differences less than 1% when changing the length.

Simple and reliable theoretical models were presented by Raoof and Davies in their work [8], based on correlations developed by Raoof and Kraincanic. These models are useful to characterize independent wire rope core (IWRC) or fiber core (FC) large-diameter multi-strand wire ropes. The analytical expressions presented by the authors allow the prediction of the axial stiffness with the hypothesis of no-slip or full-slip. These simple models were established based on a very large experimental third-party database which sustains the reliability of the approaches.

The Stomer–Verlet method presented in [9] is very useful in the motion analysis of a multi-body system with wire rope connections between bodies. This method was derived from a discrete Euler–Lagrange (DEL) equation. One main advantage of this approach comes from the fact that the stretching of wire ropes can be mathematically modeled as a constraint for stability. During computational investigations, the DEL equation, with wire ropes as constraints, provides relatively stable solutions.

Within their studies [10,11], Kaczmarczyk and Ostachowicz proposed an interesting distributed parameter model for the dynamic analysis of the behavior of a deep mine hoisting cable. They approached the non-stationary aspects of the problem due to the coupled transversal-longitudinal responses, involving non-linear partial differential equations in a mathematical model. A discrete formulation of this model was obtained using a Rayleigh–Ritz procedure, finally using a non-linear, non-stationary, coupled second-order ODE system applicable for computational-based analyses [10].

The authors treated the problem both in fast and slow time scales in order to investigate the complete response of the system. Equivalent proportional damping was considered for the solving procedure for the stiff problem. The results indicate various transient resonances within non-linear evolutions of the system. An important finding is that small changes in excitation characteristics are able to induce large changes in dynamic response due to the shifting of the resonances [11].

Computational approaches regarding the behaviors of the wire ropes and cables during intensive and variable dynamic exploitation regimes can be consulted in various research reports [12–14]. Hereby, extensive analyses of the problem regarding cable dynamics during the lifting process have been provided within studies [12,13], where the simulations were based on linear and non-linear, lumped or continuous mass distribution models. In addition, the dynamics of wire ropes, such as in the main hoisting component of technological equipment, have been treated in computational research [14].

Hobbs and Raoof presented a relevant analysis procedure based on an exhaustive literature review inside their research paper [15]. The contact between the wires within the rope strands, the axial stiffness, the hysteresis variability with the charge characteristics, the loaded cable bending problem, and the axial cyclic charging of strand-based wire ropes were clearly presented and extensively discussed within this article. In addition, particular attention was given to the aspects regarding the localized and repeated bending phenomena that frequently occurs close to the cable terminations.

An interesting approach successfully combining computational-based simulation with experimental investigations is provided within [16]. In addition, it has to be mentioned that the second edition of Costello's study [17] details a comprehensive analysis of wire behavior dynamics and provides useful information for computational approaches based on extensive experimental investigations. In the same area of experimental-based analyses of lifting process dynamics, the research presented in [18] must be mentioned.

Demšić and Raduka [19] proposed an analytical model developed for inhomogeneous boundary conditions which takes into account the system nonlinearities in terms of quadratic and cubic formulations. The authors used both a numerical integration procedure and multiple scales of a perturbation method in order to solve the reduced mathematical model and the parametric oscillations, respectively.

A comprehensive study regarding special cables used at cable-supported bridges was provided within [20]. The authors investigated a wide palette of parameters that are able to characterize the bridge behavior under dynamic loads. Among these, they analyzed the anomalies of damping properties and nonlinear parametric vibration. One has to note the experimental developments presented within this study.

Elata et al. proposed an interesting computational model for the dynamic response analysis of wire ropes with an independent core that are axially charged by a load and a torque [21]. The model fully supposed a double-helix configuration for component wires and assumed a fiber elastic response of individual wires, thus it becomes able to provide the values of the stress at the wire level.

The previous briefly introduction does not have the intention to be exhaustive but has the role of revealing the available literature related to theoretical and applicable research studies into the field of wire ropes. These studies are useful through their gains in dynamics modeling, simulation, and practical implementation. Considering the wide applications of strands wire ropes, this clearly results in the necessity of large and various categories for knowing the characteristic parameters in order to carry out reliable computational analyses. However, the available studies individually treat each type of problem regarding the wire rope dynamic behavior. Operational tests are rather nonexistent. Experimental investigations have been strictly developed for particular cases of simulation models and they do not supply generalizable characteristics. Computational models have plenty of representations, both from the point of view of behavioral schematizations and also solving techniques. Nevertheless, these approaches require many parameters to accurately perform the analysis.

Hereby, the present study presents some useful assessments based on experimental dynamics investigation which are able to provide a flexible and adaptive set of data according to a large category of numerical models. Flexible means that the post-processing procedure can be extended in respect to the requirements (in terms of parameter type and precision). This is a qualitative approach. Adaptive means that the number of evaluated parameters can be permanently harmonized with the model order. This is a quantitative approach. The main advantages are supplied by the reduced experimental investigations volume, and, in addition, by the capability of operational implementation (where the exploitation regime implies dynamic evolution or the system supports external dynamic perturbation). Thus, the goal of this research is the estimation of wire rope characteristics, closely related to the model complexity, where these parameters will be assumed.

2. Materials and Methods

This research was primarily based on experimental investigations which have developed using a dedicated laboratory setup. Thus, the first stage of analysis features a set of experimental tests developed on strands wire rope samples subjected to both static and dynamic loads. The second stage of analysis features the acquired data post-processing, which were used to evaluate the characteristic parameters derived from the wire rope transitory dynamics.

2.1. Experimental Setup

The experimental laboratory setup used for the first stage of analysis is presented in Figure 1. It consists of a rigid latticed tower that sustains the inspected wire rope sample. The free end of the rope can be charged by a permanent static load and, additionally, by a dynamic excitation. For analysis, a short dynamic pulse was used, in terms of singular acting and strong enough hammer impact [22]. The shock was axially applied downwards of the suspended mass (static load) in a vertical direction, thus, the secondary effects regarding the transversal dynamics of the wire rope were minimized.

Dynamic evolution of the entire ensemble was monitored by a computer-based analysis system that is able to acquire signals from both accelerometers and force transducers (see Figure 1a). The acquisition was assured by the cDAQ-9174© chassis (National Instruments, USA), supplied with NI-USB-9233© (National Instruments, USA) for accelerometers and NI-USB-9237© (National Instruments, USA) for the strain–bridge-based force transducers.

The main parameters regarding the motion of the loading mass hung from the free end of the tested wire rope were monitored through the mass vertical direction oscillations and, respectively, the force within the rope sample (see Figure 1c). These jobs were carried out with the help of a CCP-ICP© 320C34 (PCB, USA) uniaxial accelerometer with a magnetic mounting base and a S9-20kN (HBM, Germany) strain-bridge based traction force transducer.

The unavoidable residual transitory dynamics of the tower were directly monitored at its upper side, nearby the device that sustains the inspected rope (see Figure 1b). This operation was carried out by two accelerometers (the same type as that mounted on the loading mass), mounted for both the vertical and the horizontal directions of possible motion provided by the free end of the tower. It was assumed that the lateral oscillations of the tower during the tests could be neglected (there were no external excitations applied in that direction and the experimental site does not supply any essential basement displacements).

(a) (b) (c)

Figure 1. The laboratory experimental setup: (**a**) General view with a latticed tower, wire rope sample, static charging device, and compute-based acquisition system; (**b**) detailed view of the upper side of the tower, with the hung rope device and residual motion monitoring transducers; (**c**) detailed view of the wire rope charging device with an additional mass applied and transducers for monitoring the forces and oscillations, respectively.

2.2. Tested Wire Rope Sample Characteristics

The strands wire rope considered in this analysis was a regular 6(12+*FC*)+*FC* commercial cable, type CA1AA072A© (*CABLERO, Romania*), 6 mm diameter, based on 6 strands with 12 steel wires of 0.4 mm diameter, designed for a 288-kg maximum service load. The symbol *FC* within its code means that the rope has a fiber core, both for the strands and for the rope. The unitary length mass is 0.11 kg/m and the theoretical failure load is 9.82 kN.

According to [23] (pp. 40–45), the fill factor f and the rope mass factor w can be used in order to characterize the nominal metallic cross-section area S from the circumscribed area S_u of the cable, and, respectively, the nominal rope length-related mass m.

Taking into account the diameters ratio $\delta/d = 0.4/6 = 0.067$, the fill factor f yields a value of 0.32. Thus, the nominal metallic cross-section area yields $S = (\pi/4)\, fd^2 = 9.05\ \text{mm}^2$. Respectively, with a rope mass factor $w = 0.306$, the nominal rope length-related mass, evaluated with widely accepted empirical expression $m = 0.01\ wd^2$, acquires 0.11016 kg/m, which accurately approximates the initially provided value from the producer.

The axial stiffness E_{wr} of a wire rope was evaluated using Hruska's approach [8]:

$$H = \frac{E_{wr}}{E_{steel}} = \frac{\sum\limits_{j=1}^{m}\left(\sum\limits_{i=1}^{n} S_i \cos^3\alpha_i\right)\cos^3\beta_j}{\sum\limits_{j=1}^{m}\left(\sum\limits_{i=1}^{n} S_i/\cos\alpha_i\right)\Big/\cos\beta_j}, \tag{1}$$

where E_{steel} denotes the steel Young's modulus, m is the total number of the rope strands, n denotes the total number of wire layers in each strand, including the king wire that is considered the first layer, and S_i is the cross-sectional area of wire within layer i of a strand j. The angles α_i and β_j respectively denote the lay angle of layer i in strand j and the lay angle of the strand j of the wire rope. Analyzing the denominator of the right-hand-side term in Equation (1), it can obviously be seen that it represents the total net steel area of the normal cross-section of the wire rope. The equation uses the hypotheses of a king wire having $\alpha_i = 0$ and the elliptical shapes of wires and strands within the normal cross-section of the rope.

The parameters in Equation (1) were evaluated based on the schematization depicted in Figure 2a, where $2r_s$ denotes the diameter of the strand circular distribution, D is the wire rope diameter, L is the cable sample length, h_s denotes the strand step, and β is the lay angle of the strand. The evaluation of the lay angle of the wires within the ropes, α_i, can be obviously obtained based on the same procedure.

The images in Figure 2b contain a region of the cable sample and a sectional view, respectively, and this helps to identify the geometrical parameters of the wire rope sample in terms of the helical step, diameter of circular distribution, and the lay angles of the wires and the strands.

Taking into account that the area S_i and angles α_i and β_i have constant values, Equation (1) acquires a simple expression in respect to these angles. For the inspected rope type, the strand diameter is 2 mm, the strands step is 46 mm, and the wires step (within the strand) is 20 mm. Hereby, $H = 0.8152$ and the axial stiffness $E_{wr} = 1.712 \times 10^5$ MPa (assuming $E_{steel} = 2.1 \times 10^5$ MPa), which is a value that is framed by the available literature data [23]. In addition, supposing the usual elongation of $\varepsilon \cong 1.7\%$, the stress yield $\sigma \cong 2910$ MPa (also in the range of available values for this kind of small diameter FC steel wire ropes).

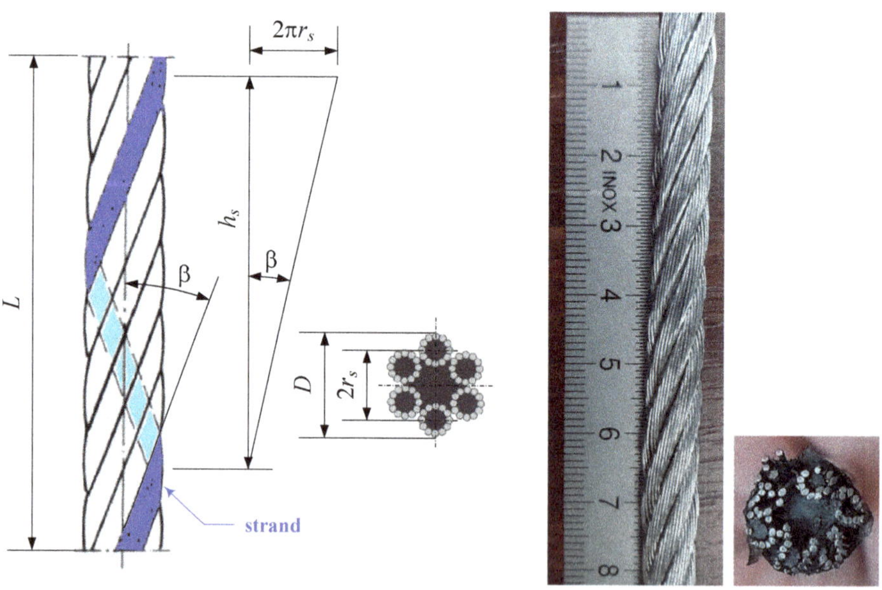

(a) (b)

Figure 2. The evaluation of the wire rope geometrical parameters: (**a**) The model of the strands wire rope, used for evaluation of the parameters involved in the lay angle computation of the strand and the rope, respectively [23] (p. 25); (**b**) detailed images of the wire rope used for the experiments, showing longitudinal and sectional views.

2.3. Post-Processing Techniques

The second stage of analysis featured the numerical investigations of the acquired signals (in terms of forces and accelerations) in order to accurately estimate the main parameters of the dynamic evolution of the experimental ensemble and, finally, to obtain the characteristics of the wire rope sample. It was initially assumed that the characteristics might be required not as simple basic values, but as extended sets of data, being able to accurately approximate the nonlinear dynamics of such a system.

Available methods for harmonics investigations include Fast Fourier Transform (FFT) techniques, adaptive filter applications, artificial neural networks, singular value decompositions (SVD), and higher-order spectra (HOSA) [22,24]. Most of these methods are able to work properly for moderate noise levels within analyzed signals and for a narrow domain of frequencies. These methods suppose that the signal only contains harmonics with fixed periodicity intervals. However, real signals might contain (and this is usually the case) inter-harmonics when the periodicity provides variable and very long intervals.

The Prony method, which is intended for use here, assumes that a sampled signal can be re-formulated as a linear combination of exponential functions [25–28]. This approach is not a spectral estimation technique, but it has a close relationship to squares linear prediction algorithms, which are usually used in auto-regressive (AR) or auto-regressive moving average (ARMA) parameter estimation methods. Opposite to AR/ARMA techniques, which try to fit a random model to the second-order data statistics, the Prony method seeks to fit a deterministic exponential-based model to the analyzed signal.

According to the previously mentioned arguments, in order to extract sinusoidal and/or exponential signals from time series data, the Prony method provides a feasible tool. This method allows the decomposition of a complex signal to a sum of exponential functions and transforms this entire computational process into solving a set of linear equations [25–28].

Hereby, assuming a signal $x(t)$, available through n samples, where $x[1]$, $x[2]$, ..., $x[n]$, it can be approximated by m terms [25–28]:

$$y[i] = \sum_{k=1}^{m} A_k \exp[(b_k + j\omega_k)(i-1)T_s + j\psi_k], \tag{2}$$

where $i = \overline{1, n}$, A_k denotes the amplitude, T_s denotes the sampling period, α_k denotes the damping factor, ω_k denotes the angular velocity, ψ_k denotes the initial phase, and $j^2 = -1$. This discrete-time expression can be concisely formulated in the form of the following:

$$y[i] = \sum_{k=1}^{m} h_k z_k^{i-1}, \tag{3}$$

where, based on exponentials, the following can be expressed:

$$h_k = A_k \exp(j\psi_k),\ z_k = \exp[(b_k + j\omega_k)T_s]. \tag{4}$$

Considering Equation (3), the estimation problem results from the minimization of the squared error over the n initial signal values (initial data) as follows [25–28]:

$$\Delta = \sum_{i=1}^{n} \left|\varepsilon[i]\right|^2, \tag{5}$$

where

$$\varepsilon[i] = x[i] - y[i] = x[i] - \sum_{k=1}^{m} h_k z_k^{i-1}. \tag{6}$$

Taking into account this nonlinear problem, we adopted the Prony method, which uses linear equation solutions [26–28]. In fact, this method (also known as a polynomial Prony method), assumes an autoregressive model of order p, which supposes that the value of the sampled signal $x[i]$ linearly depends on the preceding p values of x. In order to apply this method, one may use one of the available techniques for solving. Hereby, the original method proposed by Prony (a two-step method which additionally imposes an $n = 2p$ condition and, in some cases, can become unstable due to ill-conditioned matrix equation solving) was considered. In addition, it can be taken into account the matrix pencil method (MPM), which solves an eigenvalues problem (with the main advantage that is less sensitive to the noises within the signal) [25,28].

In this study, the authors have chosen the MPM. This method starts from two matrices, namely, $\mathbf{Y}_1 \in C^{mxn}$ and $\mathbf{Y}_2 \in C^{mxn}$. Generally, the set of matrices, with the expressions $(\mathbf{Y}_2 - \lambda\mathbf{Y}_1)$ and $(\lambda \in C)$, form a matrix pencil. For the case of sampled initial signal x, a rectangular Henkel matrix $\mathbf{Y}$ is considered, with the pencil parameter p, in the following form [25,28]:

$$\mathbf{Y} = \begin{pmatrix} x[1] & x[2] & \cdots & x[p] & x[p+1] \\ x[2] & x[3] & \cdots & x[p+1] & x[p+2] \\ \vdots & \vdots & \ddots & \vdots & \vdots \\ x[N-p] & x[N-p+1] & \cdots & x[N-1] & x[N] \end{pmatrix}_{(N-p)\times(p+1)}. \tag{7}$$

The matrix $\mathbf{Y}$ helps to obtain the matrices $\mathbf{Y}_1$ and $\mathbf{Y}_2$. Thus, $\mathbf{Y}_1$ results from $\mathbf{Y}$ without the last column and, respectively, $\mathbf{Y}_2$ results from $\mathbf{Y}$, but without the first column [25–28]:

$$\mathbf{Y}_1 = \begin{pmatrix} x[1] & x[2] & \cdots & x[p] \\ x[2] & x[3] & \cdots & x[p+1] \\ \vdots & \vdots & \ddots & \vdots \\ x[N-p] & x[N-p+1] & \cdots & x[N-1] \end{pmatrix}_{(N-p)\times p}, \mathbf{Y}_2 = \begin{pmatrix} x[2] & \cdots & x[p] & x[p+1] \\ x[3] & \cdots & x[p+1] & x[p+2] \\ \vdots & \ddots & \vdots & \vdots \\ x[N-p+1] & \cdots & x[N-1] & x[N] \end{pmatrix}_{(N-p)\times p}. \tag{8}$$

Supposing that m denotes the real number of poles of the function $x[i]$, with the conditioning $m \leq p \leq n - m$, it can be seen that z_k, where $(k = 1...p)$, are the generalized eigenvalues of the matrix pencil $(\mathbf{Y}_2 - \lambda\mathbf{Y}_1)$. Taking into account that matrices $\mathbf{Y}_1$ and $\mathbf{Y}_2$ are ill-conditioned, the QZ algorithm is not stable enough to provide adequate results. In this case, it is more accurate to use the following expression:

$$z_k = eig\left(\mathbf{Y}_1^+ \mathbf{Y}_2\right), \tag{9}$$

where $\mathbf{Y}_1{}^+$ denotes the Moore–Penrose pseudo-inverse matrix of $\mathbf{Y}_1$. In such conditions, the damping b_k and the frequency $f_k = 2\pi\omega_k$ results from the eigenvalues of z_k [25,28].

$$b_k = \frac{\ln|z_k|}{T_s},\ f_k = \frac{1}{2\pi T_s}\arctan\left[\frac{\mathrm{Im}(z_k)}{\mathrm{Re}(z_k)}\right]. \tag{10}$$

Finally, by solving the system [25,28]

$$\mathbf{Z}_{p\times p}\,\mathbf{h}_{p\times 1} = \mathbf{x}_{p\times 1}, \tag{11}$$

the amplitude A_k and the phase ψ_k [25,28] may be found:

$$A_k = |h_k|,\ \psi_k = \arctan\left[\frac{\mathrm{Im}(h_k)}{\mathrm{Re}(h_k)}\right]. \tag{12}$$

3. Results

The experimental investigations considered many tests based on five wire rope samples with the characteristics presented in the previous chapter of this paper. The number of rope samples was initially stipulated by the research theme. Within this process, the variables were (1) the static loading mass, (2) the application mode of the initial shock, and (3) the acquisition procedure in terms of analog data filtering, taking into account or not prior sampling of the signals. Based on the initial evaluation of the results, the authors adopted two significant cases for presentation and discussion within this paper.

Hereby, it was considered that the static loading masses should be 19 and 8 kg, respectively. The same rope sample was used, with an effective length $L = 2.200$ m between the hook devices. It was assumed that the analog filtered signal (low-pass filter with $f_{cut} = 400$ Hz) was provided by the force transducer. For both cases, multiple tests were performed. Two situations for each case were adopted for graphical presentation.

The raw acceleration signals are presented in Figure 3, acquired onto the loading mass and nearby the upper hanging point (in both the vertical and horizontal directions). Each set of diagrams contain the three timed signals and the corresponding FFT spectral magnitude, respectively.

The signals provided by the force transducer (raw digital and analog filtered signals) are depicted in Figure 4, where each group of diagrams represent both the timed evolution and spectral composition (in terms of the FFT magnitude).

The diagrams in Figure 5 present the results of a joint time-frequency analysis [22] of the absolute motion of the loading mass in terms of acceleration. The exclusive motion of the mass results through elimination of the parasitic motions, recorded on the upper hanging device, from the signal directly acquired at the loading point (see Figure 1). The absolute acceleration of the mass, as a function of time, is presented in Figure 6, together with its spectral composition. Obviously, the diagrams in Figure 6 are related to each testing situation. The essential magnitudes were evaluated and red circles were marked on the spectral diagrams in order to identify the minimum set of eigenfrequencies in the system evolution. The evaluation of these points was performed with the help of a computational routine which identifies the peaks within the signal [22] that simultaneously satisfy the following conditions: Minimum height raised above 10% of maximum magnitude and prominence greater than 0.03 (value adopted based on the initial analysis of all signals spectra). The procedure only takes into account the first five peaks, because the analysis of high frequency range does not meet the purposes of this research. These aspects have to be mentioned because of the various numbers of marked peaks on the spectra graphs in Figure 6.

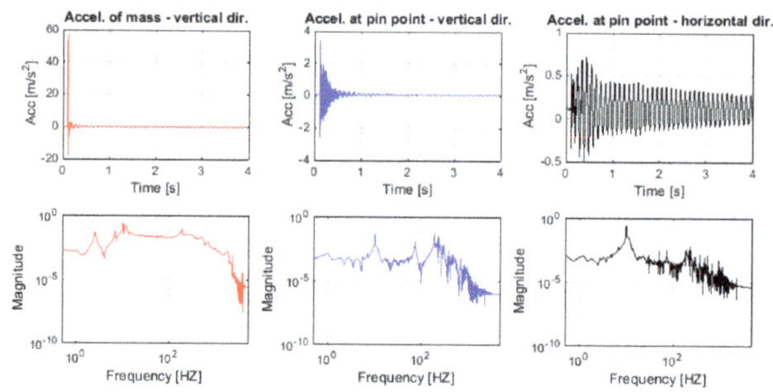

(a)

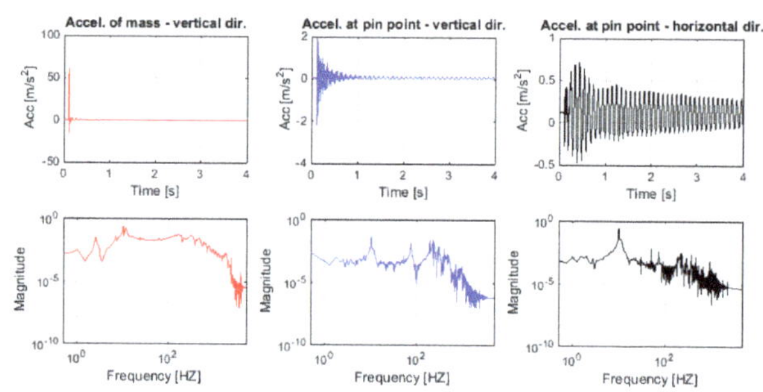

(b)

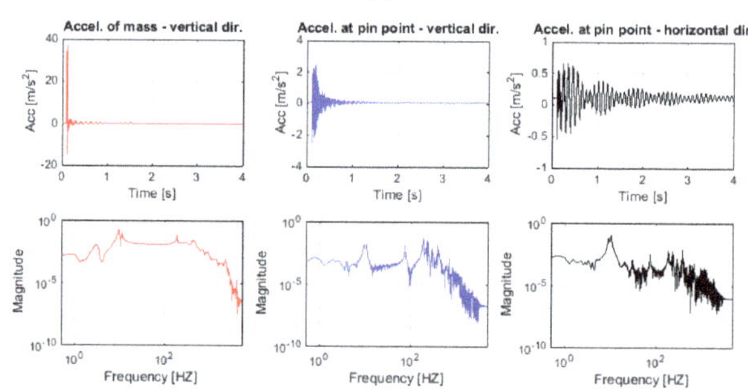

(c)

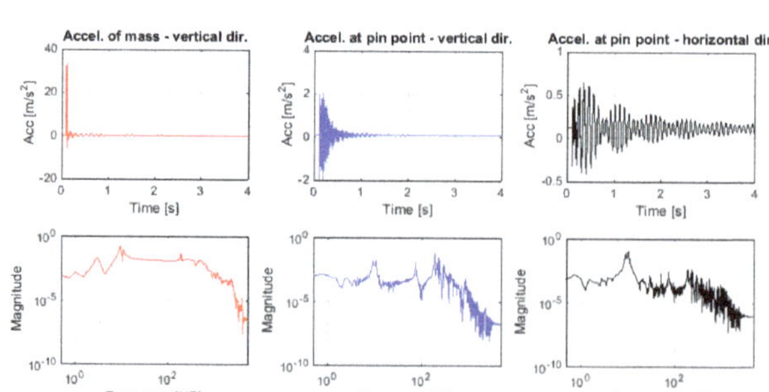

(d)

Figure 3. The raw signals of acceleration in terms of both the timed evolution and the spectral magnitude: (**a**) First situation within the case of m = 8 kg; (**b**) second situation within the case of m = 8 kg; (**c**) first situation within the case of m = 19 kg; (**d**) second situation within case of m = 19 kg.

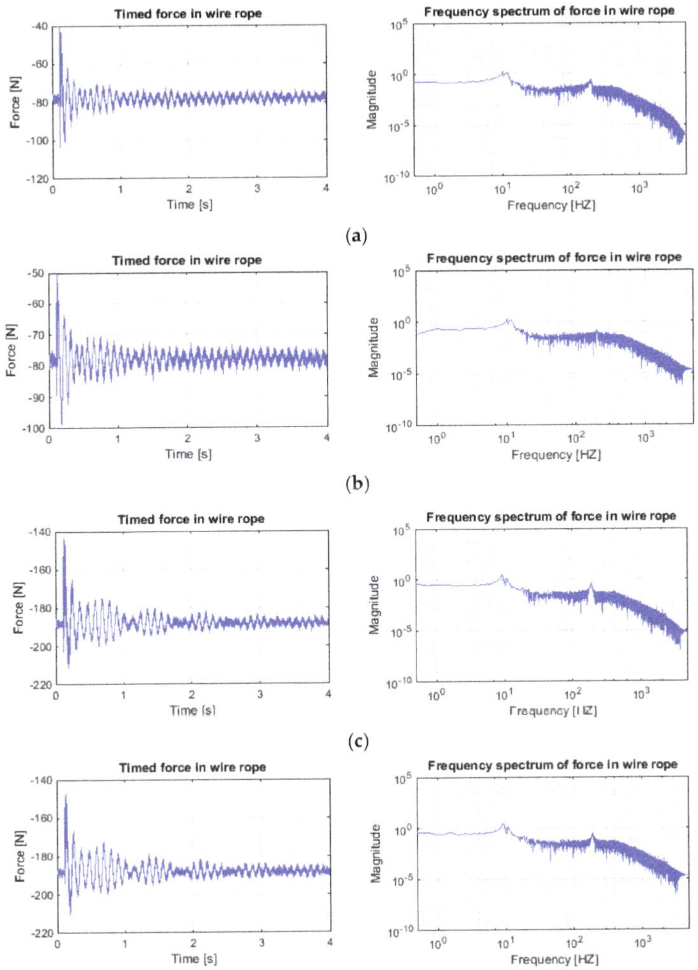

Figure 4. The raw signals from the force transducer in terms of the timed evolution and spectral magnitude: (**a**) First situation within the case of $m = 8$ kg; (**b**) second situation within the case of $m = 8$ kg; (**c**) first situation within the case of $m = 19$ kg; (**d**) second situation within the case of $m = 19$ kg.

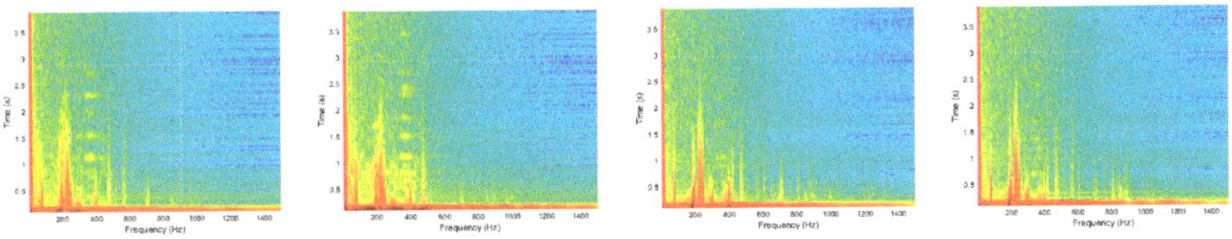

(a) (b) (c) (d)

Figure 5. Joint time-frequency analysis of the absolute motion of the loading mass in terms of the acceleration signal: (**a**) First situation within the case of $m = 8$ kg; (**b**) second situation within the case of $m = 8$ kg; (**c**) first situation within the case of $m = 19$ kg; (**d**) second situation within the case of $m = 19$ kg.

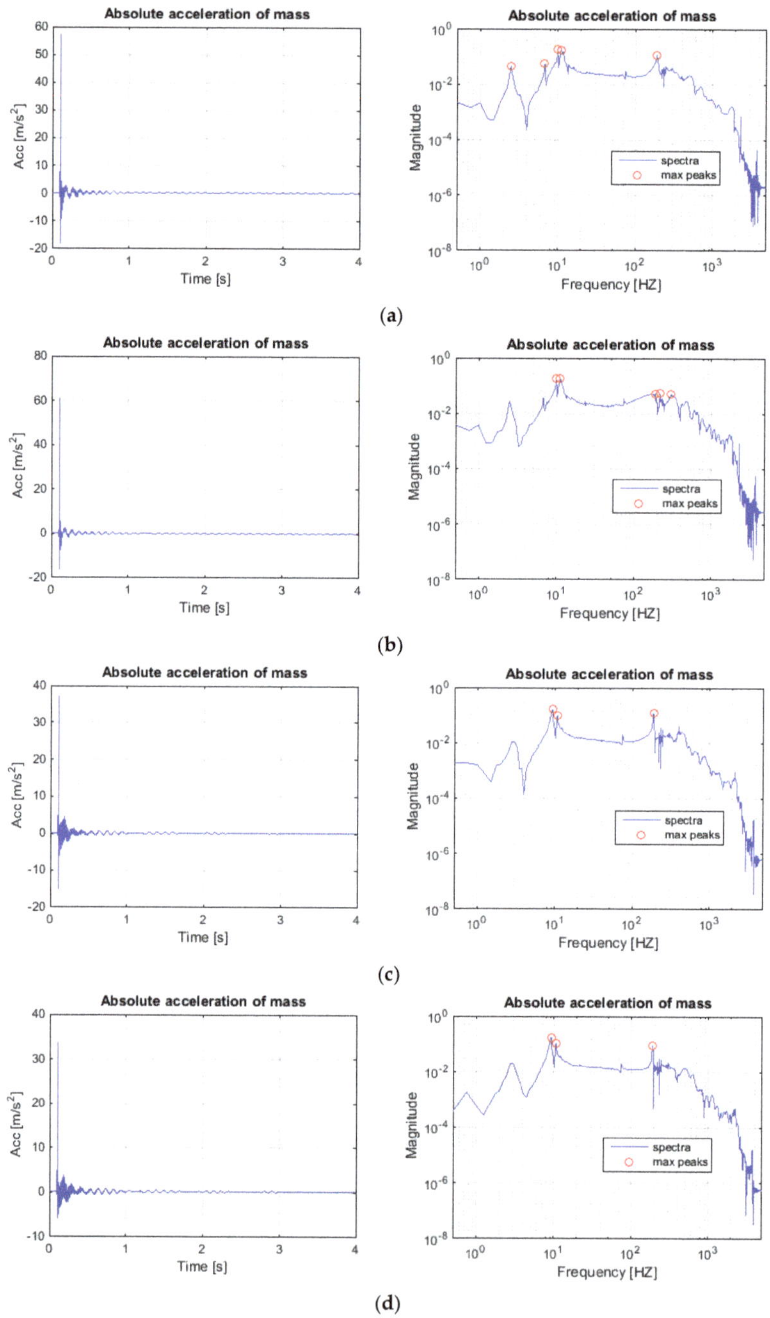

Figure 6. Timed evolution and related spectral magnitude of the absolute motion in terms of acceleration, recorded at the loading mass: (**a**) First situation within the case of m = 8 kg; (**b**) second situation within the case of m = 8 kg; (**c**) first situation within the case of m = 19 kg; (**d**) second situation within the case of m = 19 kg. Note: Red circles on the right-side diagrams denote the maximum peaks satisfying the imposed conditions (see text for details).

The authors have evaluated the transfer function, taking into account the force signal, as the input to the system, and the loading mass acceleration signal, as the output of system [22,24,29]. In order to facilitate a comparative analysis between excitation, the transfer function, and system response, Figure 7 simultaneously presents the three spectra (in terms of the FFT magnitude). The four groups of diagrams in Figure 7 correspond to each considered test.

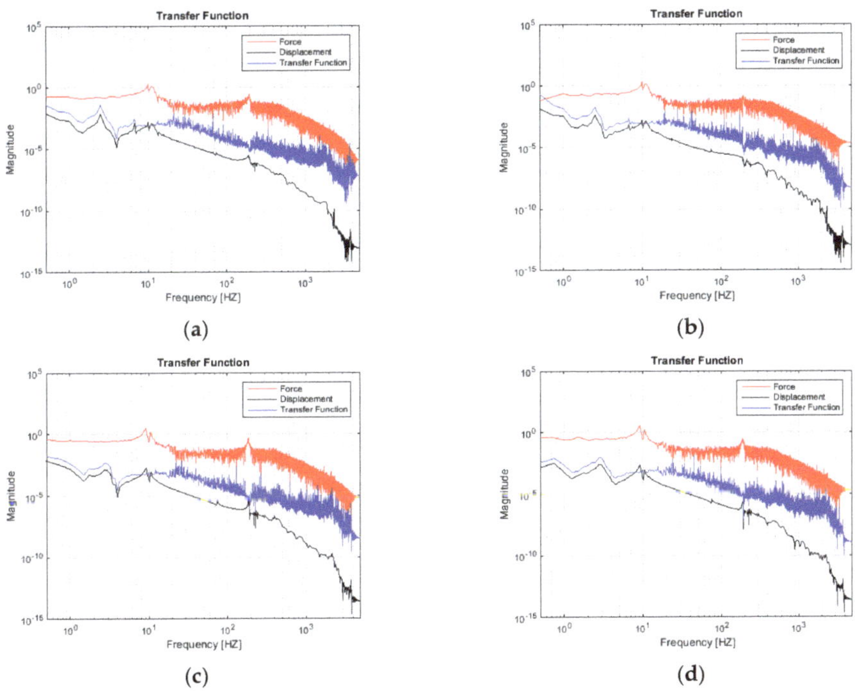

Figure 7. Transfer function of the tested ensemble, comparatively presented with the input and output spectra (magnitudes): (**a**) First situation within the case of $m = 8$ kg; (**b**) second situation within the case of $m = 8$ kg; (**c**) first situation within the case of $m = 19$ kg; (**d**) second situation within the case of $m = 19$ kg.

One interesting parameter that is able to characterize the system behavior under dynamic excitations is the dynamic rigidity [30]. This parameter relates to the perturbing force with the system response (in terms of displacement) and provides information regarding the dynamic evolution of this ratio in respect to the perturbation frequency. The spectra of dynamic rigidity, according to the four considered cases, are presented in Figure 8.

Additionally, in the second stage of analysis, the Prony method was used in order to estimate an extended range of spectral parameters, such as the amplitude, frequencies, damping, and phase. Practical implementation of this procedure uses MPM. During the analysis, different values for the pencil parameter were tested, finally adopting the value of 512. This value assures an adequate ratio between the additional parameters and computational resource requirements.

Hereby, the diagrams in Figure 9 show the amplitude, damping factor, and damping ratio respectively, in respect to the frequency. These diagrams were grouped according to the four presented experimental cases. Supposing that the Prony algorithm directly provided amplitudes and damping factors, the damping ratio was subsequently evaluated based on the available data from the Prony/MPM implementations. In addition, an equivalent value of the damping ratio was estimated using the hypothesis of the same energy values supplied for both the initial system (simulated by the finite sum of exponential functions) and for an equivalent single-degree-of-freedom (SDoF) system with a natural frequency identical to the dominant frequency within the signal spectra [22,30–32]. Taking into account the five maximum dominant spectral peaks previously evaluated, this results in the five maximum values of the equivalent damping ratio. These equivalent values are depicted in the damping ratio graphs within Figure 9 with dashed red lines.

Supposing that each frequency provided by the Prony/MPM algorithms can be considered the natural frequency of a SDoF linear model, with a correspondent amplitude, damping, and phase (also available from the same algorithm), it becomes simple to evaluate a theoretical response of a multi-degree-of-freedom (MDoF) linear model as a finite sum of all components supplied by the Prony method.

These cumulative responses have a graphical representation, such as a function of two variables, e.g., the perturbation and the natural frequencies (see Figures 10a, 11a, 12a and 13a). The correlated overlapping spectral diagrams, following the SDoF natural frequency parameter, are presented in Figures 10b, 11b, 12b and 13b. The global response of the equivalent MDoF model was depicted using a blue continuous thick line on each diagram within Figure 10, Figure 11, Figure 12, Figure 13 according to the assumed tests/cases for presentation.

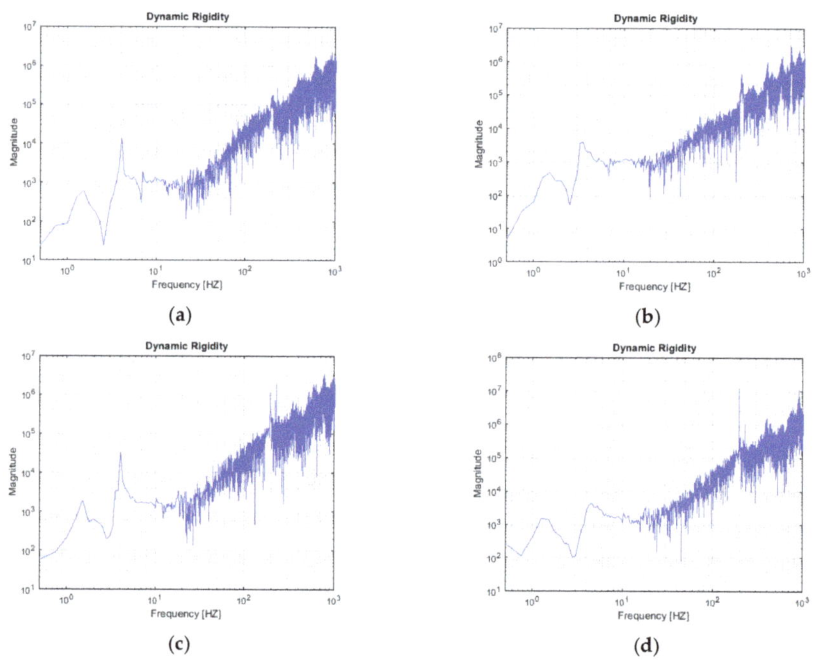

Figure 8. Dynamic rigidity spectra: (**a**) First situation within the case of $m = 8$ kg; (**b**) second situation within the case of $m = 8$ kg; (**c**) first situation within the case of $m = 19$ kg; (**d**) second situation within the case of $m = 19$ kg.

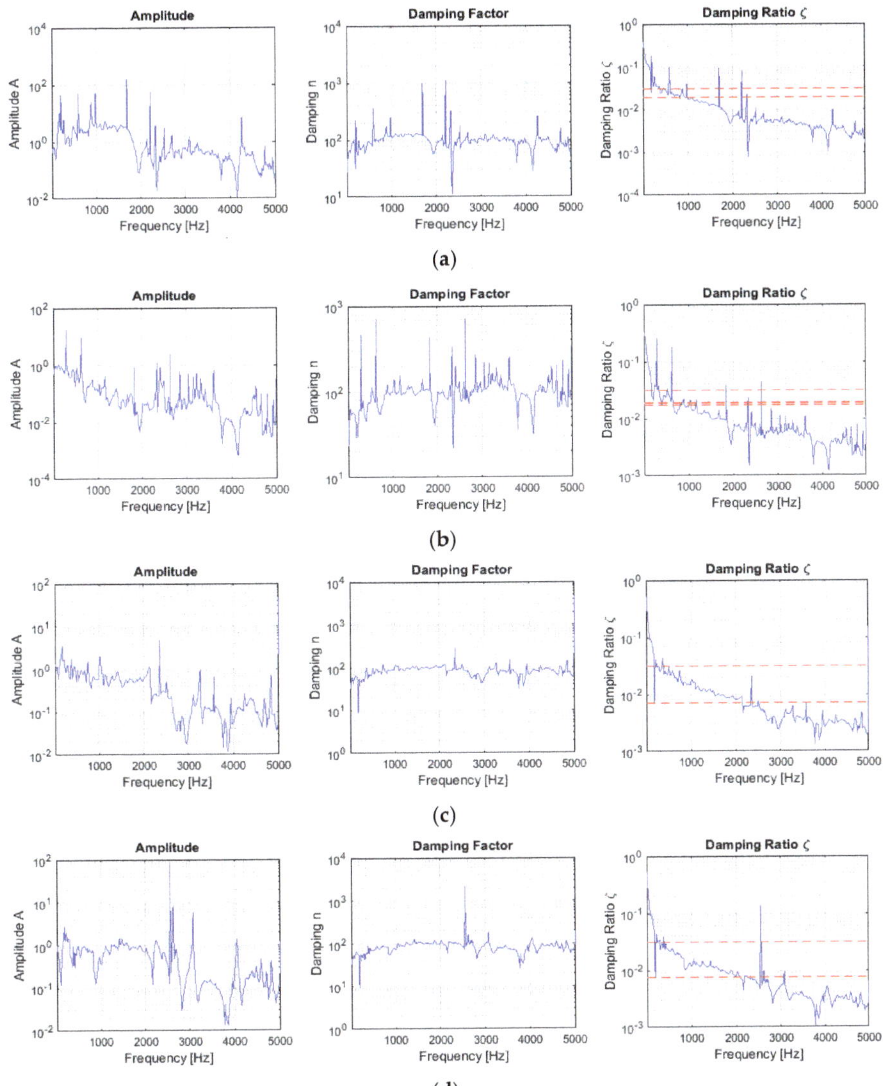

Figure 9. Amplitudes, damping factor, and damping ratio provided by the Prony method as functions of the modal frequencies (according to the first 512 terms of signal decomposition): (**a**) First situation within the case of $m = 8$ kg; (**b**) second situation within the case of $m = 8$ kg; (**c**) first situation within the case of $m = 19$ kg; (**d**) second situation within the case of $m = 19$ kg. Red dashed lines within the damping ratio diagrams denote an equivalent damping ratio (see text for details).

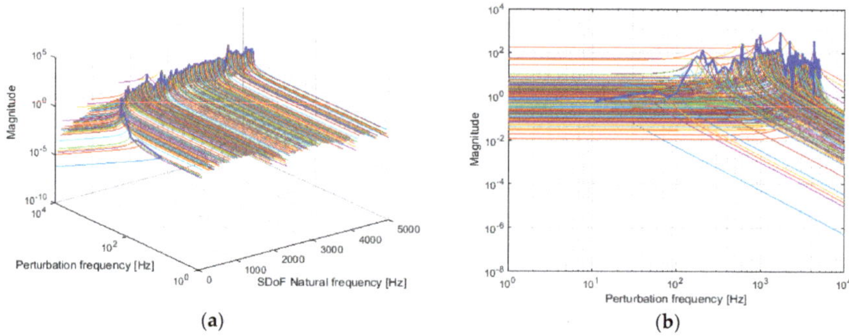

Figure 10. Double-spectra evolution of the first 512 terms in the exponential function decomposition according to the Prony method with the first test within the case of $m = 8$ kg: (**a**) Behavior of each component in respect to the perturbation frequency in the range of interest; (**b**) overlapped spectral diagrams. The blue continuous thick lines on the graphs denote the response spectra of the linear system, assuming the available terms.

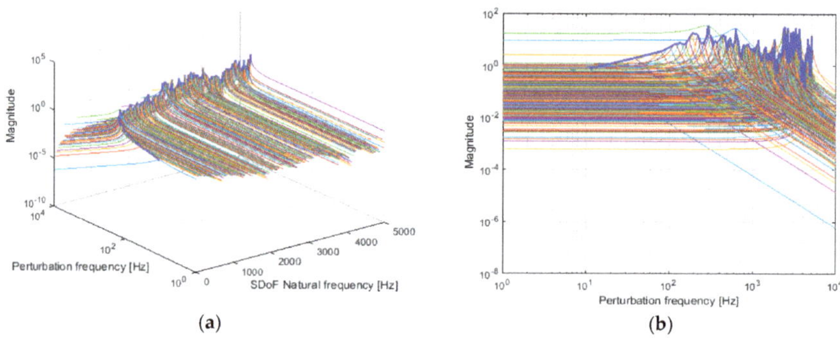

Figure 11. Double-spectra evolution of the first 512 terms in the exponential function decomposition according to the Prony method for the second test within the case of $m = 8$ kg: (**a**) Behavior of each component in respect to the perturbation frequency in the range of interest; (**b**) overlapped spectral diagrams. The blue continuous thick lines on graphs denote the response spectra of the linear system, assuming the available terms.

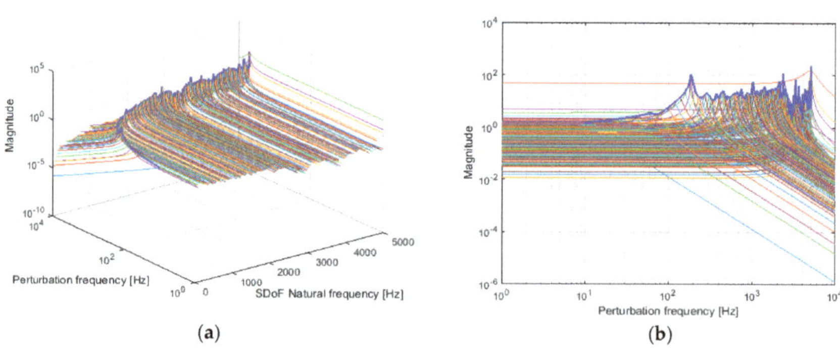

Figure 12. Double-spectra evolution of the first 512 terms in the exponential function decomposition according to the Prony method for the first test within the case of $m = 19$ kg: (**a**) Behavior of each component in respect to the perturbation frequency in the range of interest; (**b**) overlapped spectral diagrams. The blue continuous thick lines on graphs denote the response spectra of the linear system, assuming the available terms.

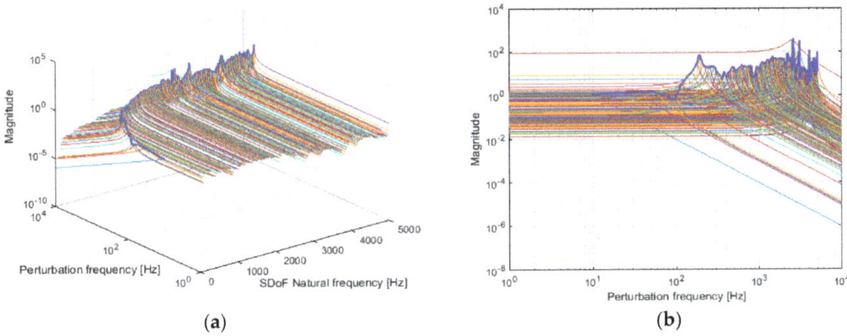

Figure 13. Double-spectra evolution of the first 512 terms in the exponential function decomposition according to the Prony method for the second test within the case of m = 19 kg: (**a**) Behavior of each component in respect to the perturbation frequency in the range of interest; (**b**) overlapped spectral diagrams. The blue continuous thick lines on graphs denote the response spectra of a cumulative linear system, assuming the available terms.

4. Discussion

A comparative analysis of the raw acceleration signals reveals dominant frequencies within the spectra. A pertinent evaluation of these values must take into account the force signal spectra. Thus, the spectral peaks in the range of interest becomes quite visible around 10 Hz and in the domain of 90–400 Hz. The loading mass acceleration spectra presents a rather wide plate evolution from 10 to 1000 Hz, containing a maximum of three essential peaks, comparative with the other two acceleration signals, which provide many of distinctive peaks (see Figure 3). This qualitative difference is due to the initial shock directly applied to the mass and, respectively, to the dissipative and elastic characteristics of the rope mounted between the two points of acceleration monitoring. The spectral force diagrams, which are even as noisy as the acceleration diagrams, provide two clear main peaks around the values of 10 and 200 Hz, respectively (see Figure 4). An interesting signal is that acquired nearby the upper hanging device in the horizontal direction, because this preponderantly contains the effects due to the support tower bending motion. Hereby, the spectrum of this signal contains a very clear maximum in the range of 9.25 to 10.75 Hz, depending on the test case, which represents the first modal frequency of the tower. This preliminary analysis offers the primary qualitative findings, which are useful for the next evaluations.

The effective (absolute) acceleration of the loading mass is able to characterize the dynamics of the wire rope used to suspend the mass oscillator. This signal was obtained from the initial acceleration recorded on the mass, through elimination of the perturbing signals due to additional unavoidable motion of the support tower. Timed evolutions of these signals were firstly evaluated using a joint time-frequency algorithm and the results are depicted in Figure 5. This investigation helps to correctly identify the time-frequency areas within the effective mass acceleration signal, showing the transitory regimes of each dominant frequency in terms of both the initial time and during the timed evolution. The results show the two relevant spectral components supplied by the support structure (low frequency area) and, respectively, by the rope–mass-vibrating ensemble (with most relevant values centered around 200 Hz).

Following the previous qualitative results, the acquired signals were quantitatively analyzed using, firstly, a classical FFT algorithm. The magnitude diagrams were computationally scanned for the five maximum peaks (this value was settled after the preliminary analysis that had indicated the average number of peaks provided by the entire test set).

As it can be seen in Figure 6, the peaks were grouped around 10 and 200 Hz. For the cases/situations within this paper, the frequency values of marked peaks in Figure 6 are presented in Table 1. For the first case (m = 8kg), the scanning procedure identified five values in each situation. For the second case, the procedure yielded only three values, where all attended the domains of interest. Additionally, identified components at 2.5, 6.75, 220, and 302 Hz appeared in a certain singular situation. However, following the graphs, it can be clearly seen that these dominant frequencies were also included by all signals, but with various amplitudes and prominences, thus that they were not constantly pointed out by the algorithm. Low values (2.5 and 6.75 Hz) result from the cable transversal parasitic oscillations and high values (220 and 302 Hz) results from the dynamics of the hooking devices.

Table 1. Peaks frequencies related to the diagrams in Figure 8.

Case/Situation of Analysis	Frequency (Hz)
m = 8 kg/situation I	2.499938, 6.749831, 9.99975, 11.24972, **190.9952**
m = 8 kg/situation II	9.99975, 11.24972, **188.7453**, 219.9945, 301.9925
m = 19 kg/situation I	9.249769, 10.74973, **186.7453**
m = 19 kg/situation II	9.249769, 10.74973, **189.4953**

Force and acceleration can be assumed as the input and the output, respectively, of the rope–load ensemble. Hereby, the transfer function, in terms of driving point function, becomes easily estimated for each inspected case (see Figure 7). The effective acceleration of loading mass (Figure 6), double integrated during the analysis time, was considered in order to obtain the displacement signal. The driving point transfer function represents the system admittance viewed through the input point, thus it is able to provide useful information related to the evolution of dynamic rigidity with respect to the excitation frequency (Figure 8). Targeted frequencies around 190 Hz were pointed out in the dynamic rigidity graphs, but low frequencies (9 to 11 Hz) were not present because this parameter exclusively characterizes the wire rope dynamics.

Starting from the hypothesis that the product $kL = ES = const.$, where k denotes the rigidity, L is the wire rope sample effective length, E is the rope Young's modulus, and S denotes the effective sectional area of the rope [7,14,30], it can be seen that the sample rigidity features a constant value for all assumed tests because the same cable sample with a constant length was used. According to initial evaluation of the rope parameters (see the values within Section 2.2), the rigidity was 7.04×10^5 Nm^{-1}. The experimental results provided rather different values for the cable rigidity because of dynamic effects (in fact, we discussed the dynamic rigidity). Thus, the values of this parameter acquired the following values: 9.8×10^5 Nm^{-1} for the case of m = 8 kg and 24×10^5 Nm^{-1} for the case of m = 19 kg.

The secondary stage of analysis was related to the dissipative characteristics provided by the experiments. The dissipation of the wire rope-loading mass ensemble could be globally characterized through an overall damping factor, evaluated based on timed effective acceleration diagrams (see Figure 6). The simplest evaluation procedure consists of an exponential function-based interpolation of the magnitude of the Hilbert Transform applied to the inspected function [22,31]. This method, such as the classical procedure of the logarithm decrement, is affected by certain operational errors regarding the initial parameters, the time length of the processed signal, and the nonlinearities within the system evolution. However, it was adopted because it avoids the uncertainness of the maximum point picking procedure.

The results according to the Hilbert Transform-based method are presented in Table 2. It has to be noted that there is a relative wide spread of the values, which may produce certain difficulties in the selection of the parametrical values for some MDoF computational models.

In order to accurately simulate the behavior of a real ensemble, an equivalent linear MDoF theoretical system is suitable for use. The dissipative characteristics of such models require supplementary parameters. Thus, an additional computational procedure was adopted, based on Prony approximation and MPM, which were briefly presented in Section 2.3. The results of these analyses were comparatively presented in Figure 9, and the MDoF system approximations were, respectively, depicted in Figures 10–13. The evaluation had supposed the value of 512 for the pencil parameter. Thus, 256 serviceable values were yielded, respectively, for the amplitude, damping, and phase parameters, in respect to the frequency. These values can be taken into account for any linear computational model intended to simulate the dynamic behavior of a wire rope.

Within some practical situations, a computational approximation is often required with the simplest models, such as the SDoF schematization [16,33]. An equivalent value of the dissipative characteristic is required for these cases. Using an algorithm that supposes a SDoF system, enabling the same energy with the initial MDoF system, it becomes easy to evaluate the equivalent damping parameter (in terms of the damping ratio) [22,30–32]. Successively, we assumed the natural frequencies for the SDoF system (see values in Table 1) and the equivalent damping ratios were computed. In addition, the damping factors were also evaluated. The results are given in Table 3. It has to be noted that there are consistent values within each testing case, but a wide spread between the cases, which shows the rope dissipation dependence with respect to the loading conditions.

Table 2. Overall damping factors related to the diagrams in Figure 8.

Case/Situation of Analysis	Overall Damping Factor (s^{-1})
m = 8 kg/situation I	24.5810
m = 8 kg/situation II	23.8685
m = 19 kg/situation I	12.7252
m = 19 kg/situation II	14.2641

Table 3. Equivalent damping factors and ratios related to the diagrams in Figure 8.

Case/Situation of Analysis	Equivalent SDoF Damping Ratio (-)	Equivalent SDoF Damping Factor (s^{-1})
m = 8 kg/sit. I	0.0312, 0.0312, 0.0312, 0.0312, **0.0192**	0.4908, 1.3233, 1.9589, 2.2047, **23.0028**
m = 8 kg/sit. II	0.0311, 0.0312, **0.0186**, 0.0174, 0.0193	1.9567, 2.2024, **22.0632**, 24.1166, 36.6487
m = 19 kg/sit. I	0.0312, 0.0312, **0.0072**	1.8117, 2.1062, **8.3916**
m = 19 kg/sit. II	0.0312, 0.0312, **0.0077**	1.8152, 2.1102, **9.1820**

Comparative analysis of values in Tables 2 and 3 reveals certain differences between the damping factors for the two evaluative algorithms. Basic justification of this fact results from the appreciative qualitative character of the first proposed algorithm. However, this technique was presented because it is easily implemented and requires reduced computing resources (comparative to the second technique). Regarding the differences between the values of the two cases according to the second algorithm (significant values in Table 3 are denoted in boldface), the damping factor depends on the excitation. Thus, in Table 4, the damping coefficient was exclusively provided as that evaluated for the significant values. The average values of this parameter were 360.528 and 333.906 Nsm^{-1}, which presents a 2.1% and 4.7% maximum deviation related to the computed values.

The investigation of the diagrams in Figures 10–13 shows that, for any considered case/situation, the spectra of the equivalent linear MDoF system does not enable the low frequencies related to the dynamic behavior of the tower structure. This approach supplies the frequencies of interest. The signal spectrum additionally contains high frequencies due to the wave propagation within the wire rope. Hereby, Prony method/MPM assessments are able to provide the required data for any linear MDoF system approximation.

Table 4. Equivalent damping coefficients related to the data in Table 3.

Case/Situation of Analysis	Equivalent SDoF Damping Coefficient (Nsm^{-1})
m = 8 kg/situation I	368.048
m = 8 kg/situation II	353.008
m = 19 kg/situation I	318.896
m = 19 kg/situation II	348.916

5. Conclusions

The discussion section contains findings related to both the conservative and dissipative aspects within dynamic behaviors of strands wire ropes. It was shown that the methods used for investigation of dynamic data are able to provide suitable information, depending on the numerical model requirement. The findings within this study were systematically presented, starting with a SDoF model (completely characterized by the global damping and stiffness) and progressing to complex MDoF approaches (which usually require a variable range frequency, depending on the dissipations and rigidities). Hereby, the presented assessments have demonstrated the flexibility and adaptability of the proposed techniques and, additionally, with the advantage of reduced experimental investigations. This study only includes laboratory tests based on an experimental setup, according to the initial research theme. This limitation does not restrict the applicability of the findings, as per the study objective. However, the operational mode of application was not completely proven. This research sustains the feasibility of the proposed post-processing assessments. Future developments will be focused on operational test applications using the same analysis technique but instead applied to operationally acquired signals. Thus, it will be able to provide suitable arguments for enlarging the utilization range of the authors' assessments.

Author Contributions: Conceptualization, G.D.M.(A.) and C.D.; methodology, C.D. and S.V.; software, C.D.; validation, G.D.M.(A.), C.D. and S.V.; formal analysis, C.D. and S.V.; investigation, G.D.M.(A.), C.D. and S.V.; resources, G.D.M.(A.); data curation, C.D.; writing—original draft preparation, G.D.M.(A.) and C.D.; writing—review and editing, G.D.M.(A.) and C.D.; supervision, S.V.; project administration, G.D.M.(A.) and C.D. All authors have read and agreed to the published version of the manuscript.

Funding: This research received no external funding.

Acknowledgments: The authors feel grateful to the editor and thank the referees for the valuable comments that helped us to improve the manuscript.

Conflicts of Interest: The authors declare no conflict of interest.

References

1. Huang, Q.; Li, Z.; Xue, H.-Q. Multi-body dynamics co-simulation of hoisting wire rope. *J. Strain Anal.* **2018**, *53*, 36–45. [CrossRef]
2. Pu, H.; Xie, X.; Liang, G.; Yun, X.; Pan, H. Analysis for dynamic characteristics in load-lifting system of the crane. *Procedia Eng.* **2011**, *16*, 586–593. [CrossRef]
3. Haniszewski, T. Preliminary modelling studies of an experimental test stand of a crane, for investigation of its dynamic phenomena of lifting and driving mechanism. *Transp. Probl.* **2017**, *12*, 115–126. [CrossRef]
4. Haniszewski, T. Modeling the dynamics of cargo lifting process by overhead crane for dynamic overload factor estimation. *J. Vibroeng.* **2017**, *19*, 75–86. [CrossRef]
5. Argatov, I. Response of a wire rope strand to axial and torsional loads: Asymptotic modeling of the effect of interwire contact deformations. *Int. J. Solids Struct.* **2011**, *48*, 1413–1423. [CrossRef]
6. Usabiaga, H.; Pagalday, J. Analytical procedure for modelling recursively and wire by wire stranded ropes subjected to traction and torsion loads. *Int. J. Solids Struct.* **2008**, *45*, 5503–5520. [CrossRef]
7. Hamilton, J. Vibration-based techniques for measuring the elastic properties of ropes and the added mass of submerged objects. *J. Atmos. Ocean. Technol.* **1999**, *17*, 688–697. [CrossRef]
8. Raoof, M.; Davies, T.J. Simple determination of the axial stiffness for largediameter independent wire rope core or fibre core wire ropes. *J. Strain Anal.* **2003**, *38*, 577–586. [CrossRef]
9. Ham, S.-H.; Roh, M.-I.; Lee, H.; Ha, S. Multibody dynamic analysis of a heavy load suspended by a floating crane with constraint-based wire rope. *Ocean Eng.* **2015**, *109*, 145–160. [CrossRef]
10. Kaczmarczyk, S.; Ostachowicz, W. Transient vibration phenomena in deep mine hoisting cables. Part 1: Mathematical model. *J. Sound Vib.* **2003**, *262*, 219–244. [CrossRef]
11. Kaczmarczyk, S.; Ostachowicz, W. Transient vibration phenomena in deep mine hoisting cables. Part 2: Numerical simulation of the dynamic response. *J. Sound Vib.* **2003**, *262*, 245–289. [CrossRef]

12. Anghelache, G.D.; Nastac, S. Computational analysis of nonlinearities within dynamics of cable-based driving systems. *IOP Conf. Series: Mater. Sci. Eng.* **2017**, *227*, 012007. [CrossRef]
13. Gąska, D.; Margielewicz, J. Numerical modeling of dynamics of lifting the load. *Transport* **2008**, *1*, 2–5.
14. Debeleac, C.; Nastac, S. Computational Assessments on Dragline Bucket Oscillations. *Rom. J. Acoust. Vib.* **2019**, *15*, 149–157.
15. Hobbs, R.; Raoof, M. Behaviour of cables under dynamic or repeated loading. *J. Constr. Steel Res.* **1996**, *39*, 31–50. [CrossRef]
16. Oproescu, G. *Wires Dynamics. Simulation and Experiment*; (Dinamica firelor. Modelare si experiment); Impuls Publishing House: Bucharest, Romania, 2015. (In Romanian)
17. Costello, G.A. *Theory of Wire Rope*, 2nd ed.; Springer: New York, NY, USA, 1997.
18. Haniszewski, T. Preliminary modeling studies of sudden release of a part of the hoist load with using experimental miniature test crane. *Vibroeng. PROCEDIA* **2017**, *13*, 193–198. [CrossRef]
19. Demšić, M.; Raduka, V. Cable oscillations due indirect excitation. *Građevinar* **2015**, *67*, 829–841. [CrossRef]
20. Xie, X.; Li, X.; Shen, Y. Static and Dynamic Characteristics of a Long-Span Cable-Stayed Bridge with CFRP Cables. *Materials* **2014**, *7*, 4854–4877. [CrossRef]
21. Elata, D.; Eshkenazy, R.; Weiss, M.P. The mechanical behavior of a wire rope with an independent wire rope core. *Int. J. Solids Struct.* **2004**, *41*, 1157–1172. [CrossRef]
22. Nastac, S. *Experimental Procedures. Evaluative Setups and Tests*; (Proceduri experimentale. Standuri si incercari evaluative); Galati University Press: Galati, Romania, 2019. (In Romanian)
23. Feyrer, K. *Wire Ropes. Tension, Endurance, Reliability*; Springer: Berlin/Heidelberg, Germany, 2007.
24. Grigorescu, L.; Oproescu, G.; Nastac, S. *The Theory of Signals and Automatization*; (Teoria semnalelor si automatizari); Impuls Publishing House: Bucharest, Romania, 2002. (In Romanian)
25. Hua, Y.; Sarkar, T. Matrix pencil method for estimating parameters of exponentially damped/undamped sinusoids in noise. *IEEE Trans. Acoust. Speech, Signal Process.* **1990**, *38*, 814–824. [CrossRef]
26. Potts, D.; Tasche, M. Parameter estimation for nonincreasing exponential sums by Prony-like methods. *Linear Algebra Appl.* **2013**, *439*, 1024–1039. [CrossRef]
27. Peter, T.; Plonka, G. A generalized Prony method for reconstruction of sparse sums of eigenfunctions of linear operators. *Inverse Probl.* **2013**, *29*, 025001. [CrossRef]
28. Rodríguez, A.F.; De Santiago, L.; Guillén, E.L.; Ascariz, J.M.R.; Miguel, J.M.; Boquete, L. Coding Prony's method in MATLAB and applying it to biomedical signal filtering. *BMC Bioinform.* **2018**, *19*, 451. [CrossRef]
29. Debeleac, C. *Equipment-Terrain Interaction and Dynamic Simulation*; (Interactiunea echipament-mediu si modelare dinamica); Galati University Press: Galati, Romania, 2015. (In Romanian)
30. Bratu, P. *Vibration of Elastic Systems*; (Vibratiile sistemelor elastice); Tehnica: Bucuresti, Romania, 2000.
31. Nastac, S. *Introduction to Nonlinear and Random Vibrations*; (Introducere in teoria vibratiilor neliniare si aleatoare); Impuls Publishing House: Bucharest, Romania, 2012. (In Romanian)
32. Zeveleanu, C.; Bratu, P. *Nonlinear Vibrations*; (Vibratii neliniare); Impuls Publishing House: Bucharest, Romania, 2001. (In Romanian)
33. Oproescu, G.; Nastac, S. *Elements of Computational Analysis*; (Elemente de modelare numerica); Libertatea: Braila, Romania, 2000. (In Romanian)

symmetry

MDPI

Article

Kane's Formalism Used to the Vibration Analysis of a Wind Water Pump

Gabriel Leonard Mitu [1], Eliza Chircan [2], Maria Luminita Scutaru [2,*] and Sorin Vlase [2,3]

1 COMAT, SA, str. Zizinului, nr.111, 500002 Brasov, Romania; leonard.mitu@unitbv.ro
2 Department of Mechanical Engineering, Transilvania University of Brașov, B-dulEroilor 20, 500036 Brașov, Romania; chircan.eliza@unitbv.ro (E.C.); svlase@unitbv.ro (S.V.)
3 Technical Sciences Academy of Romania, B-dul Dacia 26, 030167 Bucharest, Romania
* Correspondence: lscutaru@unitbv.ro

Received: 20 May 2020; Accepted: 17 June 2020; Published: 19 June 2020

Abstract: The paper uses Kane's formalism to study two degrees of freedom (DOF) mechanisms with elastic elements = employed in a wind water pump. This formalism represents an alternative, in our opinion, that is simpler and more economical to Lagrange's equation, used mainly by researchers in this type of application. In the problems where the finite element method (FEM) is applied, Kane's equations were not used at all. The automated computation causes it to be reconsidered in the case of mechanical systems with a high DOF. Analyzing the planar transmission mechanism, these equations were applied for the study of an elastic element. An analysis was then made of the results obtained for this type of water pump. The matrices coefficients of the obtained equations were symmetric or skew-symmetric.

Keywords: Kane's equations; planar mechanism; Lagrange's equations; dynamics; finite element method (FEM); multibody system (MBS); wind water pump

1. Introduction

The analytical mechanics gave, from the beginning, several equivalent formalisms useful in the analysis of mechanical systems with a large number of degrees of freedom (DOF). The advantage of these approaches lies in the fact that they are suitable for generalizations. However, in practice, despite a relatively large number of analysis possibilities, the Lagrange's equation was used. The success of this method is based on the fact that it operates with classical and well-known notions of mechanics. However, due to the significant progress of the last five decades in the field of numerical computing methods and the use of computers in all domains, a reconsideration of these methods is required. Some of these formalisms can bring advantages in terms of modeling, writing algorithms, and computational time. The advantages and the drawbacks are summarized at the end of this section. By using a convenient method, it is possible to reduce the computing operations. For example, Hamilton's method leads us to a system of first-order differential equations; therefore, there is no need to transform the second-order system of differential equations into a first-order system of differential equations.

The last decade shows us that we have begun to reconsider all the formalisms of analytical mechanics. This is a consequence of the desire to facilitate the representation of mathematical models and the need to reduce the amount of equations and steps that are required. Worldwide, there are different studies that have applied and analyzed alternative ways of obtaining the equations of motion. In the following, we show a summary of some of the most significant research studies on the subject.

Even if alternative methods were used in the analysis of multi body systems (MBSs), such as Gibbs–Appell equations, Hamilton equations, Maggi's equations, Jacobi's equations, Kane's equations, etc., for problems that required the application of the finite element method (FEM) method, Lagrange's equations were used almost exclusively. The disadvantage is the high level of calculus due to the large number of DOF. It is natural, therefore, to identify the most advantageous methods in terms of volume and calculus time. We mention that, even though different methods were used, the resulting system of differential equations was the same.

Using the results presented in [1–16], the following conclusions can be formulated regarding the advantages and the drawbacks of modeling an elastic mechanical system using the equivalent formalism of analytical mechanics.

As a method currently used, the Lagrange's equations (LE) method, preferred in the case of mechanical systems with a high DOF, has proven that it can be successfully applied to a wide range of situations. The main advantage of the method is that basic and well-known notions are used. In the finite element analysis (FEA) of MBS systems with elastic elements, this method has been used by almost all researchers. Mechanical systems modeled with various types of finite elements have been studied using this method. In the first step, the equations of motion are computed. The next step is the assembly of these equations, which implies the writing of the constraint conditions between the finite elements, expressed generally by non-holonomic conditions. This step is actually the elimination of Lagrange's multipliers from the obtained algebraic differential system (DAE). It has been proven that the method of LE is convenient for many applications [1–7]. The use of other equivalent formulations in analytical mechanics has been only sporadically investigated in relation to these problems.

The most used method by researchers to study such systems is the Lagrange's method. The main reason is the degree of generality and abstraction that this method allows (as opposed to the Euler method, where every liaison force must be introduced and determined), but also the fact that researchers are accustomed to it. A major disadvantage is the fact that one must determine the multipliers, which, for large systems, can become a difficult thing.

Gibbs–Appell (GA) equations, although rarely used in the study of mechanical systems, have obvious advantages in terms of the representation form to be used. Compared with LE, it has advantages regarding the required computation time [8].

The disadvantage is that researchers are less familiar with some notions, such as energy of accelerations. The GA formalism is useful for non-holonomic systems. Advantages of the GA method have been presented in studies related to the calculus of MBSs with rigid elements [9]. The final equations obtained by applying this method are identical to those obtained with LE, but the computational effort is lower [10]. The major advantage of the method is that Lagrange's multipliers do not appear in the written equations; the method eliminates them directly and, as a result, the number of unknowns is lower [11]. These advantages have been observed by the researchers and, lately, the method tends to become a [12] procedure. Gibbs–Appell equations have the advantage that the level of calculus involved in the modeling process is smaller. On the other hand, this method involves the introduction of a notion that researchers are less accustomed to operate with, namely the energy of accelerations.

Most analytical methods lead to a system of second-order differential equations. To solve this kind of system using commercial software, it is necessary to transform it by introducing additional variables to a system of first-order differential equations. Hamiltonian mechanics allows obtaining a system of first order and thus shorter time is needed to solve it. In this way, a time-consuming stage in the process of transforming from second-order differential equation systems to first-order differential systems is avoided. However, the method has a disadvantage in the need to perform numerous preliminary calculations [16]. We do not know a concrete study of all the aspects involved in the use of this method.

The last years have revealed increasing interest from researchers for Maggi's equations (ME), which were little applied and which are an alternative formalism to the other formalisms in analytical mechanics [13,14]. The interest is also generated by the development of modern mechanical manufacturing systems (robots and manipulators) [15]. In the case of the study of non-linear systems, the feedback linearization technique involves the application of ME. However, this method is rarely used, despite all these advantages, and is less familiar to researchers. The use of Maggi's and Kane's equations (two equivalent formulations of the same principles) has advantages in the analysis of non-holonomic systems. The liaison conditions, expressed linearly, can be used immediately in the equations of motion. It also has the advantage that the obtained equations no longer contain Lagrange multipliers or the liaison forces from the Newton–Euler equations.

The paper aims to analyze the possibility of using Kane's equations to derive the equations of motion for a finite element and also to identify the advantages of this approach for the researchers.

In order to test the possibility of applying the method and to find the potential problems that might appear during its use, we made an application based on a planar mechanism.

2. Finite Element Kinematics

In our application, we used a one-dimensional finite element to study a planar mechanism with two degrees of freedom. To apply Kane's equation (KE), it is necessary to know the kinematic of the element. In such analysis, the velocity and the acceleration of a point are of significant importance and are expressed in terms of nodal coordinates. An analysis of this problem can be found in [2–6]. The shape function that was used helped determine the displacements field in terms of independent nodal coordinates. (Figure 1)

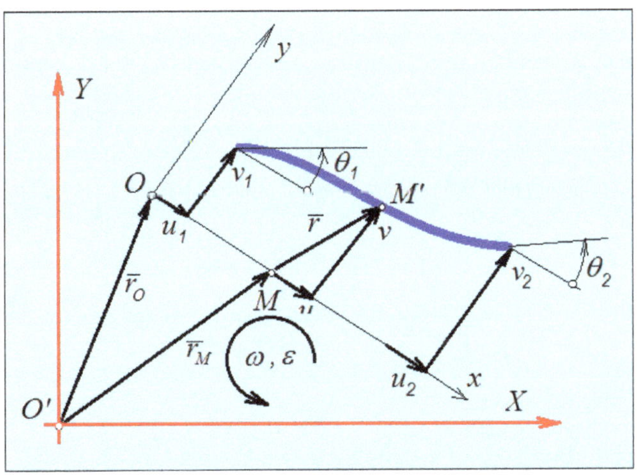

Figure 1. Plane finite element.

The new position vector M' of an arbitrary chosen point M becomes [7]:

$$\{r_{M'}\}_G = \{r_M\}_G + \{f\}_G = \{r_M\}_G + [R]\left\{\begin{array}{c} u \\ v \end{array}\right\} = \{r_o\}_G + [R]\left\{\begin{array}{c} x \\ 0 \end{array}\right\} + [R][N]\{\delta\} \tag{1}$$

where $\{f(u,v)\}$ is the displacement vector; the shape functions are in $[N] = \left[\begin{array}{c} N_{(u)} \\ N_{(v)} \end{array}\right] = \left[\begin{array}{c} N_{(1)} \\ N_{(2)} \end{array}\right]$, and $\{\delta\} = \left\{\begin{array}{c} \delta_1 \\ \delta_2 \end{array}\right\}$ and $\{\delta_1\}$, $\{\delta_2\}$ are the displacement vectors. The slope is [6]:

$$\beta = \frac{dv}{dx} = \frac{d}{dx}\left(\left[N_{(v)}\right]\{\delta_e\}\right) = \left[N'_{(v)}\right]\{\delta_e\} \tag{2}$$

With G index, we denote a vector or a matrix with components in the global coordinate system, and with L index, we denote a size with components in the local system.

From Equation (1), the velocity of a point results in:

$$\{v_{M'}\}_G = \left\{\dot{r}_{M'}\right\}_G = \{v_O\}_G + \left[\dot{R}\right]\left\{\begin{array}{c} x \\ 0 \end{array}\right\}_L + \left[\dot{R}\right][N]\{\delta\}_L + [R][N]\left\{\dot{\delta}\right\}_L \tag{3}$$

Calculated in [6,7], where $\{v_O\}_G = \left\{\dot{r}_O\right\}_G$ is the velocity vector of the origin of the mobile coordinate system.

The transformation of a vector is performed by the matrix $[R]$:

$$\{t\}_G = [R]\{t\}_L \tag{4}$$

In Appendix A are presented more results used in the paper.

3. Kane's Equations Used in Conjunction with FEA

Kane's method and equations have represented, during the last decades, a method that was successfully applied in the study of MBS systems. The KE method can be used for both holonomic and non-holonomic systems and can eliminate some of the disadvantages of classical methods in analytical mechanics (Newton–Euler and Lagrange). From a theoretical point of view, Maggi and Kane equations are equivalent, representing the same method very well suited for systems with non-holonomic constraints [17]. The application of Kane's equations was determined by the need to study complex flexible structures such as robots and manipulators [18]. It has been found that applying this method allows efficient calculation for these systems [19–22]. The advantage of these equations is that they do not contain constraints forces [23]. The paper presents the necessary formalism to derive the motion equation for a finite element. A planar mechanism is studied, but the method can be applied for every type of an elastic MBS. Of course, the specific form of the equations must be calculated each time considering the shape functions required by the specific finite element chosen.

A short presentation of the Kane's equations is presented in Appendix B. For an elastic finite element, considered as a solid, the Kane's equation can be written as:

$$\sum_{i=1}^{N} \overline{F}_i \frac{\partial \overline{v}_i}{\partial \dot{q}_k} = \int_V \overline{a} \frac{\partial \overline{v}}{\partial \dot{q}_k} dm \quad k = \overline{1,n} \tag{5}$$

where by $k = \overline{1,n}$ is meant $k = 1,2,\ldots \ \ldots,n$.

Differentiating the expression of velocity Equation (3), we obtain:

$$\{a_{M'}\}_G = \{a_o\}_G + \left[\ddot{R}\right]\left\{\begin{array}{c} x \\ 0 \end{array}\right\} + \left[\ddot{R}\right][N]\{\delta\}_L + 2[R][N]\left\{\dot{\delta}\right\}_L + [R][N]\left\{\ddot{\delta}\right\}_L \tag{6}$$

From Equation (1), we obtain:

$$\frac{\partial\{v'_M\}_G}{\partial\left\{\dot{\delta}\right\}_L} = [R][N] \quad \text{and} \left(\frac{\partial\{v'_M\}_G}{\partial\left\{\dot{\delta}\right\}_L}\right)^T = [N]^T[R]^T \tag{7}$$

The expression $\frac{\partial\{T\}}{\partial\{\delta\}}$, if $\{T\} = [A]\{\delta\}$, is $[A]$. It is possible to write:

$$\left(\frac{\partial\{v_{M'}\}_G}{\partial\left\{\dot{\delta}\right\}_G}\right)^T \{a_{M'}\}_G = [N]^T[R]^T\{a_{M'}\}_G \tag{8}$$

$$\begin{aligned}&\int_V\left(\frac{\partial\{v_{M'}\}_G}{\partial\{\dot{\delta}\}_G}\right)^T\{a_{M'}\}_G=\left(\int_V\rho[N]^TdV\right)\{a_O\}_L+\int_V[N]^T([\varepsilon]_L+[\omega]_L[\omega]_L)\{r\}_L\rho dV+\\&+\left(\int_V[N]^T([\varepsilon]_L+[\omega]_L[\omega]_L)[N]\rho dV\right)\{\delta\}_L+2\left(\int_V\rho[N]^T[\omega]_L[N]dV\right)\{\dot{\delta}\}_L+\left(\int_V[N]^T[N]\rho dV\right)\{\ddot{\delta}\}_L\\&=-\left[m_O^i\right]\{a_O\}_L-\left\{q^i(\varepsilon)\right\}-\left\{q^i(\omega)\right\}+[k(\varepsilon)]\{\delta\}+[k(\omega)]\{\delta\}+[c(\omega)]\{\dot{\delta}\}_L+[m]\{\ddot{\delta}\}_L\end{aligned}\tag{9}$$

The following notations were made:

$$[m]=\int_V[N]^T[N]\rho dV=[m_{11}]+[m_{22}]\quad;\quad\left[m_O^i\right]=\int_V\rho[N]dV;\tag{10}$$

$$[c(\omega)]=\int_0^L[N]^T[\omega_L][N]\rho Adx=\omega([m_{21}]-[m_{12}])\quad;\quad[k(\varepsilon)]=\int_0^L[N]^T[\varepsilon_L][N]\rho Adx=\varepsilon([m_{21}]-[m_{12}]);\tag{11}$$

$$[k(\omega)]=-\int_0^L[N]^T\left[\dot{R}\right]^T\left[\dot{R}\right][N]\rho Adx=\int_0^L[N]^T[\omega_L][\omega_L][N]\rho Adx=-\omega^2([m_{11}]+[m_{22}])\tag{12}$$

$$\left\{q^i(\varepsilon)\right\}=\int_0^L[N]^T[\varepsilon_L]\left\{\begin{array}{c}x\\0\end{array}\right\}\rho Adx=\varepsilon\{m_{2x}\}\quad;\quad\left\{q^i(\omega)\right\}=\int_0^L[N]^T[\omega_L][\omega_L]\left\{\begin{array}{c}x\\0\end{array}\right\}\rho Adx=\omega^2\{m_{1x}\}\tag{13}$$

We obtain:

$$\begin{aligned}[m]\{\ddot{\delta}\}+\ &2\omega([m_{21}]-[m_{12}])\{\dot{\delta}\}+\left([k_b]+[k_a]+\varepsilon([m_{21}]-[m_{12}])-\omega^2([m_{11}]+[m_{22}])+[k^g]\right)\{\delta\}=\\&=\left\{q^{ext}\right\}+\left\{q^{liaison}\right\}-\varepsilon\{m_{2x}\}-\omega^2\{m_{1x}\}-\left[m_o^i\right]\{\ddot{r}_o\}_L\end{aligned}\tag{14}$$

In Equation (14), Equation (7) was applied. Thus, in a nodal point, k acts as generalized forces $\{F_k\}$ composed by the external forces (concentrated and distributed), the liaison forces, and the forces deriving from a potential. The corresponding velocities of these points can be found inside the velocities vector $\{\dot{\delta}\}_G$. We obtain:

$$\left(\frac{\partial\{v_{M'}\}_G}{\partial\{\dot{\delta}\}_G}\right)^T\{F_k\}\rho dV=([k]+[k(\varepsilon)]+[k(\omega)]+[k^g])\{\delta\}_L-\left\{q^{ext}\right\}_L-\left\{q^{liaison}\right\}_L\tag{15}$$

$\left\{q^{ext}\right\}_L$ is the generalized external nodal forces vector, $([k]+[k(\varepsilon)]+[k(\omega)]+[k^g])\{\delta\}_L$—is the vector of forces deriving from a potential, $\left\{q^{liaison}\right\}_L$—is the liaison vector [6], and $\{v_{M'}\}$ is, in this representation, the velocity of the nodal point k.

4. Study of a Planar Mechanism Used in a Wind Water Pump

The mechanism of the transmission of the wind water pump is presented in Figure 2 [24]. The elements of the mechanism are rigid despite the B C′ beam, which is considered linear elastic.

To determine the constraint forces in the connection point, we consider all the elements of the mechanism being rigid (it is a first approximation, and we consider that the deformability of the beam does not influence the rigid body motion of the mechanism) [25,26]. In this case, if we know the moment offered by the electric engine and the friction in the joint after the integration of the equations, we can compute the liaison forces in the hinges B and C′. These forces can be introduced now in the finite element model.

However, due to the many parameters that we have to take into account and which are difficult to determine or approximate, we used another method of determining the liaison forces. Recordings were made of the movement of some characteristic points of the mechanism, their trajectories were obtained, and, based on them, the speed and the acceleration fields of the mechanism were determined. Knowing the accelerations and the equations of motion, we could determine the forces that appeared in the two hinges (by solving a linear system) and use them in the finite element model.

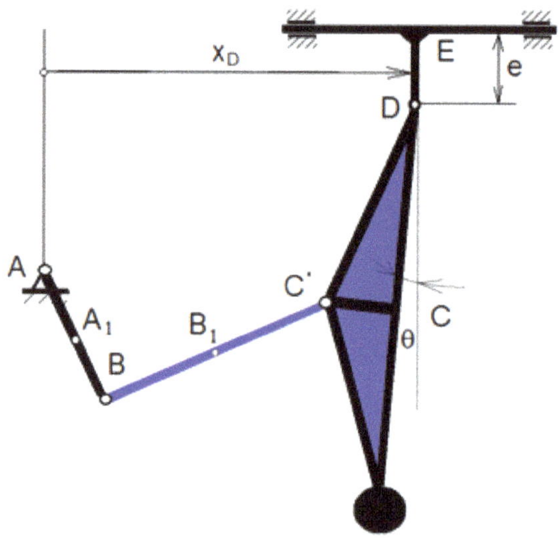

Figure 2. Kinematic mechanism of the transmission of a wind water pump.

In Figure 3 are presented the points where the markers were glued. These forces are necessary for writing the equations of motion of the BC′ elastic beam. They were determined previously using Kane's equations. After this step, we computed the free vibrations of the equation system that was obtained.

Figure 3. Position of the markers.

The markers placed on the bar were tracked using a video analysis and modeling program. Following the recording of the markers' positions with the help of the analysis program, the speeds and the accelerations of the points were determined as well as the trajectories traveled by them.

In Figure 4 are presented the trajectories of each chosen point on the bar using the data taken by the video analysis program. These are represented according to the positions given by the main coordinate axes of the chosen system, XOY, OX being the horizontal axis and OY the vertical axis.

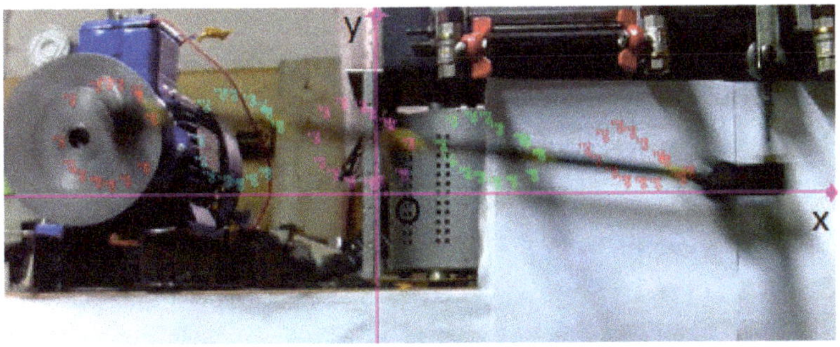

Figure 4. Test bench.

For analysis and processing of the results as well as for the modeling with finite elements, simple codes written in MATLAB were used. This was done due to the simplicity and the versatility of this programming environment. Some of its advantages would be that it allows the easy implementation of algorithms, it is possible to develop codes easily, corrections can be made quickly, it allows the use of relatively large data sizes of the studied system, and it shows fast graphical results. The trajectories were presented in Figure 5.

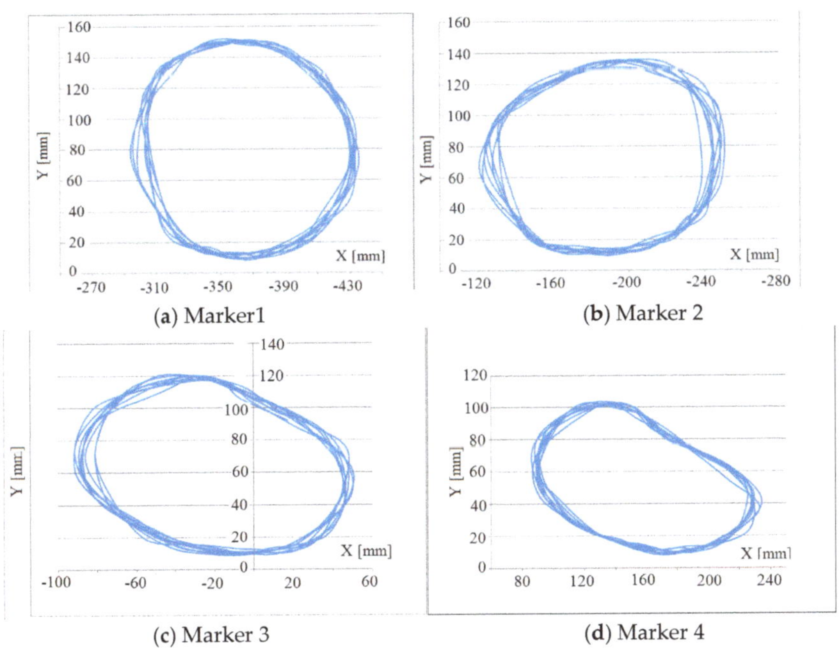

Figure 5. *Cont.*

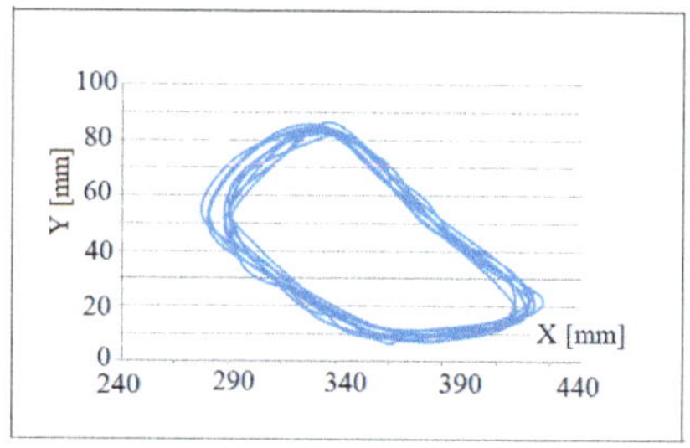

(**e**) Marker 5

Figure 5. The trajectories of the five points analyzed at an engine angular velocity of 140 rpm.

The velocity and the acceleration fields (Figure 6) were determined using the aforementioned optical method. The acceleration fields are presented in the figures below for the five nodes studied.

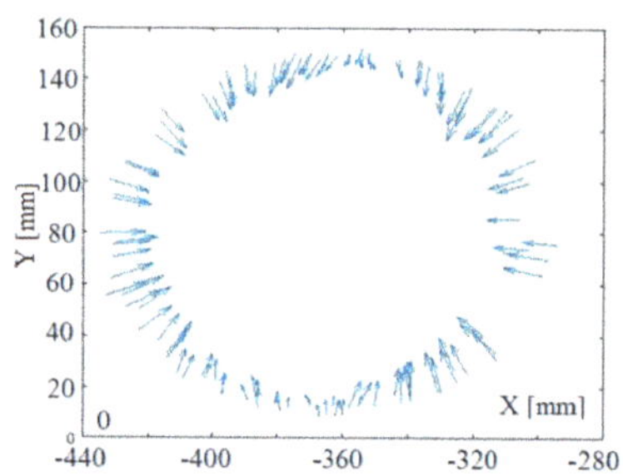

(**a**) Marker1

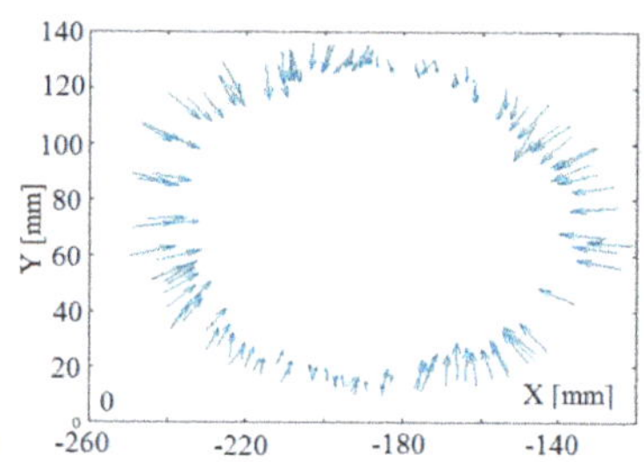

(**b**) Marker 2

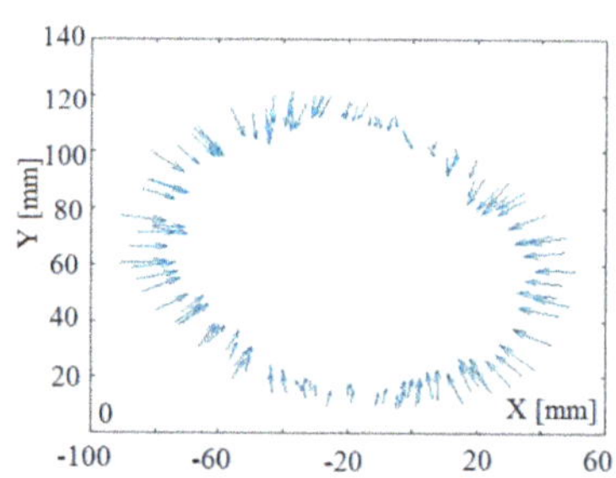

(**c**) Marker 3

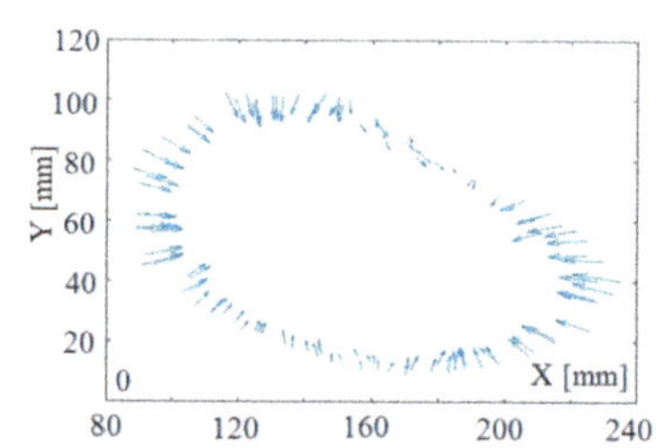

(**d**) Marker 4

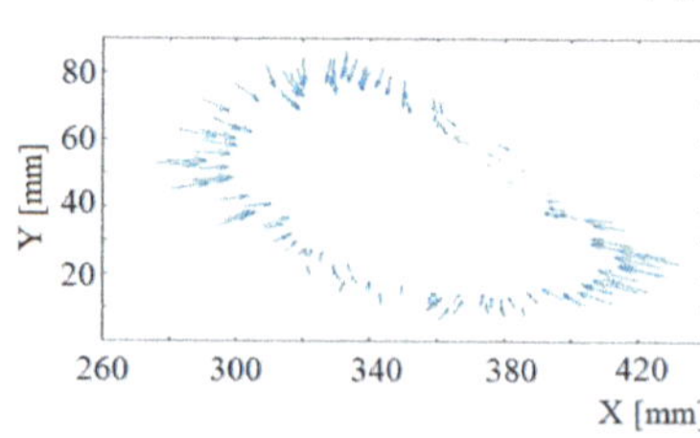

(**e**) Marker 5

Figure 6. Acceleration field of the five points analyzed at an engine angular velocity of 140 rpm. The value of the maximum acceleration was 14.33 m/s^2.

The model of the beam divided in finite elements is presented in Figure 7. In the hinge B and C, we could introduce the reactions previously determined using the optical measurement and solving the linear system of equation containing the reaction (liaison forces). Then, it was possible to perform a modal analysis of the beam. The results presented in the graphs below were taken for the same operating angular velocity, 140 rpm, this being the maximum speed at which experimental measurements were made (Figure 8).

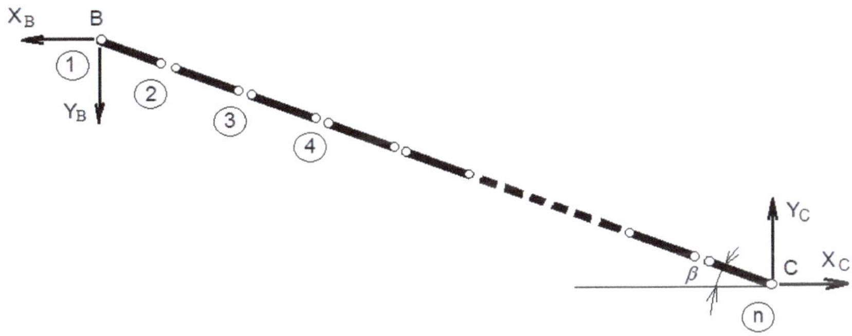

Figure 7. Elastic lever BC′ divided in finite elements.

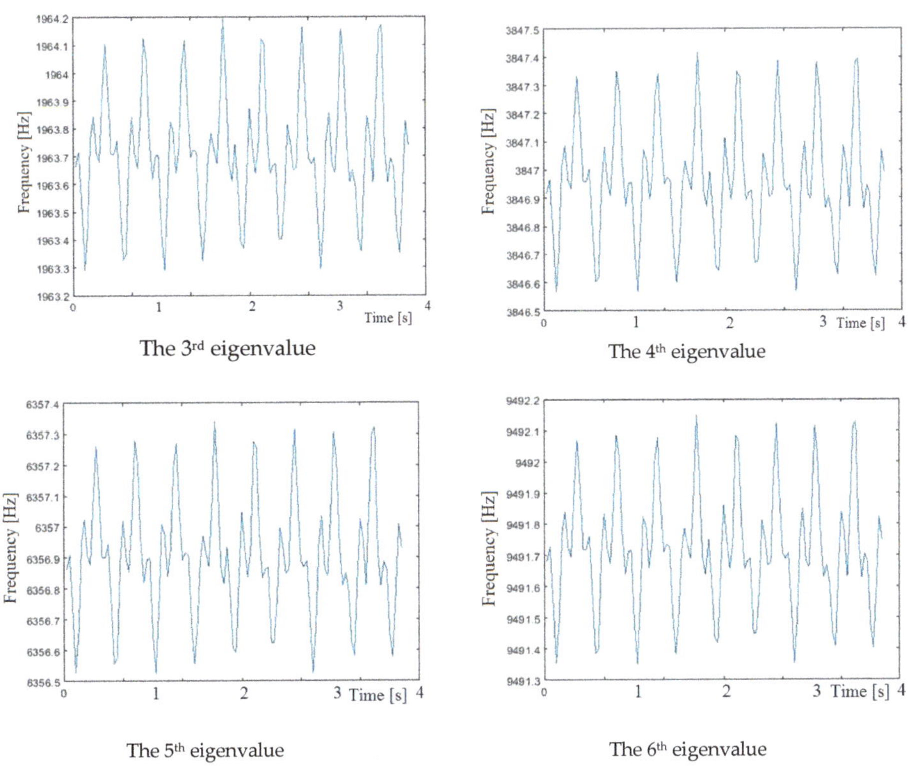

Figure 8. Four eigenfrequencies (eigenfrequencies 3, 4, 5, and 6) of the beam discretized in 10 finite elements for an angular velocity at 140 rpm.

In the following, we try a comparison between the use of Kane's and Lagrange's equations based on reference books [27,28]. The procedure follows a series of similar steps. The differences occur when Lagrange's formulas and Kane's formulas are actually applied (Figures 9 and 10). The number of distinctive operations in the two cases is studied. There are still elementary matrix operations, but we do not make the comparison there, considering that the number of terms in the case of Lagrange's method is significantly higher than in the case of applying Kane's method (Table 1).

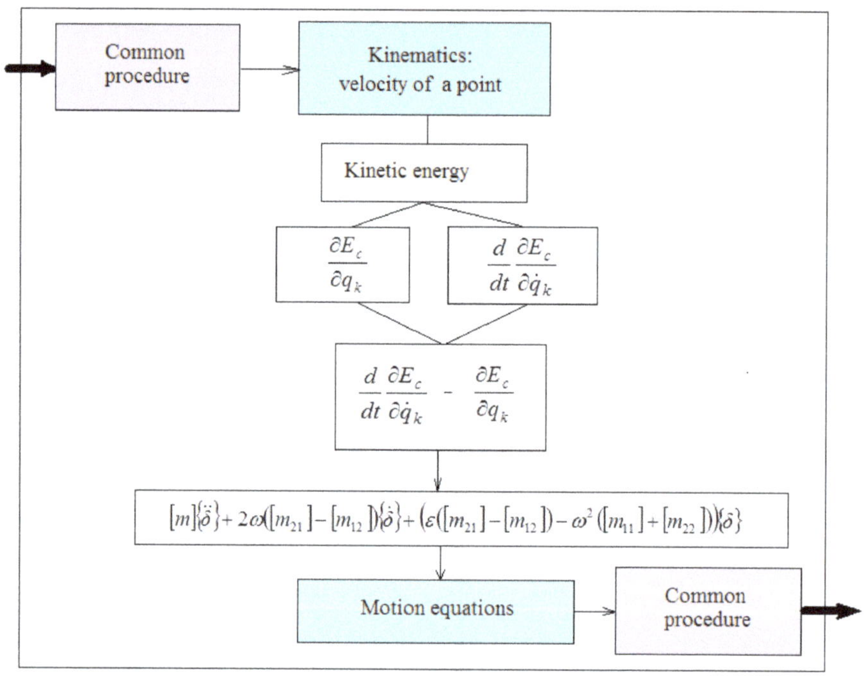

Figure 9. The section from the procedure of deriving the motion equations differences using Kane's or Lagrange's formalism. The use of Lagrange's formalism.

Table 1. Comparison between the numbers of operations for the two methods.

	$\frac{\partial E_c}{\partial \dot q_k}, \frac{\partial E_c}{\partial q_k}$			$\left(\frac{\partial \{v_{M'}\}_G}{\partial \{\dot\delta\}_G}\right)$	
	Number of elements			Number of elements	
1	10	30	1	10	30
		Number of operations			
78	1008	9486	6	33	93

From the analysis of the presented calculus, it was found that there was an obvious difference between the numbers of differences implied by the two methods. Apparently, the difference was very large and should have led to significant time savings if the method of Kane's equations was applied. In reality, this step represented only a fraction of the multitude of operations necessary to perform. Consequently, the economy achieved was much smaller. For example, in the case of writing a calculus program in MATLAB for the calculation of the beam studied in the paper, which was discretized in 30 finite elements, the comparison led to a decrease in calculation time of only 7%.

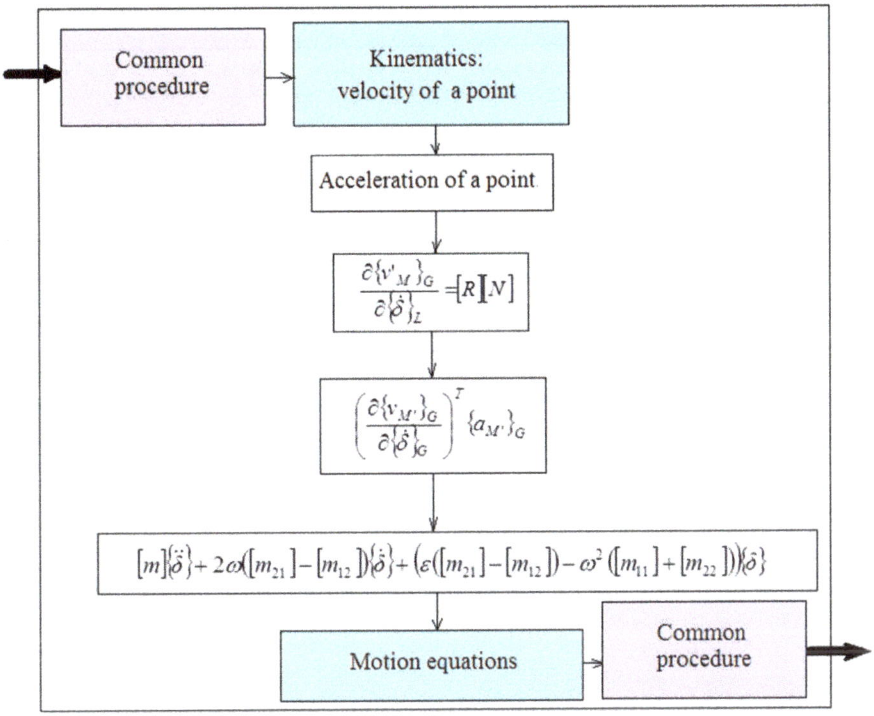

Figure 10. The section from the procedure of deriving the motion equations differences using Kane's or Lagrange's formalism. The use of Kane's formalism.

5. Conclusions and Discussions

The literature studied suggests that LE presents some advantages for the researchers. The main disadvantage is the need to determine the Lagrange multipliers from the DAE obtained, which is difficult to achieve in the case of large DOF systems. This involves numerous calculations and large computational times. By comparison, the GA method seems to eliminate this disadvantage but introduces a new notion, energy of the accelerations. An alternative is ME, very useful when the constraints are non-holonomic. The last years have indicated an interest in researchers towards this method in the context of the need to study complex systems such as robots and manipulators applied at the present time on a large scale in the manufacturing industry. Hamiltonian formalism has the advantage of obtaining a system of first-order differential equations, which can become a plus in the case of using numerical calculation. However, the complexity of the intermediate calculations makes the method inconvenient for the researchers. The KE method, which is equivalent to the ME method, has begun to be used more widely in the last decade, with studies also led by the automation industry and industrial robots. In the paper, the KE formalism was applied to determine the dynamic response of an elastic MBS. The application considered was that of the transmission of a water pump driven by the wind. The finite element chosen was a one-dimensional element where classical third and fifth degree interpolation polynomials were used. It can be seen that KE can be a successful alternative in determining the equations of motion with the advantage of being economical. The uses of these equations present a natural alternative for non-holonomic mechanical systems. A comparison between the two methods was made in the paper. For the example highlighted in the paper, which is of small size, this difference may seem minor, but for large systems with high DOF, the difference may become significant.

Kane's equations can be an economical and a simple alternative to the problem. The mentioned methods and the other known ones from the analytical mechanics with their equivalents will be re-evaluated in the context of the development of the modern industry, characterized by working mechanisms with high speeds and high loads. This implies the fact that new technologies and software development on these new methods are needed.

Author Contributions: Conceptualization, G.L.M., E.C., M.L.S., and S.V.; methodology, E.C., M.L.S.; writing—original draft preparation, M.L.S. and S.V.; writing—review and editing G.L.M., E.C., M.L.S., and S.V.; visualization, M.L.S. and S.V.; supervision, M.L.S., S.V.; funding acquisition, G.L.M. and M.L.S. All authors have read and agreed to the published version of the manuscript.

Funding: This research received no external funding.

Conflicts of Interest: The authors declare no conflict of interest.

Appendix A

Matrix $[R]$ is:

$$[R] = \begin{bmatrix} \cos\theta & -\sin\theta & 0 \\ \sin\theta & \cos\theta & 0 \\ 0 & 0 & 1 \end{bmatrix} \tag{A1}$$

It being orthonormal, results:

$$[R][R]^T = [R]^T[R] = [E] = \begin{bmatrix} 1 & 0 & 0 \\ 0 & 1 & 0 \\ 0 & 0 & 1 \end{bmatrix} \tag{A2}$$

We use the notations:

$$\left[\dot{R}\right] = \omega \begin{bmatrix} -\sin\theta & -\cos\theta & 0 \\ \cos\theta & -\sin\theta & 0 \\ 0 & 0 & 0 \end{bmatrix} \tag{A3}$$

$$[\omega]_G = \left[\dot{R}\right][R]^T = -[R]\left[\dot{R}\right]^T = \begin{bmatrix} 0 & -\omega & 0 \\ \omega & 0 & 0 \\ 0 & 0 & 0 \end{bmatrix} \tag{A4}$$

and:

$$[\omega]_L = [R]^T[\omega]_G[R] = \begin{bmatrix} 0 & -\omega & 0 \\ \omega & 0 & 0 \\ 0 & 0 & 0 \end{bmatrix} \tag{A5}$$

The skew symmetric operator angular velocity, expressed in global reference frame is $[\omega]_G$, and in the local reference frame $[\omega]_L$. The angular velocity vector of the solid is $\overline{\omega}$. There are the relations:

$$[\omega]_G = -[\omega]_G^T \text{and} [\omega]_L = -[\omega]_L^T \tag{A6}$$

Differentiating Equation (A5) it obtains:

$$\left[\ddot{R}\right] = \varepsilon \begin{bmatrix} -\sin\theta & -\cos\theta & 0 \\ \cos\theta & -\sin\theta & 0 \\ 0 & 0 & 0 \end{bmatrix} - \omega^2 \begin{bmatrix} \cos\theta & -\sin\theta & 0 \\ \sin\theta & \cos\theta & 0 \\ 0 & 0 & 0 \end{bmatrix} \tag{A7}$$

and the angular acceleration operator:

$$[\varepsilon]_G = \left[\dot{\omega}\right]_G = \begin{bmatrix} 0 & -\varepsilon & 0 \\ \varepsilon & 0 & 0 \\ 0 & 0 & 0 \end{bmatrix} \tag{A8}$$

in the global reference frame and:

$$[\varepsilon]_L = \left[\dot{\omega}\right]_L = \begin{bmatrix} 0 & -\varepsilon & 0 \\ \varepsilon & 0 & 0 \\ 0 & 0 & 0 \end{bmatrix} \tag{A9}$$

in a local reference frame. After some elementary calculus, results are:

$$\left[\ddot{R}\right][R]^T = \begin{bmatrix} -\omega^2 & -\varepsilon & 0 \\ \varepsilon & -\omega^2 & 0 \\ 0 & 0 & 0 \end{bmatrix} \tag{A10}$$

$$[R]^T\left[\ddot{R}\right] = \begin{bmatrix} -\omega^2 & -\varepsilon & 0 \\ \varepsilon & -\omega^2 & 0 \\ 0 & 0 & 0 \end{bmatrix} \tag{A11}$$

Since we are dealing with a plane problem, we can reduce the size of the matrices used. Thus, the rotation matrix is:

$$[R^*] = \begin{bmatrix} \cos\theta & -\sin\theta \\ \sin\theta & \cos\theta \end{bmatrix} \tag{A12}$$

And all previously written formulas can be reduced to relations between two-dimensional squared matrices.

Appendix B

Kane's equations [29].

It starts from the relationship [1,16]:

$$\sum_{i=1}^{N}\left(\overline{F}_i - m_i\bar{a}_i\right)\frac{\partial\bar{r}_i}{\partial q_k} = 0 \quad ; \quad k = \overline{1,n} \tag{A13}$$

In analytical mechanics, it is shown that we have the relation:

$$\frac{\partial\bar{r}_i}{\partial q_k} = \frac{\partial\bar{v}_i}{\partial\dot{q}_k} \quad ; \quad k = \overline{1,n} \quad ; \quad i = \overline{1,N} \tag{A14}$$

Substituting in Equation (A13) we obtain:

$$\sum_{i=1}^{N}\left(\overline{F}_i - m_i\bar{a}_i\right)\frac{\partial\bar{v}_i}{\partial\dot{q}_k} = 0 \quad ; \quad k = \overline{1,n} \tag{A15}$$

If noted:

$$\frac{\partial\bar{v}_i}{\partial\dot{q}_k} = \frac{\partial\bar{v}_i}{\partial u_k} = \bar{v}_i^{(k)} \quad ; \quad k = \overline{1,n} \quad ; \quad i = \overline{1,N} \tag{A16}$$

It results in:

$$\sum_{i=1}^{N}\overline{F}_i\frac{\partial\bar{v}_i}{\partial u_k} = \sum_{i=1}^{N}m_i\bar{a}_i\frac{\partial\bar{v}_i}{\partial u_k} \quad ; \quad k = \overline{1,n} \quad ; \quad i = \overline{1,N} \tag{A17}$$

where, with $\overline{F}_i$, the external forces acting in nodes were noted.

References

1. Erdman, A.G.; Sandor, G.N.; Oakberg, A. A General Method for Kineto-Elastodynamic Analysis and Synthesis of Mechanisms. *J. Eng. Ind.* **1972**, *94*, 1193. [CrossRef]

2. Nath, P.K.; Ghosh, A. Kineto-Elastodynamic Analysis of Mechanisms by Finite Element Method. *Mech. Mach. Theory* **1980**, *15*, 179. [CrossRef]
3. Thompson, B.S.; Sung, C.K. A survey of Finite Element Techniques for Mechanism Design. *Mech. Mach. Theory* **1986**, *21*, 351–359. [CrossRef]
4. Gerstmayr, J.; Schöberl, J. A 3D Finite Element Method for Flexible Multibody Systems. *Multibody Syst. Dyn.* **2006**, *15*, 305–320. [CrossRef]
5. Vlase, S.; Marin, M.; Öchsner, A.; Scutaru, M.L. Motion equation for a flexible one-dimensional element used in the dynamical analysis of a multibody system. *Contin. Mech.* **2019**, *31*, 715–724. [CrossRef]
6. Vlase, S. Dynamical Response of a Multibody System with Flexible Elements with a General Three-Dimensional Motion. *Rom. J. Phys.* **2012**, *57*, 676–693.
7. Vlase, S.; Dănăşel, C.; Scutaru, M.L.; Mihălcică, M. Finite Element Analysis of a Two-Dimensional Linear Elastic Systems with a Plane "rigid Motion". *Rom. J. Phys.* **2014**, *59*, 476–487.
8. Negrean, I.; Crișan, A.-D.; Vlase, S. A New Approach in Analytical Dynamics of Mechanical Systems. *Symmetry* **2020**, *12*, 95. [CrossRef]
9. Vlase, S.; Marin, M.; Scutaru, M.L. Maggi's Equations Used in the Finite Element Analysis of the Multibody Systems with Elastic Elements. *Mathematics* **2020**, *8*, 399. [CrossRef]
10. Mehrjooee, O.; Dehkordi, S.F.; Korayem, M.H. Dynamic modeling and extended bifurcation analysis of flexible-link manipulator. *Mech. Based Des. Struct. Mach.* **2019**. [CrossRef]
11. Korayem, M.H.; Dehkordi, S.F. Motion equations of cooperative multi flexible mobile manipulator via recursive Gibbs-Appell formulation. *Appl. Math. Model.* **2019**, *65*, 443–463. [CrossRef]
12. Korayem, M.H.; Dehkordi, S.F. Derivation of dynamic equation of viscoelastic manipulator with revolute-prismatic joint using recursive Gibbs-Appell formulation. *Nonlinear Dyn.* **2017**, *89*, 2041–2064. [CrossRef]
13. Haug, E.J. Extension of Maggi and Kane Equations to Holonomic Dynamic Systems. *J. Comput. Nonlinear Dyn.* **2018**, *13*. [CrossRef]
14. Vlase, S.; Negrean, I.; Marin, M.; Scutaru, M.L. Energy of Accelerations Used to Obtain the Motion Equations of a Three-Dimensional Finite Element. *Symmetry* **2020**, *12*, 321. [CrossRef]
15. Ladyzynska-Kozdras, E. Application of the Maggi equations to mathematical modeling of a robotic underwater vehicle as an object with superimposed non-holonomic constraints treated as control laws. *Mechatron. Syst. Mech. Mater.* **2012**, *180*, 152–159. [CrossRef]
16. Malvezzi, F.; Matarazzo Orsino, R.M.; Hess Coelho, T.A. Lagrange's, Maggi's and Kane's equations to the dynamic modelling of serial manipulator. In Proceedings of the DINAME 2017-XVII International Symposium on Dynamic Problems of Mechanics, ABCM, São Sebastião, Brazil, 5–10 March 2017.
17. Wang, J.T.; Huston, R.L. Kane's equations with undetermined multipiers-application to constrained multibody systems. *ASME J. Appl. Mech.* **1987**, *54*, 424–429. [CrossRef]
18. Kane, T.R.; Levinson, D.A. Multibody dynamics. *ASME J. Appl. Mech.* **1983**, *50*, 1071–1078. [CrossRef]
19. Rosenthal, D. An order n formulation for robotic systems. *J. Astronaut. Sci.* **1990**, *38*, 511–529.
20. Banerjee, A. Block-diagonal equations for multibodyelastodynamics with geometric stiffness and constraints. *J. Guid. Control Dyn.* **1993**, *16*, 1092–1100. [CrossRef]
21. Anderson, K.; Critchley, J. Improved order-N performance algorithm for the simulation of constrained multi-rigid-body dynamic systems. *Multibody Syst. Dyn.* **2003**, *9*, 185–212. [CrossRef]
22. Kane, T.R.; Levinson, D.A. Use of Kane's Dynamical Equations in Robotics. *Int. J. Robot. Res.* **1983**, *2*, 3–21. [CrossRef]
23. Bajodah, A.H.; Hodges, D.H.; Chen, Y.-H. New Form of Kane's Equations of Motion for Constrained Systems. *J. Guid. Control Dyn.* **2003**, *26*, 79–88. [CrossRef]
24. Scutaru, M.L. Models for the Study of Mechanical Response of the Solids and Systems of Solids. Ph.D. Thesis, Transylvania University, Lexington, KY, USA, 2014.
25. Itu, C.; Oechsner, A.; Vlase, S.; Marin, M. Improved rigidity of composite circular plates through radial ribs. *Proc. Inst. Mech. Eng. Part L J. Mater. Des. Appl.* **2019**, *233*, 1585–1593. [CrossRef]
26. Nastac, S.; Debeleac, C.; Vlase, S. Hysteretically Symmetrical Evolution of Elastomers-Based Vibration Isolators within alpha-Fractional Nonlinear Computational Dynamics. *Symmetry* **2019**, *11*, 924. [CrossRef]
27. Papastavridis, J.G. *Analytical Mechanics: A Comprehensive Treatise on the Dynamics of Constrained Systems; For Engineers, Physicists, and Mathematicians*; Oxford University Press: Oxford, UK, 2002.

28. Roithmayr, C.M.; Hodges, D.H. *Dynamics: Theory and Application of Kane's Method*, 1st ed.; Cambridge University Press: Cambridge, UK, 2016.
29. Ursu-Fisher, N. *Elements of Analitical Mechanics*; House of Science Book Press: Cluj-Napoca, Romania, 2015.

symmetry

MDPI

Article

The Effect of Fractional Time Derivative of Bioheat Model in Skin Tissue Induced to Laser Irradiation

Aatef Hobiny [1], Faris Alzahrani [1], Ibrahim Abbas [1,2,*] and Marin Marin [3]

1 Nonlinear Analysis and Applied Mathematics Research Group (NAAM), Mathematics Department, King Abdulaziz University, Jeddah 21521, Saudi Arabia; ahobany@kau.edu.sa (A.H.); falzahrani1@kau.edu.sa (F.A.)
2 Mathematics Department, Faculty of Science, Sohag University, Sohag 82524, Egypt
3 Department of Mathematics and Computer Science, Transilvania University of Brasov, 500093 Brasov, Romania; m.marin@unitbv.ro
* Correspondence: ibrabbas7@science.sohag.edu.eg

Received: 6 March 2020; Accepted: 18 March 2020; Published: 10 April 2020

Abstract: This work uses the "fractional order bio-heat model" (Fob) model of heat conduction to offer a new interpretation to study the thermal damages in a skin tissue caused by laser irradiation. The influences of fractional order and the thermal relaxation time parameters on the temperature of skin tissue and the resulting thermal damage are studied. In the Laplace domain, the analytical solutions of temperature are obtained. Using the equation of Arrhenius, the resulting thermal injury to the tissues is assessed by the denatured protein ranges. The numerical results of the thermal damages and temperature are presented graphically. A parametric analysis is dedicated to the identifications of suitable procedures for the selection of significant design variables to achieve an effective thermal in the therapy of hyperthermia.

Keywords: fractional derivative; skin tissues; thermal damages; Laplace transforms

1. Introduction

Recently, the development of studies shows that the problem of heat transfer in living tissues introduces many complications. However, there are various discussions and findings in this field. Approaches of heating operations have been used in modernistic medicine, such as hyperthermia [1], laser tissue soldering [2], and laser surgery [3]. Some applications of bioheat models in clinical hyperthermia were discussed, especially regarding large, thermally unequilibrated blood vessels that have a worthy impact on the temperature distribution and cause serious temperature inhomogeneities. Through this clinical procedure, the applications of moving heat source on the skin tissue considering the perfusions rate is seen in some plastic surgery processes, such as in the use of laser radiation for spots or tattoos or in the heat actions of the cornea, using the laser for correcting hyperopia and removing moles.

In 1948, Pennes [4] presented the temperature distribution in the forearm skin temperature. The formulation can be analyzed by various approaches to get the solution of the heat transfer model for infinite thermal wave propagations, which is based on classical Fourier thermal conductions. In point of fact, it is still found that heat spreads at a finite rate in the biological tissue because they have highly nonhomogeneous inner structures. To dissolve the paradox that occurred in Penne's bio-heat formulation, thermal wave theory of bio-heat transfer is presented, which is based on heat waves constitutive relations, as given in [5,6]. Abbas and Hobiny [7] investigated the analysis of thermal damage in skin tissue subject to moving heat sources. The homotopy perturbation method [8] and the finite-decomposition method have also been applied [9]. Esneault and Dillenseger [10] investigated the temperature increment over times in hypothermia by the finite difference approach. Zhu et al. [11] estimated the depositions of light energy in tissues and the rate operation model for the thermal damage resulting from the use of the theorem of diffusions.

Many current models of physical processes have been successfully modified using fractional computation. We can say that the totality of integral theories and fractional derivatives was created in the last half of the last century. Definitions and different methods of fractional derivatives have become the main focus of numerous investigations. Using the fractional time derivative, many physical models' processes have been successfully modified. Ezzat et al. [12,13] introduced a new fractional bio-heat model using the fractional heat conduction equation. Ghanmi and Abbas [14] studied the fractional transient heating within the skin tissue owing to a moving heat source. Mondal et al. [15] investigated the transient heating within skin tissue owing to time-dependent thermal therapy in the context of memory-dependent heat transport law.

The analytical solutions are very interesting owing to their lower expense and accurate estimation compared with experimental and numerical calculations. Using the finite element approach, Diaz et al. [16] introduced the solutions of thermo-diffusions model in the tissue to study the resulting thermal damage. When a real phenomenon regarding heat transfer in a bounded media is studied, the nonlinear models and linear models of thermal transfer were developed and their numerical or analytical solutions are presented. Abbas and Zenkour [17] used the Green–Naghdi model to study the effect of rotation and initial stress on thermal shock problem for a fiber-reinforced anisotropic half-space. Abbas et al. [18] studied the effect of thermal dispersion on free convection in a fluid-saturated porous medium. Abbas [19] investigated the effects of relaxation times and a moving heat source on a two-temperature generalized thermoelastic thin slim strip. El-Naggar et al. [20] studied the effects of the initial stress, magnetic field, voids, and rotation on plane waves in generalized thermoelasticity. Marin and Marin et al. [21–23] used various models to study the dipolar bodies. Abbas [24] studied the nonlinear transient thermal stress analysis of a thick-walled FGM cylinder with temperature-dependent material properties. Zenkour and Abbas [25] used the finite element method to study the magneto-thermoelastic response of an infinite functionally graded cylinder.

This paper explores the effect of fractional order derivative on the thermal damage of living tissue using a bioheat model. The numerical results can be used as a substantiations division for living tissue interaction such as continual scanning laser interaction. The comparisons are made with the calculations obtained in the cases of the absence of the fractional time derivative and the thermal relaxation time parameters.

2. Mathematical Model

A semi-infinite biological tissue under thermal isolation is considered. On the basis of Cattaneo [5] and Ezzat et al. [12], the fractional bio-heat formulation in skin tissue can be expressed by

$$k\nabla^2 T = \left(1 + \frac{\tau_o^\alpha}{\Gamma(\alpha+1)}\frac{\partial^\alpha}{\partial t^\alpha}\right)\left(\rho c\frac{\partial T}{\partial t} + \omega_b \rho_b c_b (T - T_b) - Q_m - Q_{ext}\right),\ 0 < \alpha \leq 1, \tag{1}$$

Taking into consideration the above definition, it is possible to write the following:

$$\frac{\partial^\alpha h(\boldsymbol{r},t)}{\partial t^\alpha} = \begin{cases} h(\boldsymbol{r},t) - h(\boldsymbol{r},0), & \alpha \to 0, \\ I^{\alpha-1}\frac{\partial h(\boldsymbol{r},t)}{\partial t}, & 0 < \alpha < 1, \\ \frac{\partial h(\boldsymbol{r},t)}{\partial t}, & \alpha = 1, \end{cases} \tag{2}$$

$$I^\nu h(\boldsymbol{r},t) = \int_0^t \frac{(t-s)^\nu}{\Gamma(\nu)} h(\boldsymbol{r},s)ds,\ \nu > 0, \tag{3}$$

$$\lim_{\nu \to 1}\frac{\partial^\nu h(\boldsymbol{r},t)}{\partial t^\nu} = \frac{\partial h(\boldsymbol{r},t)}{\partial t}. \tag{4}$$

The full spectrum of the local thermal condition can be described through the standard thermal condition using the definition of the fractional derivative given in Equation (2). The different values of fractional parameter $0 < \alpha \leq 1$ cover two types of conductivity: $0 < \alpha < 1$ for low conductivity and $\alpha = 1$ for normal conductivity. Here, k is the thermal conductivity of tissue, ω_b is the rate of blood perfusion, t is the time, ρ_b is the blood mass density, ρ is the tissue mass density, T_b is the blood temperature, T is the tissue temperature, τ_o is the thermal relaxation time, c is the specific heat of tissues, c_b is the specific heat of blood, Q_m is the metabolic heat generations in living tissue, and Q_{ext} refers to the heat generated per unit volume of tissues. Gardner et al. [26] suggested the laser thermal source form by the following:

$$Q_{ext}(x,\ t) = I_o\mu_a\Big[U(t) - U(t-\tau_p)\Big]\Big[C_1 e^{-\frac{k_1}{\delta}x} - C_2 e^{-\frac{k_2}{\delta}x}\Big], \tag{5}$$

where $U(t)$ is the unit step function, μ_a is the coefficient of absorption, I_o is the intensity of the laser, τ_p is the exposure time of the laser, and δ is the penetration depth. $C_1,\ C_2,\ k_1$, and k_2 are the functions of diffuse reflectance R_d and are mentioned in [26]. The penetration depth is defined by the following [26]:

$$\delta = \frac{1}{\sqrt{3\mu_a(\mu_a + \mu_s(1-g))}} \tag{6}$$

where μ_s is the scattering coefficient and g is the factor of anisotropy. Now, both the lower and upper surfaces are supposed to be thermally isolated as the boundary conditions and the reference temperature are equal to its normal temperature. So, the initial conditions and the boundary conditions are presented as

$$-k\frac{\partial T(L,t)}{\partial x} = 0,\ -k\frac{\partial T(0,t)}{\partial x} = 0 \tag{7}$$

$$T(x,0) = T_b,\ \frac{\partial T(x,0)}{\partial t} = 0.0 \tag{8}$$

For appropriateness, the non-dimensional forms can be given by

$$\begin{gathered} T' = \tfrac{T-T_o}{T_o},\ T'_b = \tfrac{T_b-T_o}{T_o},\ t' = \tfrac{k}{\rho c L^2}t,\ \tau'_o = \tfrac{k}{\rho c L^2}\tau_o,\ \tau'_p = \tfrac{k}{\rho c L^2}\tau_p,\ x' = \tfrac{x}{L}, \\ k'_1 = \tfrac{L}{\delta}k_1,\ k'_2 = \tfrac{L}{\delta}k_2,\ R_b = \tfrac{\rho_b\omega_{bo}c_b L^2}{k},\ R_m = \tfrac{L^2 Q_m}{kT_o},\ R_r = \tfrac{L^2 I_o\mu_a}{kT_o}. \end{gathered} \tag{9}$$

In terms of these non-dimensional form of physical quantities in (9), the governing Equation (1) initial and boundary conditions can be expressed as (for appropriateness, the dashes have been ignored)

$$\frac{\partial^2 T}{\partial x^2} = \left(1 + \frac{\tau_o^\alpha}{\Gamma(\alpha+1)}\frac{\partial^\alpha}{\partial t^\alpha}\right)\left(\frac{\partial T}{\partial t} - R_b(T_b - T) - R_m - R_r f(x,t)\right) \tag{10}$$

$$\frac{\partial T(0,t)}{\partial x} = 0,\ \frac{\partial T(L,t)}{\partial x} = 0 \tag{11}$$

$$T(x,0) = 0,\ \frac{\partial T(x,0)}{\partial t} = 0 \tag{12}$$

where $f(x,t) = \Big[U(t) - U(t-\tau_p)\Big]\Big[C_1 e^{-k_1 x} - C_2 e^{-k_2 x}\Big]$.

The transforms of Laplace for any function $M(x,\ t)$ can be expressed by

$$\overline{M}(x,s) = L[M(x,t)] = \int_0^\infty M(x,t)e^{-st}dt,\ s > 0, \tag{13}$$

where s is the parameter of Laplace transform. Thus, the governing equations are expressed as

$$\frac{d^2\overline{T}}{dx^2} - f_1\overline{T} = -f_2 - f_3e^{-k_1x} - f_4e^{-k_2x} \tag{14}$$

$$\frac{\partial\overline{T}(0,t)}{\partial x} = 0, \ \frac{\partial\overline{T}(L,t)}{\partial x} = 0 \tag{15}$$

where

$$\begin{pmatrix} f_1 = \left(1 + \frac{s^\alpha \tau_o^\alpha}{\Gamma(\alpha+1)}\right)(s + R_b) \\ f_2 = \frac{1}{s}(R_bT_b + R_m) \\ f_3 = \frac{R_rC_1}{s}(1 - e^{-s\tau_p}) \\ f_4 = -\frac{R_rC_2}{s}(1 - e^{-s\tau_p}) \end{pmatrix} \tag{16}$$

The exact solution of Equation (14) is written as

$$\overline{T}(x,s) = \frac{f_2}{f_1} + A_1e^{\sqrt{f_1}x} + A_2e^{-\sqrt{f_1}x} + \frac{f_3}{f_1 - k_1^2}e^{-k_1x} + \frac{f_4}{f_1 - k_1^2}e^{-k_2x} \tag{17}$$

To obtain the complete solution of (14), boundary conditions represented by Equation (11) are used to get the constants A_1 and A_2, which are written as

$$\begin{aligned} A_1 &= \frac{e^{-L(k_1+k_2)}(e^{Lk_2}(e^{Lk_1}-e^{L\sqrt{\beta_1}})k_1(k_2^2-\beta_1)\beta_3+e^{Lk_1}(e^{Lk_2}-e^{L\sqrt{\beta_1}})k_1^2k_2\beta_4-e^{Lk_1}(e^{Lk_2}-e^{L\sqrt{\beta_1}})k_2\beta_1\beta_4)}{(e^{2L\sqrt{\beta_1}}-1)\sqrt{\beta_1}(\beta_1-k_1^2)(\beta_1-k_2^2)}, \\ A_2 &= \frac{e^{L\sqrt{\beta_1}}}{(e^{2L\sqrt{\beta_1}}-1)\sqrt{\beta_1}}\left(\frac{e^{-Lk_1}(e^{L(k_1+\sqrt{\beta_1})}-1)k_1\beta_3}{k_1^2-\beta_1} + \frac{e^{-Lk_2}(e^{L(k_2+\sqrt{\beta_1})}-1)k_2\beta_4}{k_2^2-\beta_1}\right). \end{aligned} \tag{18}$$

Finally, for the solutions of the temperature increment, a numerically reversal approach was adopted depending on Stehfest [27]. In this approach, the inverse $\mathrm{M}(x,t)$ of the Laplace transforms $\overline{\mathrm{M}}(x,s)$ can be approximated as

$$M(x,t) = \frac{ln2}{t}\sum_{j=1}^{M} V_j\overline{M}(x, j\frac{ln2}{t}), \tag{19}$$

where V_j is given by the following equation:

$$V_j = (-1)^{\frac{n}{2}+1}\sum_{k=\frac{i+1}{2}}^{min(i,\frac{n}{2})} \frac{k^{\frac{n}{2}+1}(2k)!}{(\frac{n}{2}-k)!k!(i-k)!(2k-1)!} \tag{20}$$

The evaluations of burns are one of the most remarkable attributes in the bio-engineering science in a living tissue. To quantify thermal damage, the technique expanded by Henriques and Moritz [28,29] can be used.

$$\Omega = \int_0^t Be^{-\frac{E_a}{RT}}dt \tag{21}$$

where $\mathrm{R} = 8.313$ J/mol·K is the universal gas constant, $\mathrm{B} = 3.1 \times 10^{98}\ s^{-1}$ is the factor of frequency, and $E_a = 6.28 \times 10^5$ J/mol is the activation energy.

3. Results and Discussions

Several simulations were conducted to test the performance of the proposed linear thermal model based on the bio-heat transfer model. For numerical computations, the values of thermal properties for living tissue were written [30].

$$\rho_b = 1060(kg)(m^{-3}), c_b = 3860\ (J)(kg^{-1})(k^{-1}),\ \omega_b = 1.87 \times 10^{-3}(s^{-1}),\ T_b = 37\ ^\circ\mathrm{C},$$

$$c = 4187\,(J)(kg^{-1})(k^{-1}), \rho = 1000(kg)(m^{-3}), k = 0.628\,(W)(m^{-1})(k^{-1}),\ \tau_o = 5(s),$$
$$\tau_p = 10(s), Q_m = 1.19 \times 10^3(W)(m^{-3}), L = 0.03(m), I_o = 122 \times 10^3(W)(m^{-2}),\ g = 0.9,$$
$$\mu_s = 12000\,(m^{-1}), \mu_a = 40(m^{-1}), T_o = 37\ °\text{C}.$$

Using MATLAB (R2017a) software, the computations are done and the results are presented graphically. The outer thermal source impact on the surface of the skin was integrated. These mathematical models, which are based on hyperbolic bio-heat transfer, were found with the interface and appropriate boundary conditions. The perfusions, metabolic, and conducting heat source terms were utilized in the formulations. A slab of tissue is 3 cm thick and the reference temperature is equal to its normal temperature, that is, $T_b = T_o = 37$ °C. In order to study the effect of fractional parameter α, the laser exposure time τ_p and the thermal relaxation time τ_o on the temperature and the thermal damages, the numerical results were presented using the graphs as in Figures 1–9. Figure 1 displays the temperature variation along the distance x at $t = 80$ s when the laser exposure and thermal relaxation times remine to constants $\tau_p = 10s, \tau_o = 5s$. It is observed that the temperature begins from the utmost values, and then decreases constantly to $T_b = 37$ °C. The time history of surface temperature through four values of the fractional order parameter α is exhibited in Figure 2. It notices that the temperature begins from T_b and increases with the time till most values, after which decreases again to T_b. Figure 3 show the resulting thermal damage through time t. Clearly, the time history of the thermal damages obtained from the different values of the fractional order parameter α is very different. As expect, the fractional order parameter α has major effects on the distributions of the variables. Figures 4–6 display the influences of the laser exposure time τ_p with the fractional bioheat model ($\alpha = 0.5$) on the resulting thermal damage and the temperature distributions. As expected, the laser exposure time τ_p has a great effect on the resulting thermal damages and the distribution of temperature. As expected, the increase of laser exposure times increases the temperature, exceptionally at the peak temperature where the difference between the diagrams. On the basis of the fractional order bioheat model ($\alpha = 0.5$), the comparison of temperature response among four types of thermal relaxation time τ_o along x is shown in Figure 7. In these cases, it was observed that the thermal relaxation time decreases the temperature of the skin surface. Gradually, the temperature decreases continuously to the normal temperature. The time history of surface temperature through four values of the thermal relaxation time is shown in Figure 8. Moreover, Figure 9 displays the variations of thermal damages with time for various values of thermal relaxation time τ_o. Figures 3, 6 and 9 show the variations of thermal damage at the point $x = 0$ under various values of the fractional parameter, the laser exposure time, and the thermal relaxation time, respectively. As observed in the figures, the thermal damages are reduced with the fractional model and hyperbolic bioheat model.

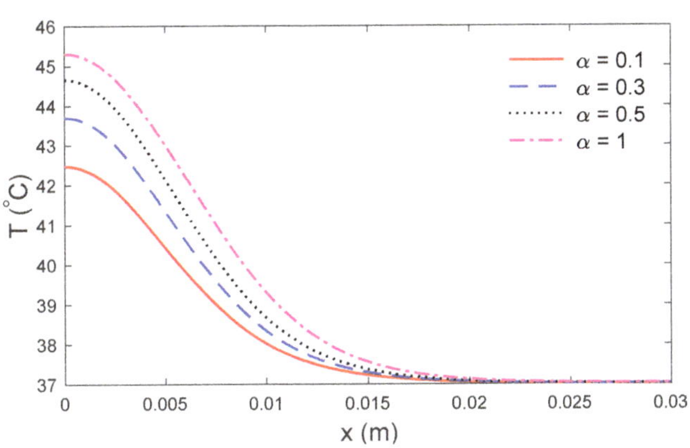

Figure 1. Temperature profile in skin tissue with and without fractional derivative.

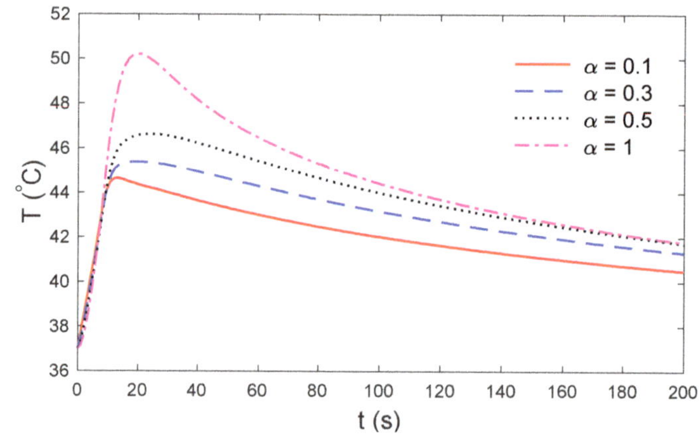

Figure 2. Temperature history at skin surface with and without fractional derivative.

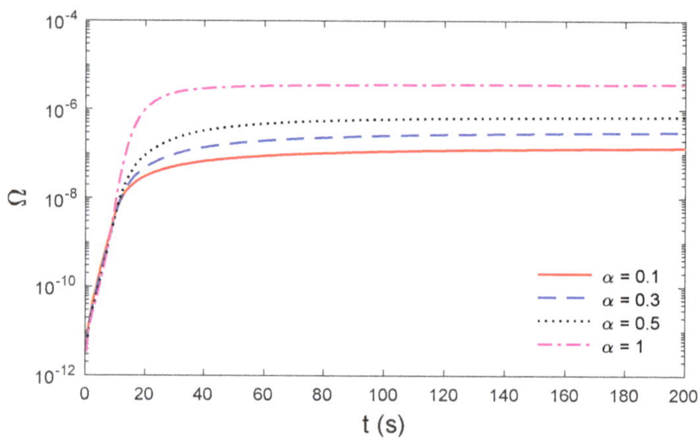

Figure 3. The variation of thermal damage at skin surface with and without fractional derivative.

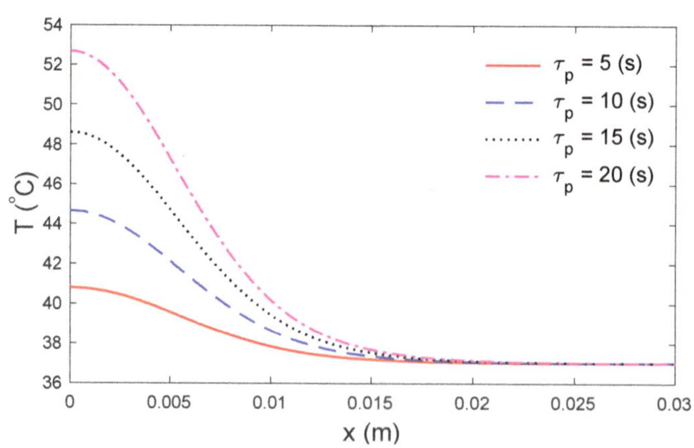

Figure 4. Temperature profile in skin tissue for different values of the laser exposure time τ_p.

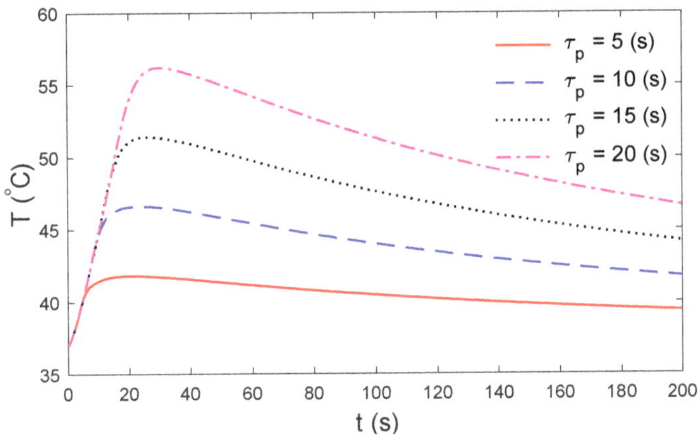

Figure 5. Temperature history at skin surface for different values of the laser exposure time τ_p.

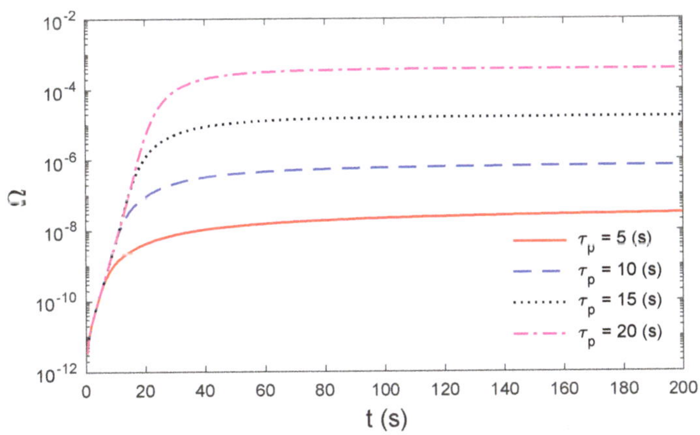

Figure 6. The variation of thermal damage with time for different values of the laser exposure time τ_p.

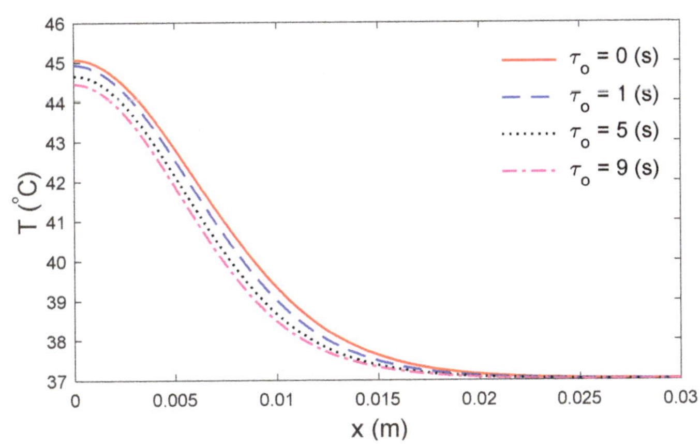

Figure 7. Temperature profile in skin tissue for different values of thermal relaxation time τ_o.

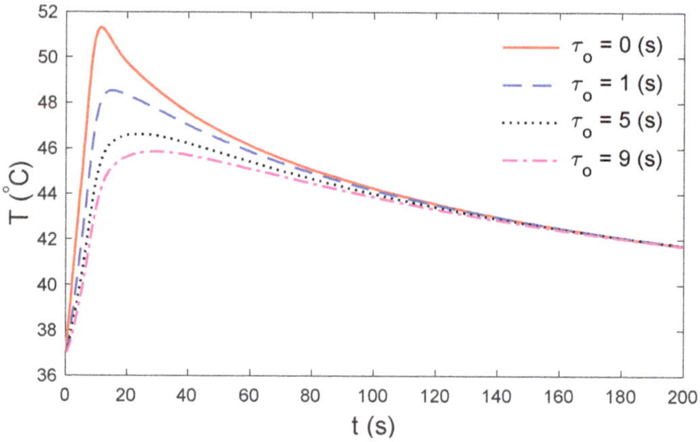

Figure 8. Temperature history at skin surface for different values of thermal relaxation time τ_o.

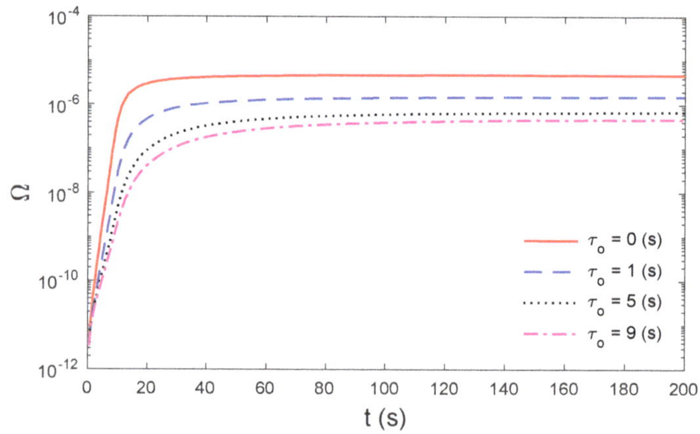

Figure 9. The variation of thermal damage with time for different values of thermal relaxation time τ_o.

4. Conclusions

On the basis of the fractional order bio-heat model in a living tissue, the generalized bio-heat equation was proposed, through the substitution of the rate of change term by a fractional time derivative. The fractional-order bio-heat transfer model is derived and its effect is given in thermotherapy application. The thermal damages are reduced with the fractional model and hyperbolic bioheat model. We believe that the analysis of the present study will be useful to understand the basic features of this new model for bioheat conduction.

Author Contributions: All four authors conceived the framework and structured the whole manuscript, checked the results, and completed the revision of the paper. The authors have equally contributed to the elaboration of this manuscript. All authors have read and approved the final form of the manuscript.

Funding: This research received no external funding.

Acknowledgments: This project was funded by the Deanship of Scientific Research (DSR) at King Abdulaziz University, Jeddah, Saudi Arabia, under grant no. (KEP-72-130-38). The authors, therefore, acknowledge with thanks DSR technical and financial support.

Conflicts of Interest: The authors declare no conflicts of interest.

References

1. Mahjoob, S.; Vafai, K. Analytical characterization of heat transport through biological media incorporating hyperthermia treatment. *Int. J. Heat Mass Transf.* **2009**, *52*, 1608–1618. [CrossRef]
2. Gabay, I.; Abergel, A.; Vasilyev, T.; Rabi, Y.; Fliss, D.M.; Katzir, A. Temperature-controlled two-wavelength laser soldering of tissues. *Lasers Surg. Med.* **2011**, *43*, 907–913. [CrossRef] [PubMed]
3. Zhou, J.; Chen, J.; Zhang, Y. Dual-phase lag effects on thermal damage to biological tissues caused by laser irradiation. *Comput. Biol. Med.* **2009**, *39*, 286–293. [CrossRef] [PubMed]
4. Pennes, H.H. Analysis of tissue and arterial blood temperatures in the resting human forearm. *J. Appl. Physiol.* **1948**, *1*, 93–122. [CrossRef] [PubMed]
5. Cattaneo, C. A form of heat conduction equation which eliminates the paradox of instantaneous propagation. *Compt. Rendu* **1958**, *247*, 431–433.
6. Vernotte, P. Les paradoxes de la theorie continue de l'equation de la chaleur. *Compt. Rendu* **1958**, *246*, 3154–3155.
7. Hobiny, A.D.; Abbas, I.A. Theoretical analysis of thermal damages in skin tissue induced by intense moving heat source. *Int. J. Heat Mass Transf.* **2018**, *124*, 1011–1014. [CrossRef]
8. Gupta, P.K.; Singh, J.; Rai, K. Numerical simulation for heat transfer in tissues during thermal therapy. *J. Therm. Biol.* **2010**, *35*, 295–301. [CrossRef]
9. Gupta, P.K.; Singh, J.; Rai, K.; Rai, S. Solution of the heat transfer problem in tissues during hyperthermia by finite difference–decomposition method. *Appl. Math. Comput.* **2013**, *219*, 6882–6892. [CrossRef]
10. Dillenseger, J.-L.; Esneault, S. Fast FFT-based bioheat transfer equation computation. *Comput. Biol. Med.* **2010**, *40*, 119–123. [CrossRef]
11. Zhu, D.; Luo, Q.; Zhu, G.; Liu, W. Kinetic thermal response and damage in laser coagulation of tissue. *Lasers Surg. Med.* **2002**, *31*, 313–321. [CrossRef] [PubMed]
12. Ezzat, M.A.; AlSowayan, N.S.; Al-Muhiameed, Z.I.; Ezzat, S.M. Fractional modelling of Pennes' bioheat transfer equation. *Heat Mass Transf.* **2014**, *50*, 907–914. [CrossRef]
13. Ezzat, M.A.; El-bary, A.A.; Al-sowayan, N.S. Tissue responses to fractional transient heating with sinusoidal heat flux condition on skin surface. *Anim. Sci. J.* **2016**, *87*, 1304–1311. [CrossRef] [PubMed]
14. Ghanmi, A.; Abbas, I.A. An analytical study on the fractional transient heating within the skin tissue during the thermal therapy. *J. Therm. Biol.* **2019**, *82*, 229–233. [CrossRef]
15. Mondal, S.; Sur, A.; Kanoria, M. Transient heating within skin tissue due to time-dependent thermal therapy in the context of memory dependent heat transport law. *Mech. Based Des. Struct. Mach.* **2019**, 1–15. [CrossRef]
16. Díaz, S.H.; Nelson, J.S.; Wong, B.J. Rate process analysis of thermal damage in cartilage. *Phys. Med. Biol.* **2002**, *48*, 19. [CrossRef]
17. Abbas, I.A.; Zenkour, A.M. The effect of rotation and initial stress on thermal shock problem for a fiber-reinforced anisotropic half-space using Green-Naghdi theory. *J. Comput. Theor. Nanosci.* **2014**, *11*, 331–338. [CrossRef]
18. Abbas, I.A.; El-Amin, M.; Salama, A. Effect of thermal dispersion on free convection in a fluid saturated porous medium. *Int. J. Heat Fluid Flow* **2009**, *30*, 229–236. [CrossRef]
19. Abbas, I.A. The effects of relaxation times and a moving heat source on a two-temperature generalized thermoelastic thin slim strip. *Can. J. Phys.* **2015**, *93*, 585–590. [CrossRef]
20. El-Naggar, A.; Kishka, Z.; Abd-Alla, A.M.; Abbas, I.A.; Abo-Dahab, S.M.; Elsagheer, M. On the initial stress, magnetic field, voids and rotation effects on plane waves in generalized thermoelasticity. *J. Comput. Theor. Nanosci.* **2013**, *10*, 1408–1417. [CrossRef]
21. Marin, M. Cesaro means in thermoelasticity of dipolar bodies. *Acta Mech.* **1997**, *122*, 155–168. [CrossRef]
22. Marin, M.; Ellahi, R.; Chirilă, A. On solutions of Saint-Venant's problem for elastic dipolar bodies with voids. *Carpathian J. Math.* **2017**, *33*, 219–232.
23. Marin, M.; Nicaise, S. Existence and stability results for thermoelastic dipolar bodies with double porosity. *Contin. Mech. Thermodyn.* **2016**, *28*, 1645–1657. [CrossRef]
24. Abbas, I.A. Nonlinear transient thermal stress analysis of thick-walled FGM cylinder with temperature-dependent material properties. *Meccanica* **2014**, *49*, 1697–1708. [CrossRef]
25. Zenkour, A.M.; Abbas, I.A. Magneto-thermoelastic response of an infinite functionally graded cylinder using the finite element method. *J. Vib. Control.* **2014**, *20*, 1907–1919. [CrossRef]

26. Gardner, C.M.; Jacques, S.L.; Welch, A. Light transport in tissue: Accurate expressions for one-dimensional fluence rate and escape function based upon Monte Carlo simulation. *Lasers Surg. Med.* **1996**, *18*, 129–138. [CrossRef]
27. Stehfest, H. Algorithm 368: Numerical inversion of Laplace transforms [D5]. *Commun. ACM* **1970**, *13*, 47–49. [CrossRef]
28. Henriques, F., Jr.; Moritz, A.R. Studies of thermal injury: I. The conduction of heat to and through skin and the temperatures attained therein. A theoretical and an experimental investigation. *Am. J. Pathol.* **1947**, *23*, 530.
29. Moritz, A.R.; Henriques, F., Jr. Studies of thermal injury: II. The relative importance of time and surface temperature in the causation of cutaneous burns. *Am. J. Pathol.* **1947**, *23*, 695.
30. Askarizadeh, H.; Ahmadikia, H. Analytical analysis of the dual-phase-lag model of bioheat transfer equation during transient heating of skin tissue. *Heat Mass Transf.* **2014**, *50*, 1673–1684. [CrossRef]

symmetry

MDPI

Article

Dynamic Hilbert-Type Inequalities with Fenchel-Legendre Transform

Ahmed A. El-Deeb [1,*], Samer D. Makharesh [1] and Dumitru Baleanu [2,3,4]

1 Department of Mathematics, Faculty of Science, Al-Azhar University, Nasr City, Cairo 11884, Egypt; sameeermakarish@yahoo.com
2 Department of Mathematics, Cankaya University, Ankara 06530, Turkey; dumitru@cankaya.edu.tr or Baleanu@mail.cmuh.org.tw
3 Institute of Space Science, 07650 Magurele-Bucharest, Romania
4 Department of Medical Research, China Medical University Hospital, China Medical University, Taichung 40402, Taiwan
* Correspondence: ahmedeldeeb@azhar.edu.eg

Received: 12 March 2020; Accepted: 26 March 2020; Published: 7 April 2020

Abstract: Our work is based on the multiple inequalities illustrated in 2020 by Hamiaz and Abuelela. With the help of a Fenchel-Legendre transform, which is used in various problems involving symmetry, we generalize a number of those inequalities to a general time scale. Besides that, in order to get new results as special cases, we will extend our results to continuous and discrete calculus.

Keywords: Hilbert's inequality; Fubini theorem; Fenchel-Legendre transform; time scale

AMS Subject Classifications: 26D10; 26D15; 26E70; 34A40

---

1. Introduction

In 2020, Hamiaz and Abuelela [1] have studied the following discrete inequalities:

Theorem 1. *Suppose $q, p \geqslant 1$, $\alpha \geqslant \beta \geqslant \frac{1}{2}$ and $(b_m)_m \geq 0$ $(a_n)_n \geq 0$ are sequences of real numbers. Define $A_n = \sum_{s=1}^{n} a_s$, $B_m = \sum_{t=1}^{m} b_t$. Then*

$$\sum_{n=1}^{k}\sum_{m=1}^{r}\frac{A_n^{2p}B_m^{2q}}{h(n)+h^*(m)} \leqslant C_1^*(p,q)\left(\sum_{n=1}^{k}(k-n+1)(a_nA_n^{p-1})^2\right)\times\left(\sum_{m=1}^{r}(r-m+1)(b_mB_m^{q-1})^2\right)$$

and

$$\sum_{n=1}^{k}\sum_{m=1}^{r}\frac{A_n^{p}B_m^{q}}{\left(|h(n)|^{\frac{1}{2\beta}}+|h^*(m)|^{\frac{1}{2\beta}}\right)^{\alpha}} \leqslant \sum_{n=1}^{k}\sum_{m=1}^{r}\frac{A_n^{p}B_m^{q}}{\sqrt{h(n)+h^*(m)}}$$
$$\leqslant C_2^*(p,q,k,r)\left(\sum_{n=1}^{k}(k-n+1)(a_nA_n^{p-1})^2\right)^{\frac{1}{2}}\times\left(\sum_{m=1}^{r}(r-m+1)(b_mB_m^{q-1})^2\right)^{\frac{1}{2}}$$

unless (a_n) *or* (b_m) *is null, where*

$$C_1^*(p,q) = (pq)^2 \text{ and } C_2^*(p,q,r,k) = pq\sqrt{kr}.$$

Hilger [2] suggested time scales theory to unify discrete and continuous analysis. More Hilbert-type inequalities and other types can be seen in [1,3–36], see also [37–53]. For more details on time scales calculus see [54].

We will need the following important relations between calculus on time scales $\mathbb{T}$ and either continuous calculus on $\mathbb{R}$ or discrete calculus on $\mathbb{Z}$. Note that:

(*i*) If $\mathbb{T} = \mathbb{R}$, then

$$\sigma(t) = t, \quad \mu(t) = 0, \quad f^{\Delta}(t) = f'(t), \quad \int_a^b f(t)\Delta t = \int_a^b f(t)dt. \tag{1}$$

(*ii*) If $\mathbb{T} = \mathbb{Z}$, then

$$\sigma(t) = t+1, \quad \mu(t) = 1, \quad f^{\Delta}(t) = f(t+1) - f(t), \quad \int_a^b f(t)\Delta t = \sum_{t=a}^{b-1} f(t). \tag{2}$$

Next is Hölder's and Jensen's inequality:

Lemma 1 ([19]). *Let* $a, b \in \mathbb{T}$ *and* $f, g \in C_{rd}([a,b]_{\mathbb{T}}, [0,\infty))$. *If* $p, q > 1$ *with* $\frac{1}{p} + \frac{1}{q} = 1$, *then*

$$\int_a^b f(t)g(t)\Delta t \le \left[\int_a^b f^p(t)\Delta t\right]^{\frac{1}{p}} \left[\int_a^b g^q(t)\Delta t\right]^{\frac{1}{q}}.$$

Lemma 2 ([19]). *Let* $a, b \in \mathbb{T}$ *and* $\check{c}, \check{d} \in \mathbb{R}$. *Assume that* $g \in C_{rd}([a,b]_{\mathbb{T}}, [\check{c}, \check{d}])$ *and* $r \in C_{rd}([a,b]_{\mathbb{T}}, \mathbb{R})$ *are nonnegative with* $\int_a^b r(t)\Delta t > 0$. *If* $\check{\Phi} \in C_{rd}((\check{c}, \check{d}), \mathbb{R})$ *be a convex function, then*

$$\check{\Phi}\left(\frac{\int_a^b g(t)r(t)\Delta t}{\int_a^b r(t)\Delta t}\right) \leqslant \frac{\int_a^b r(t)\check{\Phi}(g(t))\Delta t}{\int_a^b r(t)\Delta t}.$$

Now, we present the Fenchel-Legendre transform and refer, for example, to [11–13], for more details.

Definition 1. *Assuming* $h : \mathbb{R}^n \longrightarrow \mathbb{R} \cup \{+\infty\}$ *is a function:* $h \neq +\infty$ *i.e.,* $\text{Dom}(h) = \{x \in \mathbb{R}^n, |h(x) < \infty\} \neq \emptyset$. *Then the Fenchel-Legendre transform is defined as:*

$$h^* : \mathbb{R}^n \longrightarrow \mathbb{R} \cup \{+\infty\}, \ y \longrightarrow h^*(y) = \sup\{< y, x > -h(x), x \in \text{Dom}(h)\} \tag{3}$$

where $< .,. >$ *is the scalar product on* $\mathbb{R}^n$. *The mapping* $h \longrightarrow h^*$ *is often be called the conjugate operation.*

The domain of h^* is the set of slopes of all the affine functions minorizing the function h over $\mathbb{R}^n$. An equivalent formula for (3) is introduced as follows:

Corollary 1. *Assuming* $h : \mathbb{R}^n \longrightarrow \mathbb{R}$ *is differentiable, strictly convex and 1-coercive function. Then*

$$h^*(y) = < y, (\nabla h)^{-1}(y) > -h((\nabla h)^{-1}(y)), \tag{4}$$

$\forall\, y \in \text{Dom}(h^*)$, *where* $< .,. >$ *denotes the scalar product on* $\mathbb{R}^n$.

Lemma 3 ([13]). *Let h be a function and h^* its Fenchel-Legendre transform. Then*

$$< x, y > \leqslant h(x) + h^*(y), \tag{5}$$

for all $x \in \mathrm{Dom}(h)$, and $y \in \mathrm{Dom}(h^*)$.

In addition, we will use the following definition and lemma as we will see in the proof of our results:

Definition 2. *The function $\check{\Phi}$ is said to be a submultiplicative on $[0, \infty)$ if*

$$\check{\Phi}(xy) \leqslant \check{\Phi}(x)\check{\Phi}(y), \ \textit{for all} \ \ x, y \geqslant 0. \tag{6}$$

Lemma 4 ([20]). *Assuming $\mathbb{T}$ is a time scale with $x, a \in \mathbb{T}$ such that $x \geqslant a$. If $f \geqslant 0$ and $\tilde{\alpha} \geqslant 1$, then*

$$\left(\int_a^{\sigma(x)} f(\check{\tau}) \Delta \check{\tau} \right)^{\tilde{\alpha}} \leqslant \tilde{\alpha} \int_a^{\sigma(x)} f(\eta) \left(\int_a^{\sigma(\eta)} f(\check{\tau}) \Delta \check{\tau} \right)^{\tilde{\alpha}-1} \Delta \eta. \tag{7}$$

Next, we write Fubini's theorem on time scales.

Lemma 5 (Fubini's Theorem, see [55]). *Assume that $(X, \Sigma_1, \mu_\Delta)$ and $(Y, \Sigma_2, \nu_\Delta)$ are two finite-dimensional time scales measure spaces. Moreover, suppose that $f : X \times Y \to \mathbb{R}$ is a delta integrable function and define the functions*

$$\hat{\pi}_1(y) = \int_X f(x, y) d\mu_\Delta(x), \quad y \in Y,$$

and

$$\hat{\pi}_2(x) = \int_X f(x, y) d\nu_\Delta(y), \quad x \in X.$$

Then $\hat{\pi}_1$ is delta integrable on Y and $\hat{\pi}_2$ is delta integrable on X and

$$\int_X d\mu_\Delta(x) \int_Y f(x, y) d\nu_\Delta(y) = \int_Y d\nu_\Delta(y) \int_X f(x, y) d\mu_\Delta(x).$$

In this manuscript, by using Fubini's theorem and the Fenchel-Legendre transform, which is used in various problems involving symmetry, we extend the discrete results proved in [1] on time scales. We start from the inequalities treated in the Theorem 1. Our results can be applied to give more general forms of some previously proved inequalities through substituting h and h^* by suitable functions as we will see in the following two sections.

The following section contains our main results.

2. Main Results

We start by establishing the following useful inequality:

Lemma 6. *Assume x and $y \in \mathbb{R}$ such that $x + y \geqslant 1$, then for $\gamma > 0$, and $\alpha \geqslant \beta \geqslant \frac{1}{2}$, we get*

$$(x + y)^{\frac{1}{\gamma}} \leqslant \left(|x|^{\frac{1}{2\beta}} + |y|^{\frac{1}{2\beta}} \right)^{\frac{2\alpha}{\gamma}}. \tag{8}$$

Proof. For $x + y \geqslant 1$ and $\frac{\alpha}{\beta} \geqslant 1$, we have

$$(x + y)^{\frac{1}{2}} \leqslant \left[(x + y)^{\frac{1}{2}} \right]^{\frac{\alpha}{\beta}} = \left[(x + y)^{\frac{1}{2\beta}} \right]^{\alpha} \leqslant \left[(|x| + |y|)^{\frac{1}{2\beta}} \right]^{\alpha}. \tag{9}$$

From $\left(|x|+|y|\right)^{\frac{1}{n}} \leqslant |x|^{\frac{1}{n}}+|y|^{\frac{1}{n}}$, for all $n \geqslant 1$. Thus, from (9), and since $2\beta \geqslant 1$, we obtain:

$$(x+y)^{\frac{1}{2}} \leqslant \left[\left(|x|+|y|\right)^{\frac{1}{2\beta}}\right]^{\alpha} \leqslant \left[|x|^{\frac{1}{2\beta}}+|y|^{\frac{1}{2\beta}}\right]^{\alpha}. \tag{10}$$

Now, since $\gamma > 0$, by taking the power $1/\gamma$ for both sides of (10), we get:

$$(x+y)^{\frac{1}{\gamma}} \leqslant \left(|x|^{\frac{1}{2\beta}}+|y|^{\frac{1}{2\beta}}\right)^{\frac{2\alpha}{\gamma}}.$$

This proves our claim. □

In the next theorems, we will let $p > 1, q > 1$ and $\frac{1}{q}+\frac{1}{p}=1$.

Theorem 2. *Let $\mathbb{T}$ be a time scale with $L \geqslant 1, K \geqslant 1$ and $s, t, t_0, x, y \in \mathbb{T}$. Assume $a(\check{t}) \geq 0$ and $b(\check{t}) \geq 0$ are right-dense continuous functions on the time scales intervals $[t_0, x]_{\mathbb{T}}$ and $[t_0, y]_{\mathbb{T}}$ respectively and define*

$$A(s) := \int_{t_0}^{s} a(\check{t})\Delta\check{t}, \quad and \quad B(t) := \int_{t_0}^{t} b(\check{t})\Delta\check{t},$$

then for $\sigma(s) \in [t_0, x]_{\mathbb{T}}$ and $\sigma(t) \in [t_0, y]_{\mathbb{T}}$, we have that

$$\begin{aligned}
&\int_{t_0}^{x}\int_{t_0}^{y} \frac{A^{qK}(\sigma(s))B^{qL}(\sigma(t))}{\left(|h(\sigma(s)-t_0)|^{\frac{1}{2\beta}}+|h^*(\sigma(t)-t_0)|^{\frac{1}{2\beta}}\right)^{\frac{2q\alpha}{p}}}\Delta s\Delta t \\
&\leqslant C_1(L,K,q)\left(\int_{t_0}^{x}(\sigma(x)-\sigma(s))\left(a(s)A^{K-1}(\sigma(s)\right)^{q}\Delta s\right) \\
&\times\left(\int_{t_0}^{y}(\sigma(y)-\sigma(t))\left(b(t)B^{L-1}(\sigma(t)\right)^{q}\Delta t\right)
\end{aligned} \tag{11}$$

and

$$\begin{aligned}
&\int_{t_0}^{x}\int_{t_0}^{y} \frac{A^{K}(\sigma(s))B^{L}(\sigma(t))}{\left(|h(\sigma(s)-t_0)|^{\frac{1}{2\beta}}+|h^*(\sigma(t)-t_0)|^{\frac{1}{2\beta}}\right)^{\frac{2\alpha}{p}}}\Delta s\Delta t \\
&\leqslant C_2(L,K,p)\left(\int_{t_0}^{x}(\sigma(x)-\sigma(s))\left(A^{K-1}(\sigma(s))a(s)\right)^{q}\Delta s\right)^{\frac{1}{q}} \\
&\times\left(\int_{t_0}^{y}(\sigma(y)-\sigma(t))\left(B^{L-1}(\sigma(t))b(t)\right)^{q}\Delta t\right)^{\frac{1}{q}}
\end{aligned} \tag{12}$$

where $C_1(L,K,q)=(KL)^q$ and $C_2(L,K,p)=KL(x-t_0)^{\frac{1}{p}}(y-t_0)^{\frac{1}{p}}$.

Proof. By using the inequality (7), we obtain

$$A^{K}(\sigma(s)) \leqslant K\int_{t_0}^{\sigma(s)} a(\eta)A^{K-1}(\sigma(\eta)\Delta\eta, \tag{13}$$

$$B^{L}(\sigma(t)) \leqslant L\int_{t_0}^{\sigma(t)} b(\eta)B^{L-1}(\sigma(\eta))\Delta\eta. \tag{14}$$

We use Lemma 1. Then from (13), we get

$$A^K(\sigma(s)) \leqslant K(\sigma(s)-t_0)^{\frac{1}{p}}\left(\int_{t_0}^{\sigma(s)} (a(\eta)A^{K-1}(\sigma(\eta))^q \Delta\eta\right)^{\frac{1}{q}}. \tag{15}$$

We use Lemma 1. Then from (14), we also have

$$B^L(\sigma(t)) \leqslant L(\sigma(t)-t_0)^{\frac{1}{p}}\left(\int_{t_0}^{\sigma(t)} (b(\eta)B^{L-1}(\sigma(\eta))^q \Delta\eta\right)^{\frac{1}{q}}. \tag{16}$$

From (15) and (16), we get

$$\begin{aligned} A^K(\sigma(s))B^L(\sigma(t)) \leqslant{}& KL(\sigma(s)-t_0)^{\frac{1}{p}}(\sigma(t)-t_0)^{\frac{1}{p}} \\ &\times\left(\int_{t_0}^{\sigma(s)} (a(\eta)A^{K-1}(\sigma(\eta))^q \Delta\eta\right)^{\frac{1}{q}} \\ &\times\left(\int_{t_0}^{\sigma(t)} (b(\eta)B^{L-1}(\sigma(\eta))^q \Delta\eta\right)^{\frac{1}{q}}. \end{aligned} \tag{17}$$

From inequality (17), we have

$$\begin{aligned} A^{qK}(\sigma(s))B^{qL}(\sigma(t)) \leqslant{}& (KL)^q(\sigma(s)-t_0)^{\frac{q}{p}}(\sigma(t)-t_0)^{\frac{q}{p}} \\ &\times\left(\int_{t_0}^{\sigma(s)} (a(\eta)A^{K-1}(\sigma(\eta))^q \Delta\eta\right) \\ &\times\left(\int_{t_0}^{\sigma(t)} (b(\eta)B^{L-1}(\sigma(\eta))^q \Delta\eta\right). \end{aligned} \tag{18}$$

Using Lemma 3 in (17) and (18) gives

$$\begin{aligned} A^K(\sigma(s))B^L(\sigma(t)) \leqslant{}& KL\Big(h(\sigma(s)-t_0)+h^*(\sigma(t)-t_0)\Big)^{\frac{1}{p}} \\ &\times\left(\int_{t_0}^{\sigma(s)} (a(\eta)A^{K-1}(\sigma(\eta))^q \Delta\eta\right)^{\frac{1}{q}} \\ &\times\left(\int_{t_0}^{\sigma(t)} (b(\eta)B^{L-1}(\sigma(\eta))^q \Delta\eta\right)^{\frac{1}{q}}, \end{aligned} \tag{19}$$

$$\begin{aligned} A^{qK}(\sigma(s))B^{qL}(\sigma(t)) \leqslant{}& (KL)^q\Big(h(\sigma(s)-t_0)+h^*(\sigma(t)-t_0)\Big)^{\frac{q}{p}} \\ &\times\left(\int_{t_0}^{\sigma(s)} (a(\eta)A^{K-1}(\sigma(\eta))^q \Delta\eta\right) \\ &\times\left(\int_{t_0}^{\sigma(t)} (b(\eta)B^{L-1}(\sigma(\eta))^q \Delta\eta\right). \end{aligned} \tag{20}$$

Using Lemma 6 in (19) and (20) gives

$$A^{K}(\sigma(s))B^{L}(\sigma(t)) \leqslant KL\Big(|h(\sigma(s)-t_0)|^{\frac{1}{2\beta}}+|h^{*}(\sigma(t)-t_0)|^{\frac{1}{2\beta}}\Big)^{\frac{2\alpha}{p}} \times\Big(\int_{t_0}^{\sigma(s)}(a(\eta)A^{K-1}(\sigma(\eta))^{q}\Delta\eta\Big)^{\frac{1}{q}} \times\Big(\int_{t_0}^{\sigma(t)}(b(\eta)B^{L-1}(\sigma(\eta))^{q}\Delta\eta\Big)^{\frac{1}{q}}, \quad (21)$$

$$A^{qK}(\sigma(s))B^{qL}(\sigma(t)) \leqslant (KL)^{q}\Big(|h(\sigma(s)-t_0)|^{\frac{1}{2\beta}}+|h^{*}(\sigma(t)-t_0)|^{\frac{1}{2\beta}}\Big)^{\frac{2q\alpha}{p}} \times\Big(\int_{t_0}^{\sigma(s)}(a(\eta)A^{K-1}(\sigma(\eta))^{q}\Delta\eta\Big) \times\Big(\int_{t_0}^{\sigma(t)}(b(\eta)B^{L-1}(\sigma(\eta))^{q}\Delta\eta\Big). \quad (22)$$

Dividing both sides of (21) and (22) by $\Big(|h(\sigma(s)-t_0)|^{\frac{1}{2\beta}}+|h^{*}(\sigma(t)-t_0)|^{\frac{1}{2\beta}}\Big)^{\frac{2\alpha}{p}}$ and $\Big(|h(\sigma(s)-t_0)|^{\frac{1}{2\beta}}+|h^{*}(\sigma(t)-t_0)|^{\frac{1}{2\beta}}\Big)^{\frac{2q\alpha}{p}}$ respectively, we get that

$$\frac{A^{K}(\sigma(s))B^{L}(\sigma(t))}{\Big(|h(\sigma(s)-t_0)|^{\frac{1}{2\beta}}+|h^{*}(\sigma(t)-t_0)|^{\frac{1}{2\beta}}\Big)^{\frac{2\alpha}{p}}} \leqslant KL\Big(\int_{t_0}^{\sigma(s)}(a(\eta)A^{K-1}(\sigma(\eta))^{q}\Delta\eta\Big)^{\frac{1}{q}} \times\Big(\int_{t_0}^{\sigma(t)}(b(\eta)B^{L-1}(\sigma(\eta))^{q}\Delta\eta\Big)^{\frac{1}{q}}, \quad (23)$$

$$\frac{A^{qK}(\sigma(s))B^{qL}(\sigma(t))}{\Big(|h(\sigma(s)-t_0)|^{\frac{1}{2\beta}}+|h^{*}(\sigma(t)-t_0)|^{\frac{1}{2\beta}}\Big)^{\frac{2q\alpha}{p}}} \leqslant (KL)^{q}\Big(\int_{t_0}^{\sigma(s)}(a(\eta)A^{K-1}(\sigma(\eta))^{q}\Delta\eta\Big) \times\Big(\int_{t_0}^{\sigma(t)}(b(\eta)B^{L-1}(\sigma(\eta))^{q}\Delta\eta\Big). \quad (24)$$

From (23) by using Lemma 1 we obtain

$$\int_{t_0}^{x}\int_{t_0}^{y}\frac{A^{K}(\sigma(s))B^{L}(\sigma(t))}{\Big(|h(\sigma(s)-t_0)|^{\frac{1}{2\beta}}+|h^{*}(\sigma(t)-t_0)|^{\frac{1}{2\beta}}\Big)^{\frac{2\alpha}{p}}}\Delta s\Delta t \leqslant KL(x-t_0)^{\frac{1}{p}}(y-t_0)^{\frac{1}{p}}\int_{t_0}^{x}\Big(\int_{t_0}^{\sigma(s)}(a(\eta)A^{K-1}(\sigma(\eta))^{q}\Delta\eta\Big)\Delta s\Big)^{\frac{1}{q}} \times\int_{t_0}^{y}\Big(\int_{t_0}^{\sigma(t)}(b(\eta)B^{L-1}(\sigma(\eta))^{q}\Delta\eta\Big)\Delta t\Big)^{\frac{1}{q}}. \quad (25)$$

From (24), we get

$$\int_{t_0}^{x}\int_{t_0}^{y}\frac{A^{qK}(\sigma(s))B^{qL}(\sigma(t))}{\left(|h(\sigma(s)-t_0)|^{\frac{1}{2\beta}}+|h^*(\sigma(t)-t_0)|^{\frac{1}{2\beta}}\right)^{\frac{2q\alpha}{p}}}\Delta s\Delta t$$
$$\leqslant (KL)^q\int_{t_0}^{x}\left(\int_{t_0}^{\sigma(s)}\left(a(\eta)A^{K-1}(\sigma(\eta))^q\Delta\eta\right)\Delta s\right)$$
$$\times\int_{t_0}^{y}\left(\int_{t_0}^{\sigma(t)}\left(b(\eta)B^{L-1}(\sigma(\eta))^q\Delta\eta\right)\Delta t\right). \tag{26}$$

Applying Fubini's Theorem on (25) and (26) gives

$$\int_{t_0}^{x}\int_{t_0}^{y}\frac{A^{K}(\sigma(s))B^{L}(\sigma(t))}{\left(|h(\sigma(s)-t_0)|^{\frac{1}{2\beta}}+|h^*(\sigma(t)-t_0)|^{\frac{1}{2\beta}}\right)^{\frac{2\alpha}{p}}}\Delta s\Delta t$$
$$\leqslant KL(x-t_0)^{\frac{1}{p}}(y-t_0)^{\frac{1}{p}}\left(\int_{t_0}^{x}(x-\sigma(s))\left(a(s)A^{K-1}(\sigma(s))^q\Delta s\right)^{\frac{1}{q}}\right.$$
$$\times\left(\int_{t_0}^{y}(y-\sigma(t))\left(b(t)B^{L-1}(\sigma(t))^q\Delta t\right)^{\frac{1}{q}},$$

$$\int_{t_0}^{x}\int_{t_0}^{y}\frac{A^{qK}(\sigma(s))B^{qL}(\sigma(t))}{\left(|h(\sigma(s)-t_0)|^{\frac{1}{2\beta}}+|h^*(\sigma(t)-t_0)|^{\frac{1}{2\beta}}\right)^{\frac{2q\alpha}{p}}}\Delta s\Delta t$$
$$\leqslant (KL)^q\left(\int_{t_0}^{x}(x-\sigma(s))\left(a(s)A^{K-1}(\sigma(s))^q\right)\Delta t\right)$$
$$\times\left(\int_{t_0}^{y}(y-\sigma(t))\left(b(t)B^{L-1}(\sigma(t))^q\Delta t\right).$$

Using the facts $\sigma(x)\geqslant x,\sigma(y)\geqslant y$ yields

$$\int_{t_0}^{x}\int_{t_0}^{y}\frac{A^{K}(\sigma(s))B^{L}(\sigma(t))}{\left(|h(\sigma(s)-t_0)|^{\frac{1}{2\beta}}+|h^*(\sigma(t)-t_0)|^{\frac{1}{2\beta}}\right)^{\frac{2\alpha}{p}}}\Delta s\Delta t$$
$$\leqslant C_2(L,K,p)\left(\int_{t_0}^{x}(\sigma(x)-\sigma(s))\left(a(s)A^{K-1}(\sigma(s))^q\Delta s\right)^{\frac{1}{q}}\right.$$
$$\times\left(\int_{t_0}^{y}(\sigma(y)-\sigma(t))\left(b(t)B^{L-1}(\sigma(t))^q\Delta t\right)^{\frac{1}{q}},$$

$$\int_{t_0}^{x}\int_{t_0}^{y}\frac{A^{qK}(\sigma(s))B^{qL}(\sigma(t))}{\left(|h(\sigma(s)-t_0)|^{\frac{1}{2\beta}}+|h^*(\sigma(t)-t_0)|^{\frac{1}{2\beta}}\right)^{\frac{2q\alpha}{p}}}\Delta s\Delta t$$
$$\leqslant C_1(L,K,q)\left(\int_{t_0}^{x}(\sigma(x)-\sigma(s))\left(a(s)A^{K-1}(\sigma(s)\right)^2\Delta s\right)$$
$$\times\left(\int_{t_0}^{y}(\sigma(y)-\sigma(t))\left(b(t)B^{L-1}(\sigma(t))^q\Delta t\right).$$

This completes the proof. □

Theorem 3. *Let $a(\check{t}), b(\eta), A(s)$ and $B(t)$ be defined as in Theorem 2, thus*

$$\int_{t_0}^{x}\int_{t_0}^{y}\frac{A^q(\sigma(s))B^q(\sigma(t))}{\left(|h(\sigma(s)-t_0)|^{\frac{1}{2\beta}}+|h^*(\sigma(t)-t_0)|^{\frac{1}{2\beta}}\right)^{\frac{2q\alpha}{p}}}\Delta s\Delta t$$
$$\leqslant\left(\int_{t_0}^{x}(\sigma(x)-\sigma(s))a^q(s)\Delta s\right)\left(\int_{t_0}^{y}(\sigma(y)-\sigma(t))b^q(t)\Delta t\right)$$

and

$$\int_{t_0}^{x}\int_{t_0}^{y}\frac{A(\sigma(s))B(\sigma(t))}{\left(|h(\sigma(s)-t_0)|^{\frac{1}{2\beta}}+|h^*(\sigma(t)-t_0)|^{\frac{1}{2\beta}}\right)^{\frac{2\alpha}{p}}}\Delta s\Delta t$$
$$\leqslant(x-t_0)^{\frac{1}{p}}(y-t_0)^{\frac{1}{p}}\left(\int_{t_0}^{x}(\sigma(x)-\sigma(s))a^q(s)\Delta s\right)^{\frac{1}{q}}\left(\int_{t_0}^{y}(\sigma(y)-\sigma(t))b^q(t)\Delta t\right)^{\frac{1}{q}}.$$

Proof. Put $K=L=1$ in (11) and (12). This completes the proof. □

In Theorem 2, if we choose $\mathbb{T}=\mathbb{R}$, then we have relation (1) and the next results:

Corollary 2. *If $a(s)\geq 0, b(t)\geq 0$. Define $A(s):=\int_0^s a(\eta)d\eta$ and $B(t):=\int_0^t b(\eta)d\eta$, then*

$$\int_0^x\int_0^y\frac{A^{qK}(s)B^{qL}(t)}{\left(|h(s)|^{\frac{1}{2\beta}}+|h^*(t)|^{\frac{1}{2\beta}}\right)^{\frac{2q\alpha}{p}}}dsdt$$
$$\leqslant C_1(L,K,q)\left(\int_0^x(x-s)\left(a(s)A^{K-1}(s)\right)^q ds\right)$$
$$\times\left(\int_0^y(y-t)\left(b(t)B^{L-1}(t)\right)^q dt\right).$$

and

$$\int_0^x\int_0^y\frac{A^{K}(s)B^{L}(t)}{\left(|h(s)|^{\frac{1}{2\beta}}+|h^*(t)|^{\frac{1}{2\beta}}\right)^{\frac{2\alpha}{p}}}dsdt$$
$$\leqslant C_3(L,K,p)\left(\int_0^x(x-s)\left(A^{K-1}(s)a(s)\right)^q ds\right)^{\frac{1}{q}}$$
$$\times\left(\int_0^y(y-t)\left(B^{L-1}(t)b(t)\right)^q dt\right)^{\frac{1}{q}}$$

where

$$C_3(L,K,p)=KL(xy)^{\frac{1}{p}}.$$

In Theorem 2, if we chose $\mathbb{T}=\mathbb{Z}$, then we get (2), and the next result:

Corollary 3. *If $a(n)\geq$ and $b(m)\geq 0$. Define*

$$A(n)=\sum_{s=0}^{n}a(s),\quad B(m)=\sum_{k=0}^{m}b(k).$$

Then

$$\sum_{n=1}^{N}\sum_{m=1}^{M}\frac{A^{qL}(n)B^{qK}(m)}{\left(|h(n+1)|^{\frac{1}{2\beta}}+|h^{*}(m+1)|^{\frac{1}{2\beta}}\right)^{\frac{2q\alpha}{p}}}\leqslant C_1(K,L,q)\left(\sum_{n=1}^{N}(N+1-(n+1))(a(n)A^{L-1}(n))^{q}\right)\times\left(\sum_{m=1}^{M}(M+1-(m+1))(b(m)B^{L-1}(m))^{q}\right)$$

and

$$\sum_{n=1}^{N}\sum_{m=1}^{M}\frac{A^{L}(n)B^{K}(m)}{\left(|h(n+1)|^{\frac{1}{2\beta}}+|h^{*}(m+1)|^{\frac{1}{2\beta}}\right)^{\frac{2\alpha}{p}}}\leqslant C_4(K,L,p)\left(\sum_{n=1}^{N}(N+1-(n+1))(a(n)A^{L-1}(n))^{q}\right)^{\frac{1}{q}}\times\left(\sum_{m=1}^{M}(M+1-(m+1))(b(m)B^{L-1}(m))^{q}\right)^{\frac{1}{q}}$$

where

$$C_4(K,L,p)=KL(NM)^{\frac{1}{p}}.$$

Remark 1. *Taking $p=q=2$ in Corollary 3 gives the result due to Hamiaz and Abuelela ([1], Theorem 3).*

Corollary 4. *With the hypotheses of Theorem 2 we have:*

$$\int_{t_0}^{x}\int_{t_0}^{y}\frac{A^{qK}(\sigma(s))B^{qL}(\sigma(t))}{\left(|h(\sigma(s)-t_0)|^{\frac{1}{2\beta}}+|h^{*}(\sigma(t)-t_0)|^{\frac{1}{2\beta}}\right)^{\frac{2q\alpha}{p}}}\Delta s\Delta t$$
$$\leqslant C_1(L,K,q)\left\{h\left(\int_{t_0}^{x}(\sigma(x)-\sigma(s))(a(s)A^{K-1}(\sigma(s))^{q}\Delta s\right)+h^{*}\left(\int_{t_0}^{y}(\sigma(y)-\sigma(t))(b(t)B^{L-1}(\sigma(t))^{q}\Delta t\right)\right\}$$

and

$$\int_{t_0}^{x}\int_{t_0}^{y}\frac{A^{K}(\sigma(s))B^{L}(\sigma(t))}{\left(|h(\sigma(s)-t_0)|^{\frac{1}{2\beta}}+|h^{*}(\sigma(t)-t_0)|^{\frac{1}{2\beta}}\right)^{\frac{2\alpha}{p}}}\Delta s\Delta t$$
$$\leqslant C_2(L,K,p)\left\{h\left(\int_{t_0}^{x}(\sigma(x)-\sigma(s))\left(A^{K-1}(\sigma(s))a(s)\right)^{q}\Delta s\right)+h^{*}\left(\int_{t_0}^{y}(\sigma(y)-\sigma(t))\left(B^{L-1}(\sigma(t))b(t)\right)^{q}\Delta t\right)\right\}^{\frac{1}{q}}.$$

Proof. Using the Fenchel-Young inequality (5) in (11) and (12). This proves the claim. □

Theorem 4. *Assuming the time scale $\mathbb{T}$ with $s,t,t_0,x,y\in\mathbb{T}$, $A(s)$ and $B(t)$ are defined as in Theorem 2. Suppose $f(\check{\tau})\geq 0$ and $g(\eta)\geq 0$ are right-dense continuous functions on $[t_0,x]_{\mathbb{T}}$ and $[t_0,y]_{\mathbb{T}}$ respectively. Suppose that $\Phi\geq 0$ and $\Psi\geq 0$ are convex, and submultiplicative functions on $[0,\infty)$. Furthermore assume that*

$$F(s):=\int_{t_0}^{s}f(\check{\tau})\Delta\check{\tau},\ \text{and}\ G(t):=\int_{t_0}^{t}g(\eta)\Delta\eta, \tag{27}$$

then for $\sigma(s) \in [t_0, x]_{\mathbb{T}}$ *and* $\sigma(t) \in [t_0, y]_{\mathbb{T}}$, *we have that*

$$\int_{t_0}^{x}\int_{t_0}^{y} \frac{\check{\Phi}(A^{\sigma}(s))\check{\Psi}(B^{\sigma}(t))}{\left(|h(\sigma(s)-t_0)|^{\frac{1}{2\beta}} + |h^{*}(\sigma(t)-t_0)|^{\frac{1}{2\beta}}\right)^{\frac{2\alpha}{p}}}\Delta s\Delta t$$

$$\leqslant M_1(p)\left(\int_{t_0}^{x}(\sigma(x)-\sigma(s))\left(f(s)\check{\Phi}\left(\frac{a(s)}{f(s)}\right)\right)^{q}\Delta s\right)^{\frac{1}{q}}$$

$$\times\left(\int_{t_0}^{y}(\sigma(y)-\sigma(t))\left(g(t)\check{\Psi}\left(\frac{b(t)}{g(t)}\right)\right)^{q}\Delta t\right)^{\frac{1}{q}} \tag{28}$$

where

$$M_1(p) = \left\{\int_{t_0}^{x}\left(\frac{\check{\Phi}(F^{\sigma}(s))}{F^{\sigma}(s)}\right)^{p}\Delta s\right\}^{\frac{1}{p}}\left\{\int_{t_0}^{y}\left(\frac{\check{\Psi}(G^{\sigma}(t))}{G^{\sigma}(t)}\right)^{p}\Delta t\right\}^{\frac{1}{p}}.$$

Proof. From the properties of $\check{\Phi}$ and using (2), we obtain

$$\begin{aligned}\check{\Phi}(A^{\sigma}(s)) &= \check{\Phi}\left(\frac{F(\sigma(s))\int_{t_0}^{\sigma(s)} f(\check{\tau})\frac{a(\check{\tau})}{f(\check{\tau})}\Delta\check{\tau}}{\int_{t_0}^{\sigma(s)} f(\check{\tau})\Delta\check{\tau}}\right)\\ &\leqslant \check{\Phi}(F(\sigma(s))\check{\Phi}\left(\frac{\int_{t_0}^{\sigma(s)} f(\check{\tau})\frac{a(\check{\tau})}{f(\check{\tau})}\Delta\check{\tau}}{\int_{t_0}^{\sigma(s)} f(\check{\tau})\Delta\check{\tau}}\right)\\ &\leqslant \frac{\check{\Phi}(F(\sigma(s))}{F(\sigma(s))}\int_{t_0}^{\sigma(s)} f(\check{\tau})\check{\Phi}\left(\frac{a(\check{\tau})}{f(\check{\tau})}\right)\Delta\check{\tau}.\end{aligned} \tag{29}$$

Using (1) in (29), we see that

$$\check{\Phi}(A^{\sigma}(s)) \leqslant \frac{\check{\Phi}(F^{\sigma}(s))}{F^{\sigma}(s)}(\sigma(s)-t_0)^{\frac{1}{p}}\left(\int_{t_0}^{\sigma(s)}\left(f(\check{\tau})\check{\Phi}\left[\frac{a(\check{\tau})}{f(\check{\tau})}\right]\right)^{q}\Delta\check{\tau}\right)^{\frac{1}{q}}. \tag{30}$$

In addition, from the convexity and submultiplicative property of $\check{\Psi}$, we get by using (2) and (1):

$$\check{\Psi}(B^{\sigma}(t)) \leqslant \frac{\check{\Psi}(G^{\sigma}(t))}{G^{\sigma}(t)}(\sigma(t)-t_0)^{\frac{1}{p}}\left(\int_{t_0}^{\sigma(t)}\left(g(\eta)\check{\Psi}\left[\frac{b(\eta)}{g(\eta)}\right]\right)^{q}\Delta\eta\right)^{\frac{1}{q}}. \tag{31}$$

From (30) and (31), we have

$$\begin{aligned}\check{\Phi}(A^{\sigma}(s))\check{\Psi}(B^{\sigma}(t)) &\leqslant (\sigma(s)-t_0)^{\frac{1}{p}}(\sigma(t)-t_0)^{\frac{1}{p}}\left(\frac{\check{\Phi}(F^{\sigma}(s))}{F^{\sigma}(s)}\left(\int_{t_0}^{\sigma(s)}\left(f(\check{\tau})\check{\Phi}\left[\frac{a(\check{\tau})}{f(\check{\tau})}\right]\right)^{q}\Delta\check{\tau}\right)^{\frac{1}{q}}\right.\\ &\quad\times\left(\frac{\check{\Psi}(G^{\sigma}(t))}{G^{\sigma}(t)}\left(\int_{t_0}^{\sigma(t)}\left(g(\eta)\check{\Psi}\left[\frac{b(\eta)}{g(\eta)}\right]\right)^{q}\Delta\eta\right)^{\frac{1}{q}}\right.\end{aligned} \tag{32}$$

Using (5) on $(\sigma(s)-t_0)^{\frac{1}{p}}(\sigma(t)-t_0)^{\frac{1}{p}}$ gives:

$$\begin{aligned}\check{\Phi}(A^{\sigma}(s))\check{\Psi}(B^{\sigma}(t)) &\leqslant \left(h(\sigma(s)-t_0)+h^{*}(\sigma(t)-t_0)\right)^{\frac{1}{p}}\left(\frac{\check{\Phi}(F^{\sigma}(s))}{F^{\sigma}(s)}\left(\int_{t_0}^{\sigma(s)}\left(f(\check{\tau})\check{\Phi}\left[\frac{a(\check{\tau})}{f(\check{\tau})}\right]\right)^{q}\Delta\check{\tau}\right)^{\frac{1}{q}}\right.\\ &\quad\times\left(\frac{\check{\Psi}(G^{\sigma}(t))}{G^{\sigma}(t)}\left(\int_{t_0}^{\sigma(t)}\left(g(\eta)\check{\Psi}\left[\frac{b(\eta)}{g(\eta)}\right]\right)^{q}\Delta\eta\right)^{\frac{1}{q}}.\end{aligned} \tag{33}$$

Applying Lemma 6 on the right hand side of (33), we see that

$$\check{\Phi}(A^{\sigma}(s))\check{\Psi}(B^{\sigma}(t)) \leqslant \left(|h(\sigma(s)-t_0|^{\frac{1}{2\beta}}+|h^{*}(\sigma(t)-t_0)|^{\frac{1}{2\beta}}\right)^{\frac{2\alpha}{p}}$$
$$\times\left(\frac{\check{\Phi}(F^{\sigma}(s))}{F^{\sigma}(s)}\left(\int_{t_0}^{\sigma(s)}\left(f(\check{\tau})\check{\Phi}\left[\frac{a(\check{\tau})}{f(\check{\tau})}\right]\right)^{q}\Delta\check{\tau}\right)^{\frac{1}{q}}$$
$$\times\left(\frac{\check{\Psi}(G^{\sigma}(t))}{G^{\sigma}(t)}\left(\int_{t_0}^{\sigma(t)}\left(g(\eta)\check{\Psi}\left[\frac{b(\eta)}{g(\eta)}\right]\right)^{q}\Delta\eta\right)^{\frac{1}{q}}. \tag{34}$$

From (34), we have

$$\frac{\check{\Phi}(A^{\sigma}(s))\check{\Psi}(B^{\sigma}(t))}{\left(|h(\sigma(s)-t_0|^{\frac{1}{2\beta}}+|h^{*}(\sigma(t)-t_0)|^{\frac{1}{2\beta}}\right)^{\frac{2\alpha}{p}}} \leqslant \left(\frac{\check{\Phi}(F^{\sigma}(s))}{F^{\sigma}(s)}\left(\int_{t_0}^{\sigma(s)}\left(f(\check{\tau})\check{\Phi}\left[\frac{a(\check{\tau})}{f(\check{\tau})}\right]\right)^{q}\Delta\check{\tau}\right)^{\frac{1}{q}}$$
$$\times\left(\frac{\check{\Psi}(G^{\sigma}(t))}{G^{\sigma}(t)}\left(\int_{t_0}^{\sigma(t)}\left(g(\eta)\check{\Psi}\left[\frac{b(\eta)}{g(\eta)}\right]\right)^{q}\Delta\eta\right)^{\frac{1}{q}}. \tag{35}$$

From (35), we obtain

$$\int_{t_0}^{x}\int_{t_0}^{y}\frac{\check{\Phi}(A^{\sigma}(s))\check{\Psi}(B^{\sigma}(t))}{\left(|h(\sigma(s)-t_0|^{\frac{1}{2\beta}}+|h^{*}(\sigma(t)-t_0)|^{\frac{1}{2\beta}}\right)^{\frac{2\alpha}{p}}}\Delta s\Delta t$$
$$\leqslant \int_{t_0}^{x}\left(\frac{\check{\Phi}(F^{\sigma}(s))}{F^{\sigma}(s)}\left(\int_{t_0}^{\sigma(s)}\left(f(\check{\tau})\check{\Phi}\left[\frac{a(\check{\tau})}{f(\check{\tau})}\right]\right)^{q}\Delta\check{\tau}\right)^{\frac{1}{q}}\Delta s$$
$$\times\int_{t_0}^{y}\left(\frac{\check{\Psi}(G^{\sigma}(t))}{G^{\sigma}(t)}\left(\int_{t_0}^{\sigma(t)}\left(g(\eta)\check{\Psi}\left[\frac{b(\eta)}{g(\eta)}\right]\right)^{q}\Delta\eta\right)^{\frac{1}{q}}\Delta t. \tag{36}$$

From (36), by using (1), we have

$$\int_{t_0}^{x}\int_{t_0}^{y}\frac{\check{\Phi}(A^{\sigma}(s))\check{\Psi}(B^{\sigma}(t))}{\left(|h(\sigma(s)-t_0|^{\frac{1}{2\beta}}+|h^{*}(\sigma(t)-t_0)|^{\frac{1}{2\beta}}\right)^{\frac{2\alpha}{p}}}\Delta s\Delta t$$
$$\leqslant\left\{\int_{t_0}^{x}\left(\frac{\check{\Phi}(F^{\sigma}(s))}{F^{\sigma}(s)}\right)^{p}\Delta s\right\}^{\frac{1}{p}}\left(\int_{t_0}^{x}\int_{t_0}^{\sigma(s)}\left(f(\check{\tau})\check{\Phi}\left[\frac{a(\check{\tau})}{f(\check{\tau})}\right]\right)^{q}\Delta\check{\tau}\Delta s\right)^{\frac{1}{q}}$$
$$\times\left\{\int_{t_0}^{y}\left(\frac{\check{\Psi}(G^{\sigma}(t))}{G^{\sigma}(t)}\right)^{p}\Delta t\right\}^{\frac{1}{p}}\left(\int_{t_0}^{y}\int_{t_0}^{\sigma(t)}\left(g(\eta)\check{\Psi}\left[\frac{b(\eta)}{g(\eta)}\right]\right)^{q}\Delta\eta\Delta t\right)^{\frac{1}{q}}. \tag{37}$$

From (37), by using (5), we obtain

$$\int_{t_0}^{x}\int_{t_0}^{y}\frac{\check{\Phi}(A^{\sigma}(s))\check{\Psi}(B^{\sigma}(t))}{\left(|h(\sigma(s)-t_0|^{\frac{1}{2\beta}}+|h^{*}(\sigma(t)-t_0)|^{\frac{1}{2\beta}}\right)^{\frac{2\alpha}{p}}}\Delta s\Delta t$$
$$\leqslant M_1(p)\left(\int_{t_0}^{x}(x-\sigma(s))\left(f(s)\check{\Phi}\left[\frac{a(s)}{f(s)}\right]\right)^{q}\Delta s\right)^{\frac{1}{q}}$$
$$\times\left(\int_{t_0}^{y}(y-\sigma(t))\left(g(t)\check{\Psi}\left[\frac{b(t)}{g(t)}\right]\right)^{q}\Delta t\right)^{\frac{1}{q}}.$$

By using the facts $\sigma(x) \geqslant x$ and $\sigma(y) \geqslant y$, we obtain

$$\int_{t_0}^{x}\int_{t_0}^{y}\frac{\check{\Phi}(A^\sigma(s))\check{\Psi}(B^\sigma(t))}{\left(|h(\sigma(s)-t_0|^{\frac{1}{2\beta}}+|h^*(\sigma(t)-t_0)|^{\frac{1}{2\beta}}\right)^{\frac{2\alpha}{p}}}\Delta s\Delta t$$

$$\leqslant M_1(p)\left(\int_{t_0}^{x}(\sigma(x)-\sigma(s))\left(f(s)\check{\Phi}\left[\frac{a(s)}{f(s)}\right]\right)^q\Delta s\right)^{\frac{1}{q}}$$

$$\times\left(\int_{t_0}^{y}(\sigma(y)-\sigma(t))\left(g(t)\check{\Psi}\left[\frac{b(t)}{g(t)}\right]\right)^q\Delta t\right)^{\frac{1}{q}}$$

where

$$M_1(p)=\left\{\int_{t_0}^{x}\left(\frac{\check{\Phi}(F^\sigma(s))}{F^\sigma(s)}\right)^p\Delta s\right\}^{\frac{1}{p}}\left\{\int_{t_0}^{y}\left(\frac{\check{\Psi}(G^\sigma(t))}{G^\sigma(t)}\right)^p\Delta t\right\}^{\frac{1}{p}}.$$

This completes the proof. □

In Theorem 4, taking $\mathbb{T}=\mathbb{R}$, we have (1) and the result:

Corollary 5. *Assume that $a(s)\geq 0$, $b(t)\geq 0$, $f(\check{\tau})\geq 0$ and $g(\eta)\geq 0$, we define*

$$A(s):=\int_0^s a(\eta)d\eta,\quad B(t):=\int_0^t b(\eta)d\eta,\quad F(s):=\int_0^s f(\check{\tau})d\check{\tau},\ \text{and}\ G(t):=\int_0^t g(\eta)d\eta.$$

Then

$$\int_0^x\int_0^y\frac{\check{\Phi}(A(s)\check{\Psi}(B(t))}{\left(|h(s)|^{\frac{1}{2\beta}}+|h^*(t)|^{\frac{1}{2\beta}}\right)^{\frac{2\alpha}{p}}}dsdt \leqslant M_2(p)\left(\int_0^x(x-s)\left(f(s)\check{\Phi}\left(\frac{a(s)}{f(s)}\right)\right)^q ds\right)^{\frac{1}{q}}$$

$$\times\left(\int_0^y(y-t)\left(g(t)\check{\Psi}\left(\frac{b(t)}{g(t)}\right)\right)^q dt\right)^{\frac{1}{q}}$$

where

$$M_2(p)=\left\{\int_0^x\left(\frac{\check{\Phi}(F(s))}{F(s)}\right)^p ds\right\}^{\frac{1}{p}}\left\{\int_0^y\left(\frac{\check{\Psi}(G(t))}{G(t)}\right)^p dt\right\}^{\frac{1}{p}}.$$

In Theorem 4, taking $\mathbb{T}=\mathbb{Z}$, gives (2) and the result:

Corollary 6. *Assume that $a(n)\geq 0$, $b(m)\geq 0$, $f(n)\geq 0$, $g(m)\geq 0$ are sequences of real numbers. Define*

$$A(n)=\sum_{s=0}^{n}a(s),\quad B(m)=\sum_{k=0}^{m}b(k),\quad F(n)=\sum_{s=0}^{n}f(s)\ \text{and}\ G(m)=\sum_{k=0}^{m}g(k).$$

Then

$$\sum_{n=1}^{N}\sum_{m=1}^{M}\frac{\check{\Phi}(A(n))\check{\Psi}(B(m))}{\left(|h(n+1)|^{\frac{1}{2\beta}}+|h^*(m+1)|^{\frac{1}{2\beta}}\right)^{\frac{2\alpha}{p}}}\leqslant M_3(p)\left\{\sum_{n=1}^{N}(N+1-(n+1))\left(f(n)\check{\Phi}\left[\frac{a(n)}{f(n)}\right]\right)^q\right\}^{\frac{1}{q}}$$

$$\times\left\{\sum_{m=1}^{M}(M+1-(m+1))\left(g(m)\check{\Psi}\left[\frac{b(m)}{g(m)}\right]\right)^q\right\}^{\frac{1}{q}}$$

where

$$M_3(p) = \left\{ \sum_{n=1}^{N} \left(\frac{\check{\Phi}(F(n))}{F(n)} \right)^p \right\}^{\frac{1}{p}} \left\{ \sum_{m=1}^{M} \left(\frac{\Psi(G(m))}{G(m)} \right)^p \right\}^{\frac{1}{p}}$$

Remark 2. *In Corollary 6, if $p = q = 2$ we get the result due to Hamiaz and Abuelela ([1], Theorem 5).*

Corollary 7. *Under the hypotheses of Theorem 4 the following inequality hold:*

$$\begin{aligned}
&\int_{t_0}^{x} \int_{t_0}^{y} \frac{\check{\Phi}(A^{\sigma}(s))\Psi(B^{\sigma}(t))}{\left(|h(\sigma(s) - t_0)|^{\frac{1}{2\beta}} + |h^*(\sigma(t) - t_0)|^{\frac{1}{2\beta}} \right)^{\frac{2\alpha}{p}}} \Delta s \Delta t \\
&\leqslant M_1(p) \left[h\left(\int_{t_0}^{x} (\sigma(x) - \sigma(s)) \left(f(s)\check{\Phi}\left(\frac{a(s)}{f(s)} \right) \right)^q \Delta s \right) \right. \\
&\quad \left. + h^*\left(\int_{t_0}^{y} (\sigma(y) - \sigma(t)) \left(g(t)\Psi\left(\frac{b(t)}{g(t)} \right) \right)^q \Delta t \right) \right]^{\frac{1}{q}}.
\end{aligned}$$

Proof. Using (5) in (28). This proves our claim. □

Lemma 7. *With hypotheses of Theorem 4, we get:*

$$\begin{aligned}
&\int_{t_0}^{x} \int_{t_0}^{y} \frac{\check{\Phi}(A^{\sigma}(s))^2 \Psi(B^{\sigma}(t))^2}{\left(h(\sigma(s) - t_0) + h^*(\sigma(t) - t_0) \right)} \Delta s \Delta t \\
&\leqslant M_4 \left\{ \int_{t_0}^{x} (\sigma(x) - \sigma(s)) \left(f(s)\check{\Phi}\left[\frac{a(s)}{f(s)} \right] \right)^4 \Delta s \right\}^{\frac{1}{2}} \left\{ \int_{t_0}^{y} (\sigma(t) - \sigma(t)) \left(g(t)\Psi\left[\frac{b(t)}{g(t)} \right] \right)^4 \Delta t \right\}^{\frac{1}{2}} \quad (38)
\end{aligned}$$

where

$$M_4 = \left\{ \int_{t_0}^{x} \left(\frac{\check{\Phi}(F^{\sigma}(s))^4}{(F^{\sigma}(s))^4} \right) (\sigma(s) - t_0) \Delta s \right\}^{\frac{1}{2}} \left\{ \int_{t_0}^{y} \left(\frac{\Psi(G^{\sigma}(t))^4}{(G^{\sigma}(t))^4} \right) (\sigma(t) - t_0) \Delta t \right\}^{\frac{1}{2}}. \quad (39)$$

Proof. From (30) and (31) and by using Fenchel-Young inequality with $p = q = 2$ we have

$$\begin{aligned}
&\check{\Phi}(A^{\sigma}(s))^2 \Psi(B^{\sigma}(t))^2 \\
&\leqslant \left(h(\sigma(s) - t_0) + h^*(\sigma(t) - t_0) \right) \left(\frac{\check{\Phi}(F^{\sigma}(s))^2}{(F^{\sigma}(s))^2} \left(\int_{t_0}^{\sigma(s)} \left(f(\check{\tau})\check{\Phi}\left[\frac{a(\check{\tau})}{f(\check{\tau})} \right] \right)^2 \Delta \check{\tau} \right) \right. \\
&\quad \times \left(\frac{\Psi(G^{\sigma}(t))^2}{(G^{\sigma}(t))^2} \left(\int_{t_0}^{\sigma(t)} \left(g(\eta)\Psi\left[\frac{b(\eta)}{g(\eta)} \right] \right)^2 \Delta \eta \right). \quad (40)
\end{aligned}$$

From (40), by using (1) with $p = q = 2$, we obtain

$$\begin{aligned}
&\int_{t_0}^{x}\int_{t_0}^{y}\frac{\check{\Phi}(A^{\sigma}(s))^2\check{\Psi}(B^{\sigma}(t))^2}{\Big(h(\sigma(s)-t_0)+h^*(\sigma(t)-t_0)\Big)}\Delta s\Delta t \\
&\leqslant \int_{t_0}^{x}\left(\frac{\check{\Phi}(F^{\sigma}(s))^2}{(F^{\sigma}(s))^2}\left(\int_{t_0}^{\sigma(s)}\left(f(\check{\tau})\check{\Phi}\left[\frac{a(\check{\tau})}{f(\check{\tau})}\right]\right)^2\Delta\check{\tau}\right)\Delta s\right. \\
&\times \int_{t_0}^{y}\frac{\check{\Psi}(G^{\sigma}(t))^2}{(G^{\sigma}(t))^2}\left(\int_{t_0}^{\sigma(t)}\left(g(\eta)\check{\Psi}\left[\frac{b(\eta)}{g(\eta)}\right]\right)^2\Delta\eta\right)\Delta t \\
&\leqslant \int_{t_0}^{x}\left(\frac{\check{\Phi}(F^{\sigma}(s))^2}{(F^{\sigma}(s))^2}\right)(\sigma(s)-t_0)^{\frac{1}{2}}\left(\int_{t_0}^{\sigma(s)}\left(f(\check{\tau})\check{\Phi}\left[\frac{a(\check{\tau})}{f(\check{\tau})}\right]\right)^4\Delta\check{\tau}\right)^{\frac{1}{2}}\Delta s \\
&\times \int_{t_0}^{y}\frac{\check{\Psi}(G^{\sigma}(t))^2}{(G^{\sigma}(t))^2}(\sigma(t)-t_0)^{\frac{1}{2}}\left(\int_{t_0}^{\sigma(t)}\left(g(\eta)\check{\Psi}\left[\frac{b(\eta)}{g(\eta)}\right]\right)^4\Delta\eta\right)^{\frac{1}{2}}\Delta t \\
&\leqslant \left\{\int_{t_0}^{x}\left(\frac{\check{\Phi}(F^{\sigma}(s))^4}{(F^{\sigma}(s))^4}\right)(\sigma(s)-t_0)\Delta s\right\}^{\frac{1}{2}}\left\{\int_{t_0}^{x}\left(\int_{t_0}^{\sigma(s)}\left(f(\check{\tau})\check{\Phi}\left[\frac{a(\check{\tau})}{f(\check{\tau})}\right]\right)^4\Delta\check{\tau}\right)\Delta s\right\}^{\frac{1}{2}} \\
&\times\left\{\int_{t_0}^{y}\left(\frac{\check{\Psi}(G^{\sigma}(t))^4}{(G^{\sigma}(t))^4}\right)(\sigma(t)-t_0)\Delta t\right\}^{\frac{1}{2}}\left\{\int_{t_0}^{y}\left(\int_{t_0}^{\sigma(t)}\left(g(\eta)\check{\Psi}\left[\frac{b(\eta)}{g(\eta)}\right]\right)^4\Delta\eta\right)\Delta t\right\}^{\frac{1}{2}}. \qquad (41)
\end{aligned}$$

Applying (5) on (41), we obtain

$$\begin{aligned}
&\int_{t_0}^{x}\int_{t_0}^{y}\frac{\check{\Phi}(A^{\sigma}(s))^2\check{\Psi}(B^{\sigma}(t))^2}{\Big(h(\sigma(s)-t_0)+h^*(\sigma(t)-t_0)\Big)}\Delta s\Delta t \\
&\leqslant \left\{\int_{t_0}^{x}\left(\frac{\check{\Phi}(F^{\sigma}(s))^4}{(F^{\sigma}(s))^4}\right)(\sigma(s)-t_0)\Delta s\right\}^{\frac{1}{2}}\left\{\int_{t_0}^{x}(x-\sigma(s))\left(f(s)\check{\Phi}\left[\frac{a(s)}{f(s)}\right]\right)^4\Delta s\right\}^{\frac{1}{2}} \\
&\times\left\{\int_{t_0}^{y}\left(\frac{\check{\Psi}(G^{\sigma}(t))^4}{(G^{\sigma}(t))^4}\right)(\sigma(t)-t_0)\Delta t\right\}^{\frac{1}{2}}\left\{\int_{t_0}^{y}(t-\sigma(t))\left(g(t)\check{\Psi}\left[\frac{b(t)}{g(t)}\right]\right)^4\Delta t\right\}^{\frac{1}{2}} \\
&= M_4\left\{\int_{t_0}^{x}(x-\sigma(s))\left(f(s)\check{\Phi}\left[\frac{a(s)}{f(s)}\right]\right)^4\Delta s\right\}^{\frac{1}{2}}\left\{\int_{t_0}^{y}(t-\sigma(t))\left(g(t)\check{\Psi}\left[\frac{b(t)}{g(t)}\right]\right)^4\Delta t\right\}^{\frac{1}{2}}.
\end{aligned}$$

Since $\sigma(x) \geqslant x$ and $\sigma(y) \geqslant y$, from the last inequality above, we have

$$\begin{aligned}
&\int_{t_0}^{x}\int_{t_0}^{y}\frac{\check{\Phi}(A^{\sigma}(s))^2\check{\Psi}(B^{\sigma}(t))^2}{\Big(h(\sigma(s)-t_0)+h^*(\sigma(t)-t_0)\Big)}\Delta s\Delta t \\
&\leqslant M_4\left\{\int_{t_0}^{x}(\sigma(x)-\sigma(s))\left(f(s)\check{\Phi}\left[\frac{a(s)}{f(s)}\right]\right)^4\Delta s\right\}^{\frac{1}{2}}\left\{\int_{t_0}^{y}(\sigma(t)-\sigma(t))\left(g(t)\check{\Psi}\left[\frac{b(t)}{g(t)}\right]\right)^4\Delta t\right\}^{\frac{1}{2}}
\end{aligned}$$

where M_4 defined as in (39). This proves our claim.
□

Theorem 5. *Assume the time scale $\mathbb{T}$ with $t, s, x_0, t_0, y \in \mathbb{T}$. Suppose that $b(\check{\tau}) \geq 0$ and $a(\check{\tau}) \geq 0$ are right-dense continuous functions on $[t_0, y]_{\mathbb{T}}$ and $[t_0, x]_{\mathbb{T}}$. Let $G, F, g, f, \check{\Psi}$ and $\check{\Phi}$ be as assumed in Theorem 4. Furthermore assume that*

$$A(s) := \frac{1}{F(s)}\int_{t_0}^{s} a(\check{\tau})f(\check{\tau})\Delta\check{\tau}, \quad \textit{and} \quad B(t) := \frac{1}{G(t)}\int_{t_0}^{t} b(\eta)g(\eta)\Delta\eta, \qquad (42)$$

then for $\sigma(s) \in [t_0, x]_{\mathbb{T}}$ *and* $\sigma(t) \in [t_0, y]_{\mathbb{T}}$, *we have that*

$$\int_{t_0}^{x}\int_{t_0}^{y} \frac{\check{\Phi}(A^{\sigma}(s))\check{\Psi}(B^{\sigma}(t))F^{\sigma}(s)G^{\sigma}(t)}{\left(|h(\sigma(s)-t_0)|^{\frac{1}{2\beta}} + |h^{*}(\sigma(t)-t_0)|^{\frac{1}{2\beta}}\right)^{\frac{2\alpha}{p}}}\Delta s\Delta t$$
$$\leqslant M_5(p)\left(\int_{t_0}^{x}(\sigma(x)-\sigma(s))\left(f(s)\check{\Phi}\left(a(s)\right)\right)^{q}\Delta s\right)^{\frac{1}{q}}$$
$$\times\left(\int_{t_0}^{y}(\sigma(y)-\sigma(t))\left(g(t)\check{\Psi}\left(b(t)\right)\right)^{q}\Delta t\right)^{\frac{1}{q}} \tag{43}$$

where

$$M_5(p) = (x-t_0)^{\frac{1}{p}}(y-t_0)^{\frac{1}{p}}. \tag{44}$$

Proof. From (42), we see that

$$\check{\Phi}(A^{\sigma}(s)) = \check{\Phi}\left(\frac{1}{F^{\sigma}(s)}\int_{t_0}^{\sigma(s)} f(\check{\tau})a(\check{\tau})\Delta\check{\tau}\right). \tag{45}$$

Applying (1) on (45), we obtain

$$\check{\Phi}(A^{\sigma}(s)) \leqslant \frac{(\sigma(s)-t_0)^{\frac{1}{p}}}{F^{\sigma}(s)}\left(\int_{t_0}^{\sigma(s)}\left(f(\check{\tau})\check{\Phi}[a(\check{\tau})]\right)^{q}\Delta\check{\tau}\right)^{\frac{1}{q}}. \tag{46}$$

From (46), we get

$$\check{\Phi}(A^{\sigma}(s))F^{\sigma}(s) \leqslant (\sigma(s)-t_0)^{\frac{1}{p}}\left(\int_{t_0}^{\sigma(s)}\left(f(\check{\tau})\check{\Phi}[a(\check{\tau})]\right)^{q}\Delta\check{\tau}\right)^{\frac{1}{q}}. \tag{47}$$

Similarly, we obtain

$$\check{\Psi}(B^{\sigma}(t))G^{\sigma}(t) \leqslant (\sigma(t)-t_0)^{\frac{1}{p}}\left(\int_{t_0}^{\sigma(t)}\left(g(\eta)\check{\Psi}[b(\eta)]\right)^{q}\Delta\eta\right)^{\frac{1}{q}}. \tag{48}$$

From (47) and (48), we observe that

$$\check{\Phi}(A^{\sigma}(s))\check{\Psi}(B^{\sigma}(t))G^{\sigma}(t)F^{\sigma}(s) \leqslant (\sigma(s)-t_0)^{\frac{1}{p}}(\sigma(t)-t_0)^{\frac{1}{p}}$$
$$\times\left(\int_{t_0}^{\sigma(s)}\left(f(\check{\tau})\check{\Phi}[a(\check{\tau})]\right)^{q}\Delta\check{\tau}\right)^{\frac{1}{q}}\left(\int_{t_0}^{\sigma(t)}\left(g(\eta)\check{\Psi}[b(\eta)]\right)^{q}\Delta\eta\right)^{\frac{1}{q}}. \tag{49}$$

Applying the Lemma 3 on the term $(\sigma(s)-t_0)^{\frac{1}{p}}(\sigma(t)-t_0)^{\frac{1}{p}}$, gives:

$$\check{\Phi}(A^{\sigma}(s))\check{\Psi}(B^{\sigma}(t))G^{\sigma}(t)F^{\sigma}(s) \leqslant \left(h(\sigma(s)-t_0)+h^{*}(\sigma(t)-t_0)\right)^{\frac{1}{p}}\left(\int_{t_0}^{\sigma(s)}\left(f(\check{\tau})\check{\Phi}[a(\check{\tau})]\right)^{q}\Delta\check{\tau}\right)^{\frac{1}{q}}$$
$$\times\left(\int_{t_0}^{\sigma(t)}\left(g(\eta)\check{\Psi}[b(\eta)]\right)^{q}\Delta\eta\right)^{\frac{1}{q}}. \tag{50}$$

From 6 and (50), we obtain

$$\check{\Phi}(A^{\sigma}(s))\Psi(B^{\sigma}(t))G^{\sigma}(t)F^{\sigma}(s) \leqslant \left(|h(\sigma(s)-t_0)|^{\frac{1}{2\beta}} + |h^{*}(\sigma(t)-t_0)|^{\frac{1}{2\beta}}\right)^{\frac{2\alpha}{p}} \times \left(\int_{t_0}^{\sigma(s)} \left(f(\check{\tau})\check{\Phi}[a(\check{\tau})]\right)^{q} \Delta\check{\tau}\right)^{\frac{1}{q}} \left(\int_{t_0}^{\sigma(t)} \left(g(\eta)\Psi[b(\eta)]\right)^{q} \Delta\eta\right)^{\frac{1}{q}}. \tag{51}$$

Dividing both sides of (51) by $\left(|h(\sigma(s)-t_0)|^{\frac{1}{2\beta}} + |h^{*}(\sigma(t)-t_0)|^{\frac{1}{2\beta}}\right)^{\frac{2\alpha}{p}}$, we get

$$\frac{\check{\Phi}(A^{\sigma}(s))\Psi(B^{\sigma}(t))G^{\sigma}(t)F^{\sigma}(s)}{\left(|h(\sigma(s)-t_0)|^{\frac{1}{2\beta}} + |h^{*}(\sigma(t)-t_0)|^{\frac{1}{2\beta}}\right)^{\frac{2\alpha}{p}}} \leqslant \left(\int_{t_0}^{\sigma(s)} \left(f(\check{\tau})\check{\Phi}[a(\check{\tau})]\right)^{q} \Delta\check{\tau}\right)^{\frac{1}{q}} \times \left(\int_{t_0}^{\sigma(t)} \left(g(\eta)\Psi[b(\eta)]\right)^{q} \Delta\eta\right)^{\frac{1}{q}}. \tag{52}$$

Taking delta-integral for (52), yields:

$$\int_{t_0}^{x}\int_{t_0}^{y} \frac{\check{\Phi}(A^{\sigma}(s))\Psi(B^{\sigma}(t))G^{\sigma}(t)F^{\sigma}(s)}{\left(|h(\sigma(s)-t_0)|^{\frac{1}{2\beta}} + |h^{*}(\sigma(t)-t_0)|^{\frac{1}{2\beta}}\right)^{\frac{2\alpha}{p}}} \Delta s \Delta t \leqslant \left(\int_{t_0}^{x}\left(\int_{t_0}^{\sigma(s)} \left(f(\check{\tau})\check{\Phi}[a(\check{\tau})]\right)^{q} \Delta\check{\tau}\right)^{\frac{1}{q}} \Delta s\right)\left(\int_{t_0}^{y}\left(\int_{t_0}^{\sigma(t)} \left(g(\eta)\Psi[b(\eta)]\right)^{q} \Delta\eta\right)^{\frac{1}{q}} \Delta t\right). \tag{53}$$

Using (1) in (53), yield:

$$\begin{aligned}
&\int_{t_0}^{x}\int_{t_0}^{y} \frac{\check{\Phi}(A^{\sigma}(s))\Psi(B^{\sigma}(t))G^{\sigma}(t)F^{\sigma}(s)}{\left(|h(\sigma(s)-t_0)|^{\frac{1}{2\beta}} + |h^{*}(\sigma(t)-t_0)|^{\frac{1}{2\beta}}\right)^{\frac{2\alpha}{p}}} \Delta s \Delta t \\
&\leqslant (x-t_0)^{\frac{1}{p}}(y-t_0)^{\frac{1}{p}} \left(\int_{t_0}^{x}\left(\int_{t_0}^{\sigma(s)} \left(f(\check{\tau})\check{\Phi}[a(\check{\tau})]\right)^{q} \Delta\check{\tau}\right) \Delta s\right)^{\frac{1}{q}} \\
&\times \left(\int_{t_0}^{y}\left(\int_{t_0}^{\sigma(t)} \left(g(\eta)\Psi[b(\eta)]\right)^{q} \Delta\eta\right) \Delta t\right)^{\frac{1}{q}} \\
&= M_5(p)\left(\int_{t_0}^{x}\left(\int_{t_0}^{\sigma(s)} \left(f(\check{\tau})\check{\Phi}[a(\check{\tau})]\right)^{q} \Delta\check{\tau}\right) \Delta s\right)^{\frac{1}{q}} \\
&\times \left(\int_{t_0}^{y}\left(\int_{t_0}^{\sigma(t)} \left(g(\eta)\Psi[b(\eta)]\right)^{q} \Delta\eta\right) \Delta t\right)^{\frac{1}{q}},
\end{aligned} \tag{54}$$

where M_5 defined as in (44). From (5) and (54), we get:

$$\int_{t_0}^{x}\int_{t_0}^{y}\frac{\check{\Phi}(A^{\sigma}(s))\check{\Psi}(B^{\sigma}(t))G^{\sigma}(t)F^{\sigma}(s)}{\left(|h(\sigma(s)-t_0)|^{\frac{1}{2\beta}}+|h^{*}(\sigma(t)-t_0)|^{\frac{1}{2\beta}}\right)^{\frac{2\alpha}{p}}}\Delta s\Delta t$$
$$= M_5(p)\left(\int_{t_0}^{x}(x-\sigma(s))\left(f(s)\check{\Phi}[a(s)]\right)^{q}\Delta s\right)^{\frac{1}{q}}$$
$$\times\left(\int_{t_0}^{y}(y-\sigma(t))\left(g(t)\check{\Psi}[b(t)]\right)^{q}\Delta t\right)^{\frac{1}{q}}.$$

By using the fact $\sigma(x)\geqslant x$ and $\sigma(y)\geqslant y$, we obtain

$$\int_{t_0}^{x}\int_{t_0}^{y}\frac{\check{\Phi}(A^{\sigma}(s))\check{\Psi}(B^{\sigma}(t))G^{\sigma}(t)F^{\sigma}(s)}{\left(|h(\sigma(s)-t_0)|^{\frac{1}{2\beta}}+|h^{*}(\sigma(t)-t_0)|^{\frac{1}{2\beta}}\right)^{\frac{2\alpha}{p}}}\Delta s\Delta t$$
$$= M_5(p)\left(\int_{t_0}^{x}(\sigma(x)-\sigma(s))\left(f(s)\check{\Phi}[a(s)]\right)^{q}\Delta s\right)^{\frac{1}{q}}$$
$$\times\left(\int_{t_0}^{y}(\sigma(y)-\sigma(t))\left(g(t)\check{\Psi}[b(t)]\right)^{q}\Delta t\right)^{\frac{1}{q}}.$$

This completes the proof. □

Taking $\mathbb{T}=\mathbb{R}$ in Theorem 5 with relation (1), we have:

Corollary 8. *Assume $g(t)\geq 0, b(t)\geq 0, f(s)\geq 0, a(s)\geq 0$. Define*

$$A(s):=\frac{1}{F(s)}\int_0^s f(\check{\tau})a(\check{\tau})d\check{\tau} \text{ and } B(t):=\frac{1}{G(t)}\int_0^t g(\check{\tau})b(\check{\tau})d\check{\tau},$$

$$F(s):=\int_0^s f(\check{\tau})d\check{\tau} \text{ and } G(t):=\int_0^t g(\check{\tau})d\check{\tau}.$$

Then

$$\int_0^x\int_0^y\frac{\check{\Phi}(A(s))\check{\Psi}(B(t))F(s)G(t)}{\left(|h(s)|^{\frac{1}{2\beta}}+|h^{*}(t)|^{\frac{1}{2\beta}}\right)^{\frac{2\alpha}{p}}}dsdt \leqslant M_6(p)\left(\int_0^x(x-s)\left(f(s)\check{\Phi}\left(a(s)\right)\right)^{q}ds\right)^{\frac{1}{q}}.$$
$$\times\left(\int_0^y(y-t)\left(g(t)\check{\Psi}\left(b(t)\right)\right)^{q}dt\right)^{\frac{1}{q}}$$

where

$$M_6(p)=(x)^{\frac{1}{p}}(y)^{\frac{1}{p}}$$

Taking $\mathbb{T}=\mathbb{Z}$ in Theorem 5 with relation (2), gives:

Corollary 9. *Assume $g(n)\geq 0, b(n)\geq 0, f(n)\geq 0, a(n)\geq 0$. Define*

$$A(n):=\frac{1}{F(n)}\sum_{s=0}^{n}f(s)a(s) \text{ and } B(m):=\frac{1}{G(m)}\sum_{k=0}^{m}g(k)b(k).$$

$$F(n) := \sum_{s=0}^{n} f(s) \text{ and } G(m) := \sum_{k=0}^{m} g(k).$$

Then

$$\sum_{n=1}^{N}\sum_{m=1}^{M} \frac{\breve{\Phi}(A(n))\breve{\Psi}(B(m))F(n)G(m)}{\left(|h(n+1)|^{\frac{1}{2\beta}} + |h^*(m+1)|^{\frac{1}{2\beta}}\right)^{\frac{2\alpha}{p}}} \leqslant M_7(p)\left(\sum_{n=1}^{N}(N+1-(n+1))\left(f(n)\breve{\Phi}\left(a(n)\right)\right)^q\right)^{\frac{1}{q}}$$
$$\times\left(\sum_{m=1}^{M}(M+1-(m+1))\left(g(m)\breve{\Psi}\left(b(m)\right)\right)^q\right)^{\frac{1}{q}}$$

where

$$M_7(p) = (NM)^{\frac{1}{p}}.$$

Remark 3. *In Corollary 9, if $p = q = 2$ we get the result due to Hamiaz and Abuelela ([1], Theorem 7).*

Corollary 10. *With the hypotheses of Theorem 5, we get:*

$$\int_{t_0}^{x}\int_{t_0}^{y} \frac{\breve{\Phi}(A^\sigma(s))\breve{\Psi}(B^\sigma(t))F^\sigma(s)G^\sigma(t)}{\left(|h(\sigma(s)-t_0)|^{\frac{1}{2\beta}} + |h^*(\sigma(t)-t_0)|^{\frac{1}{2\beta}}\right)^{\frac{2\alpha}{p}}}\Delta s\Delta t$$
$$\leqslant M_5(p)\left\{h\left(\int_{t_0}^{x}(\sigma(x)-\sigma(s))\left(f(s)\breve{\Phi}\left(a(s)\right)\right)^q\Delta s\right)\right.$$
$$\left.+h^*\left(\int_{t_0}^{y}(\sigma(y)-\sigma(t))\left(g(t)\breve{\Psi}\left(b(t)\right)\right)^q\Delta t\right)\right\}^{\frac{1}{q}}.$$

where M_5 defined as in (44).

Proof. We apply the Fenchel-Young inequality (5) in (43). This completes the proof. □

3. Some Applications

We can apply our inequalities to obtain different formulas of Hilbert-type inequalities by suggesting $h^*(y)$ and $h(x)$ by some functions:

In (12), as a special case, if we take $h(x) = \frac{x^2}{2}$, we have $h^*(x) = \frac{x^2}{2}$ see [12], we get

$$\int_{t_0}^{x}\int_{t_0}^{y} \frac{A^K(\sigma(s))B^L(\sigma(t))}{\left(|h(\sigma(s)-t_0)|^{\frac{1}{2\beta}} + |h^*(\sigma(t)-t_0)|^{\frac{1}{2\beta}}\right)^{\frac{2\alpha}{p}}}\Delta s\Delta t$$
$$= \int_{t_0}^{x}\int_{t_0}^{y} \frac{A^K(\sigma(s))B^L(\sigma(t))}{\left((\sigma(s)-t_0))^{\frac{1}{\beta}} + (\sigma(t)-t_0)^{\frac{1}{\beta}}\right)^{\frac{2\alpha}{p}}}\Delta s\Delta t$$
$$\leqslant \left(\frac{1}{2}\right)^{\frac{\alpha}{p\beta}} C_2(L,K,p)\left(\int_{t_0}^{x}(\sigma(x)-\sigma(s))\left(A^{K-1}(\sigma(s))a(s)\right)^q\Delta s\right)^{\frac{1}{q}}$$
$$\times\left(\int_{t_0}^{y}(\sigma(y)-\sigma(t))\left(B^{L-1}(\sigma(t))b(t)\right)^q\Delta t\right)^{\frac{1}{q}}, \qquad (55)$$

where $C_2(L,K,p)$ defined as in Theorem 2. Consequently, for $\alpha=\beta=1$, inequality (55) produces

$$\int_{t_0}^{x}\int_{t_0}^{y}\frac{A^K(\sigma(s))B^L(\sigma(t))}{\Big((\sigma(s)-t_0))+(\sigma(t)-t_0)\Big)^{\frac{2}{p}}}\Delta s\Delta t$$
$$\leqslant\left(\frac{1}{2}\right)^{\frac{1}{p}}C_2(L,K,p)\Big(\int_{t_0}^{x}(\sigma(x)-\sigma(s))\Big(A^{K-1}(\sigma(s))a(s)\Big)^q\Delta s\Big)^{\frac{1}{q}}$$
$$\times\Big(\int_{t_0}^{y}(\sigma(y)-\sigma(t))\Big(B^{L-1}(\sigma(t))b(t)\Big)^q\Delta t\Big)^{\frac{1}{q}}. \tag{56}$$

On the other hand if we take $h(n)=\frac{n^r}{r}, r>1$, then $h^*(m)=\frac{m^k}{k}$ where $\frac{1}{r}+\frac{1}{k}=1$ and $n,m\ \mathbb{R}_+$, then (12) gives

$$\int_{t_0}^{x}\int_{t_0}^{y}\frac{A^K(\sigma(s))B^L(\sigma(t))}{\Big(|h(\sigma(s)-t_0)|^{\frac{1}{2\beta}}+|h^*(\sigma(t)-t_0)|^{\frac{1}{2\beta}}\Big)^{\frac{2\alpha}{p}}}\Delta s\Delta t$$
$$=\int_{t_0}^{x}\int_{t_0}^{y}\frac{A^K(\sigma(s))B^L(\sigma(t))}{\Big((k(\sigma(s)-t_0)^r)^{\frac{1}{2\beta}}+(r(\sigma(t)-t_0)^k)^{\frac{1}{2\beta}}\Big)^{\frac{2\alpha}{p}}}\Delta s\Delta t$$
$$\leqslant\left(\frac{1}{rk}\right)^{\frac{\alpha}{p\beta}}C_2(L,K,p)\Big(\int_{t_0}^{x}(\sigma(x)-\sigma(s))\Big(A^{K-1}(\sigma(s))a(s)\Big)^q\Delta s\Big)^{\frac{1}{q}}$$
$$\times\Big(\int_{t_0}^{y}(\sigma(y)-\sigma(t))\Big(B^{L-1}(\sigma(t))b(t)\Big)^q\Delta t\Big)^{\frac{1}{q}}. \tag{57}$$

Clearly, when $\beta=\frac{1}{2\alpha}$, the inequality (57) becomes

$$\int_{t_0}^{x}\int_{t_0}^{y}\frac{A^K(\sigma(s))B^L(\sigma(t))}{\Big((k(\sigma(s)-t_0)^r)^{\alpha}+(r(\sigma(t)-t_0)^k)^{\alpha}\Big)^{\frac{2\alpha}{p}}}\Delta s\Delta t$$
$$\leqslant\left(\frac{1}{rk}\right)^{\frac{2\alpha^2}{p}}C_2(L,K,p)\Big(\int_{t_0}^{x}(\sigma(x)-\sigma(s))\Big(A^{K-1}(\sigma(s))a(s)\Big)^q\Delta s\Big)^{\frac{1}{q}}$$
$$\times\Big(\int_{t_0}^{y}(\sigma(y)-\sigma(t))\Big(B^{L-1}(\sigma(t))b(t)\Big)^q\Delta t\Big)^{\frac{1}{q}}. \tag{58}$$

If $\beta=\alpha=1$. From (57), we get

$$\int_{t_0}^{x}\int_{t_0}^{y}\frac{A^K(\sigma(s))B^L(\sigma(t))}{\Big((k(\sigma(s)-t_0)^r)^{\frac{1}{2}}+(r(\sigma(t)-t_0)^k)^{\frac{1}{2}}\Big)^{\frac{2}{p}}}\Delta s\Delta t$$
$$\leqslant\left(\frac{1}{rk}\right)^{\frac{1}{p}}C_2(L,K,p)\Big(\int_{t_0}^{x}(\sigma(x)-\sigma(s))\Big(A^{K-1}(\sigma(s))a(s)\Big)^q\Delta s\Big)^{\frac{1}{q}}$$
$$\times\Big(\int_{t_0}^{y}(\sigma(y)-\sigma(t))\Big(B^{L-1}(\sigma(t))b(t)\Big)^q\Delta t\Big)^{\frac{1}{q}}.$$

4. Conclusions

In this paper, with the help of a Fenchel-Legendre transform, which is used in various problems involving symmetry, we generalized a number of Hilbert-type inequalities to a general time scale. Besides that, in order to obtain some new inequalities as special cases, we also extended our inequalities to discrete and continuous calculus. In the future, we can generalize these inequalities in a different way by using other mathematical tools.

Author Contributions: All authors have contributed equally. All authors have read and agreed to the published version of the manuscript.

Funding: The authors declare that they have received no funding from any funding body.

Conflicts of Interest: The authors declare that they have no competing interests.

References

1. Hamiaz, A.; Abuelela, W. Some new discrete Hilbert's inequalities involving Fenchel–Legendre transform. *J. Inequal. Appl.* **2020**, *2020*, 1–14. [CrossRef]
2. Hilger, S. Ein Maßkettenkalkül mit Anwendung auf Zentrumsmannigfaltigkeiten. Ph.D. Thesis, Universität Würzburg, Würzburg, Germany, 1988.
3. Hardy, G.; Littlewood, J.E.; Pólya, G. *Inequalities*; Cambridge University Press: London, UK; New York, NY, USA, 1952; Volume 2, pp. 151–218.
4. Zhong, J.; Yang, B. An extension of a multidimensional Hilbert-type inequality. *J. Inequal. Appl.* **2017**, *2017*, 1–12. [CrossRef]
5. Frazer, H. Note on Hilbert's inequality. *J. Lond. Math. Soc.* **1946**, *1*, 7–9. [CrossRef]
6. Chen, Q.; Yang, B. A survey on the study of Hilbert-type inequalities. *J. Inequal. Appl.* **2015**, *2015*, 302. [CrossRef]
7. Abuelela, W. On an inequality related to hilbert's with laplace transform. *Int. J. Pure Appl. Math.* **2015**, *101*, 87–94. [CrossRef]
8. Huang, Q.; Yang, B. A multiple Hilbert-type integral inequality with a non-homogeneous kernel. *J. Inequal. Appl.* **2013**, *2013*, 73. [CrossRef]
9. Kim, Y.H. An improvement of some inequalities similar to Hilbert's inequality. *Int. J. Math. Math. Sci.* **2001**, *28*, 211–221. [CrossRef]
10. Pachpatte, B. On some new inequalities similar to Hilbert's inequality. *J. Math. Anal. Appl.* **1998**, *226*, 166–179. [CrossRef]
11. Dong, Q.L.; Lu, Y.Y.; Yang, J.; He, S. Approximately solving multi-valued variational inequalities by using a projection and contraction algorithm. *Numer. Algorithms* **2017**, *76*, 799–812. [CrossRef]
12. Borwein, J.; Lewis, A.S. *Convex Analysis and Nonlinear Optimization: Theory and Examples*; Springer Science & Business Media: Berlin, Germany, 2010.
13. Arnold, V.I. *Lectures on Partial Differential Equations*; Springer Science & Business Media: Berlin, Germany, 2013.
14. Davies, G.; Petersen, G. On an inequality of Hardy's (II). *Q. J. Math.* **1964**, *15*, 35–40. [CrossRef]
15. Nemeth, J. Generalizations of hardy-littlewood inequality. *Acta Sci. Math.* **1971**, *32*, 295.
16. Pachpatte, B. A note on some series inequalities. *Tamkang J. Math.* **1995**, *27*, 77–80.
17. Pachpatte, B.G.; Talkies, N.A. A note on integral inequalities involving the product of two functions. *J. Inequal. Pure Appl. Math.* **2006**, *7*, 78.
18. Mitrinovic, D.S.; Vasic, P.M. *Analytic Inequalities*; Springer: Berlin, Germany, 1970; Volume 61.
19. Agarwal, R.; O'Regan, D.; Saker, S. *Dynamic Inequalities on Time Scales*; Springer: Cham, Swizerland, 2014; Volume 2014.
20. Saker, S.H.; El-Deeb, A.; Rezk, H.; Agarwal, R.P. On Hilbert's inequality on time scales. *Appl. Anal. Discret. Math.* **2017**, *11*, 399–423. [CrossRef]
21. Abdeldaim, A.; El-Deeb, A.A.; Agarwal, P.; El-Sennary, H.A. On some dynamic inequalities of Steffensen type on time scales. *Math. Methods Appl. Sci.* **2018**, *41*, 4737–4753. [CrossRef]

22. Akin-Bohner, E.; Bohner, M.; Akin, F. Pachpatte inequalities on time scales. *JIPAM. J. Inequal. Pure Appl. Math.* **2005**, *6*, 23.
23. Bohner, M.; Matthews, T. The Grüss inequality on time scales. *Commun. Math. Anal.* **2007**, *3*, 1–8.
24. Bohner, M.; Matthews, T. Ostrowski inequalities on time scales. *JIPAM. J. Inequal. Pure Appl. Math.* **2008**, *9*, 6, 8.
25. Dinu, C. Hermite-Hadamard inequality on time scales. *J. Inequal. Appl.* **2008**, 287947. [CrossRef]
26. El-Deeb, A.A. On some generalizations of nonlinear dynamic inequalities on time scales and their applications. *Appl. Anal. Discret. Math.* **2019**, *13*, 10. [CrossRef]
27. El-Deeb, A.A.; Cheung, W.S. A variety of dynamic inequalities on time scales with retardation. *J. Nonlinear Sci. Appl.* **2018**, *11*, 1185–1206. [CrossRef]
28. El-Deeb, A.A.; El-Sennary, H.A.; Khan, Z.A. Some Steffensen-type dynamic inequalities on time scales. *Adv. Differ. Equ.* **2019**, *2019*, 246. [CrossRef]
29. El-Deeb, A.A.; Elsennary, H.A.; Cheung, W.S. Some reverse Hölder inequalities with Specht's ratio on time scales. *J. Nonlinear Sci. Appl.* **2018**, *11*, 444–455. [CrossRef]
30. El-Deeb, A.A.; Elsennary, H.A.; Nwaeze, E.R. Generalized weighted Ostrowski, trapezoid and Grüss type inequalities on time scales. *Fasc. Math.* **2018**, 123–144. [CrossRef]
31. El-Deeb, A.A.; Xu, H.; Abdeldaim, A.; Wang, G. Some dynamic inequalities on time scales and their applications. *Adv. Differ. Equ.* **2019**, 130. [CrossRef]
32. El-Deeb, A.A. Some Gronwall-Bellman type inequalities on time scales for Volterra-Fredholm dynamic integral equations. *J. Egypt. Math. Soc.* **2018**, *26*, 1–17. [CrossRef]
33. Hilscher, R. A time scales version of a Wirtinger-type inequality and applications. *J. Comput. Appl. Math.* **2002**, *141*, 219–226. [CrossRef]
34. Li, W.N. Some delay integral inequalities on time scales. *Comput. Math. Appl.* **2010**, *59*, 1929–1936. [CrossRef]
35. Řehák, P. Hardy inequality on time scales and its application to half-linear dynamic equations. *J. Inequal. Appl.* **2005**, 495–507, doi:10.1155/JIA.2005.495. [CrossRef]
36. Tian, Y.; El-Deeb, A.A.; Meng, F. Some nonlinear delay Volterra-Fredholm type dynamic integral inequalities on time scales. *Discret. Dyn. Nat. Soc.* **2018**, 5841985, doi:10.1155/2018/5841985. [CrossRef]
37. El-Deeb, A.; Khan, Z.A. Certain new dynamic nonlinear inequalities in two independent variables and applications. *Bound. Value Probl.* **2020**, *2020*, 31. [CrossRef]
38. El-Deeb, A.; El-Sennary, H.; Agarwal, P. Some opial-type inequalities with higher order delta derivatives on time scales. *Rev. Real Acad. Cienc. Exactas Fís. Nat. Ser. A Mat.* **2020**, *114*, 29. [CrossRef]
39. Abdeldaim, A.; El-Deeb, A.; Ahmed, R.G. On retarded nonlinear integral inequalities of gronwall and applications. *J. Math. Inequal.* **2019**, *13*, 1023–1038. [CrossRef]
40. El-Deeb, A.A.; Kh., F.M.; Ismail, G.A.F.; Khan, Z.A. Weighted dynamic inequalities of Opial-type on time scales. *Adv. Differ. Equ.* **2019**, *2019*, 393. [CrossRef]
41. Kh., F.M.; El-Deeb, A.A.; Abdeldaim, A.; Khan, Z.A. On some generalizations of dynamic Opial-type inequalities on time scales. *Adv. Differ. Equ.* **2019**, *2019*, 1–14.
42. Abdeldaim, A.; El-Deeb, A.A. Some new retarded nonlinear integral inequalities with iterated integrals and their applications in retarded differential equations and integral equations. *J. Fract. Calc. Appl.* **2014**, *5*, 9. [CrossRef]
43. Abdeldaim, A.; El-Deeb, A.A. On generalized of certain retarded nonlinear integral inequalities and its applications in retarded integro-differential equations. *Appl. Math. Comput.* **2015**, *256*, 375–380. [CrossRef]
44. Abdeldaim, A.; El-Deeb, A.A. On some generalizations of certain retarded nonlinear integral inequalities with iterated integrals and an application in retarded differential equation. *J. Egypt. Math. Soc.* **2015**, *23*, 470–475. [CrossRef]
45. Abdeldaim, A.; El-Deeb, A.A. On some new nonlinear retarded integral inequalities with iterated integrals and their applications in integro-differential equations. *Br. J. Math. Comput. Sci.* **2015**, *5*, 479–491. [CrossRef]
46. Agarwal, R.P.; Lakshmikantham, V. *Uniqueness and Nonuniqueness Criteria for Ordinary Differential Equations*; Series in Real Analysis; World Scientific Publishing: Singapore, 1993; Volume 6.
47. El-Deeb, A.A. *On Integral Inequalities and Their Applications*; LAP Lambert Academic Publishing: Saarbrücken, 2017.
48. El-Deeb, A.A. A Variety of Nonlinear Retarded Integral Inequalities of Gronwall Type and Their Applications. In *Advances in Mathematical Inequalities and Applications*; Springer: Berlin, Germany, 2018; pp. 143–164.

49. El-Deeb, A.A.; Ahmed, R.G. On some explicit bounds on certain retarded nonlinear integral inequalities with applications. *Adv. Inequal. Appl.* **2016**, *2016*, 15.
50. El-Deeb, A.A.; Ahmed, R.G. On some generalizations of certain nonlinear retarded integral inequalities for Volterra-Fredholm integral equations and their applications in delay differential equations. *J. Egypt. Math. Soc.* **2017**, *25*, 279–285. [CrossRef]
51. El-Owaidy, H.; Abdeldaim, A.; El-Deeb, A.A. On some new retarded nonlinear integral inequalities and their applications. *Math. Sci. Lett.* **2014**, *3*, 157. [CrossRef]
52. El-Owaidy, H.M.; Ragab, A.A.; Eldeeb, A.A.; Abuelela, W.M.K. On some new nonlinear integral inequalities of Gronwall-Bellman type. *Kyungpook Math. J.* **2014**, *54*, 555–575. [CrossRef]
53. Li, J.D. Opial-type integral inequalities involving several higher order derivatives. *J. Math. Anal. Appl.* **1992**, *167*, 98–110.
54. Bohner, M.; Peterson, A. *Dynamic Equations on Time Scales: An Introduction with Applications*; Birkhauser: Boston, MA, USA, 2001.
55. Bibi, R.; Bohner, M.; Pecarić, J.; Varosanec, S. Minkowski and Beckenbach-Dresher inequalities and functionals on time scales. *J. Math. Inequal.* **2013**, *7*, 299–312. [CrossRef]

symmetry

MDPI

Article

Energy of Accelerations Used to Obtain the Motion Equations of a Three- Dimensional Finite Element

Sorin Vlase [1,2], Iuliu Negrean [2,3], Marin Marin [4] and Maria Luminița Scutaru [1,*]

1 Department of Mechanical Engineering, Faculty of Mechanical Engineering, Transilvania University of Brașov, B-dul Eroilor 29, 500036 Brașov, Romania; svlase@unitbv.ro
2 Technical Sciences Academy of Romania; B-dul Dacia 26, 030167 Bucharest, Romania; Iuliu.Negrean@mep.utcluj.ro
3 Department of Mechanical Systems Engineering, Technical University of Cluj-Napoca, Str. Memorandumului 28, 400114 Cluj-Napoca, Romania
4 Department of Mathematics and Computer Science, Transilvania University of Brașov, B-dul Eroilor 29, 500036 Brașov, Romania; m.marin@unitbv.ro
* Correspondence: lscutaru@unitbv.ro

Received: 20 January 2020; Accepted: 13 February 2020; Published: 23 February 2020

Abstract: When analyzing the dynamic behavior of multi-body elastic systems, a commonly used method is the finite element method conjunctively with Lagrange's equations. The central problem when approaching such a system is determining the equations of motion for a single finite element. The paper presents an alternative method of calculation theses using the Gibbs–Appell (GA) formulation, which requires a smaller number of calculations and, as a result, is easier to apply in practice. For this purpose, the energy of the accelerations for one single finite element is calculated, which will be used then in the GA equations. This method can have advantages in applying to the study of multi-body systems with elastic elements and in the case of robots and manipulators that have in their composition some elastic elements. The number of differentiation required when using the Gibbs–Appell method is smaller than if the Lagrange method is used which leads to a smaller number of operations to obtain the equations of motion.

Keywords: Gibbs–Appell; energy of accelerations; finite element; nonlinear system; elastic elements; analytical dynamics; robotics

1. Introduction

The common method to obtain the equations of motion for a finite element belonging to an elastic element of a multibody system is represented by Lagrange equations. In this way, the most important step in the analysis of the multi-body systems with elastic elements is solved, the rest of the procedures being the classical ones currently used in the finite element method FEA and verified by practice. So, it is possible to determine the dynamic response of a single finite element. Using the known assembly methods and introducing loads and proper boundary conditions, the set of differential equations describing the behavior of the entire elastic system is obtained.

In this process, the type of finite element chosen will determine the shape functions that will be used and which will give the final form of the coefficients of the matrix appearing in the equations of motion. In such an analysis, it is considered that the deformations are small enough to not have influence on the general, rigid motion of the multibody system. Thus, a large class of systems has been studied, having in their structure elastic elements modeled with one-dimensional, two-dimensional or three-dimensional finite elements. One-dimensional finite elements were among the first studied. We can mention here [1–7]. The method was then extended naturally to bi- and three-dimensional elements [8–10].

The basic hypothesis used was that the rigid motion of the multibody system is known and, as a consequence, the velocity and acceleration field of each solid of every element of the system. It is also considered that this rigid motion does not influence the elastic behavior of each element.

Customary Lagrange's equations were used to determine the dynamic response of a single element. For this, it is necessary to determine, as the first step, the Lagrangian composed by kinetic energy, internal energy and the mechanical work of the external loads. The studied elements were considered linear elastic. Thus, the evolution equations of a finite element are obtained. The use of Lagrange's equations proved to be a useful and practical method and was, generally, the only method used to solve this type of problem. Recent works in the field use Lagrange's equations for the study of multi-body systems with elastic elements [11–17]. Theoretically, however, other methods of solving these dynamic systems can be applied, which will ultimately lead to the same equations of motion. In this paper, we will deal with the use of Gibbs–Appell (GA) equations, which have advantages in terms of the number of operations required to obtain the final results and thus prove more economical.

One problem which led to the GA equations was to deduce the differential equations of motion, which are easier to apply in the case of non-holonomic systems but which can also be applied to the holonomic systems. This method was discovered by Gibbs in 1879 [18] and, independently, by Appell in 1899 [19]. The function of Lagrange or Hamilton is replaced, in this formalism, by the energy of accelerations and represents an application of the well-known Gauss principle of least constraint being equivalent to the other formulations currently used in classical mechanics.

The modern mechanical systems are characterized by ultra-fast movements and velocities and include serial structures of robots as well. The fast movements are generally characterized by a time variation law for accelerations and lead to the occurrence of accelerations of higher orders. The last decades are characterized by a sustained development of the use of robots in the industry. This involves studies that allow good control of such systems, especially when working with high speeds and forces. In the case of such systems, which give very high working speeds and especially important speed variations, accelerations become significant. In these cases, the GA method becomes interesting. This type of formalism is useful for a wide class of engineering problems. For example, it can be used to analyze rigid dynamical systems, considering quasi-velocities [20]. The equations of motion can be easily obtained, both for holonomic systems and for non-holonomic systems. It is thus possible to eliminate at this step Lagrange's multipliers and thus reduce the number of unknowns with which they work. The method can be applied both to linear and nonlinear problems. An example is presented in [21] where are obtained the motion equations for an N-link robot, with flexible elements. Experimental procedures validate the result obtained using the GA formalism. In the last period, this formalism tends to become a usual procedure. There are many papers presenting this kind of approach, like [22], where an application consists of an N-flexible-link manipulator used in the cooperative flexible mobile manipulator. The main advantage of recursive GA formulation is used to determine the motion equations without classical use of Lagrange multipliers. Some new methods to study the dynamic behavior of the multibody system have, as the central piece in approach, this formalism. As an example, in [23], such a system is defined using a classic set of coupled differential equations and a set of algebraic equations for expressing the liaisons (constraints). The use of Lagrange formalism asks a large number of derivatives in the governing equations. In these cases, the use of the recursive GA formalism becomes easier to apply. In this mode, it is possible to obtain the equations of motion automatically. In [24], the main advantage of this formulation used in applications is that it is not necessary to eliminate Lagrange's multipliers and, as a consequence, the computational complexity decrease. Identically the formalism is also used in [25]. Dynamical response and motion equations can be obtained via recursive GA formulation and represent the way used to reduce the computational effort, significant for this application. The main advantage is fewer mathematical calculations compared to other formulations. There are a lot of papers using this formalism in the study of the dynamical system [26]. Interesting results are presented in [27–29]. The present paper only aims at obtaining the equations of motion for a three-dimensional finite element with a general motion.

The problems regarding the accuracy of the results and the accuracy of the models used in the case of the mechanisms, which have significant displacements and deformations, were not developed in the paper. These very interesting and current study domains themselves represent a distinct field of research. In this paper, we only dealt with the presentation of an alternative method, which presents advantages of calculation, of determining the equations of motion for a single element. The other operations used then in the method of finite elements were considered the ones currently used in practice.

2. Gibbs–Appell Formalism

In the case of scleronomic liaisons, the acceleration of a point i of a system of material points having n degree of freedom has, generally, the expression:

$$\bar{a}_i = \sum_{k=1}^{n}\sum_{j=1}^{n} \frac{\partial^2 \bar{r}_i}{\partial q_k \partial q_j}\dot{q}_k\dot{q}_j + \sum_{k=1}^{n} \frac{\partial \bar{r}_i}{\partial q_k}\ddot{q}_k \tag{1}$$

where $\bar{r}_i$ is the position vector of the point and q_j, $j = \overline{1,n}$ represent the independent coordinates. In the following, the dot placed above a scalar, vector or matrix size means the derivative of the respective size with respect to time. The notion energy of accelerations for a system of material points is defined as [27]:

$$S = \frac{1}{2}\sum_{i=1}^{N} m_i a_i^2 \tag{2}$$

The form of this expression is similar to the expression of kinetic energy but the mechanical significance is different. In the case of a solid body, the definition can be extended to the whole domain of the body as:

$$S = \frac{1}{2}\int_V \rho a^2 dV \tag{3}$$

where the acceleration of a certain point of the solid is similar to Equation (1):

$$\bar{a} = \sum_{k=1}^{n}\sum_{j=1}^{n} \frac{\partial^2 \bar{r}}{\partial q_k \partial q_j}\dot{q}_k\dot{q}_j + \sum_{k=1}^{n} \frac{\partial \bar{r}}{\partial q_k}\ddot{q}_k \tag{3a}$$

Introducing Equation (3a) in Equation (3) we get:

$$\begin{aligned}
S &= \tfrac{1}{2}\int_V \rho a^2 dV = \tfrac{1}{2}\int_V \rho\left(\sum_{k=1}^{n}\sum_{j=1}^{n} \tfrac{\partial^2 \bar{r}}{\partial q_k \partial q_j}\dot{q}_k\dot{q}_j + \sum_{k=1}^{n} \tfrac{\partial \bar{r}}{\partial q_k}\ddot{q}_k\right)^2 dV = \\
&= \tfrac{1}{2}\int_V \rho\left(\sum_{k=1}^{n}\sum_{j=1}^{n} \tfrac{\partial^2 \bar{r}}{\partial q_k \partial q_j}\dot{q}_k\dot{q}_j + \sum_{k=1}^{n} \tfrac{\partial \bar{r}}{\partial q_k}\ddot{q}_k\right)^2 dV = \\
&= \tfrac{1}{2}\int_V \rho\left[\left(\sum_{k=1}^{n}\sum_{j=1}^{n} \tfrac{\partial^2 \bar{r}}{\partial q_k \partial q_j}\dot{q}_k\dot{q}_j\right)^2 + 2\left(\sum_{k=1}^{n}\sum_{j=1}^{n} \tfrac{\partial^2 \bar{r}}{\partial q_k \partial q_j}\dot{q}_k\dot{q}_j\right)\left(\sum_{k=1}^{n} \tfrac{\partial \bar{r}}{\partial q_k}\ddot{q}_k\right) + \left(\sum_{k=1}^{n} \tfrac{\partial \bar{r}}{\partial q_k}\ddot{q}_k\right)^2\right]^2 dV = \\
&= \tfrac{1}{2}\int_V \rho\left[\sum_{k=1}^{n}\sum_{j=1}^{n}\sum_{l=1}^{n}\sum_{m=1}^{n} \tfrac{\partial^2 \bar{r}}{\partial q_k \partial q_j}\tfrac{\partial^2 \bar{r}}{\partial q_l \partial q_m}\dot{q}_k\dot{q}_j\dot{q}_l\dot{q}_m + \sum_{k=1}^{n}\sum_{j=1}^{n}\sum_{l=1}^{n} \tfrac{\partial^2 \bar{r}}{\partial q_k \partial q_j}\tfrac{\partial \bar{r}}{\partial q_l}\dot{q}_k\dot{q}_j\ddot{q}_l + \sum_{k=1}^{n}\sum_{j=1}^{n} \tfrac{\partial \bar{r}}{\partial q_k}\tfrac{\partial \bar{r}}{\partial q_j}\ddot{q}_k\ddot{q}_j\right] dV = \\
&= \tfrac{1}{2}\int_V \rho\left(\sum_{k=1}^{n}\sum_{j=1}^{n}\sum_{l=1}^{n}\sum_{m=1}^{n} \tfrac{\partial^2 \bar{r}}{\partial q_k \partial q_j}\tfrac{\partial^2 \bar{r}}{\partial q_l \partial q_m}\dot{q}_k\dot{q}_j\dot{q}_l\dot{q}_m\right) dV + \\
&+ \tfrac{1}{2}\int_V \rho\left(\sum_{k=1}^{n}\sum_{j=1}^{n}\sum_{l=1}^{n} \tfrac{\partial^2 \bar{r}}{\partial q_k \partial q_j}\tfrac{\partial \bar{r}}{\partial q_l}\dot{q}_k\dot{q}_j\ddot{q}_l\right) dV + \tfrac{1}{2}\int_V \rho\left(\sum_{k=1}^{n}\sum_{j=1}^{n} \tfrac{\partial \bar{r}}{\partial q_k}\tfrac{\partial \bar{r}}{\partial q_j}\ddot{q}_k\ddot{q}_j\right) dV = \\
&= \tfrac{1}{2}\left(\int_V \rho \sum_{k=1}^{n}\sum_{j=1}^{n}\sum_{l=1}^{n}\sum_{m=1}^{n} \tfrac{\partial^2 \bar{r}}{\partial q_k \partial q_j}\tfrac{\partial^2 \bar{r}}{\partial q_l \partial q_m} dV\right)\dot{q}_k\dot{q}_j\dot{q}_l\dot{q}_m + \tfrac{1}{2}\left(\int_V \rho \sum_{k=1}^{n}\sum_{j=1}^{n}\sum_{l=1}^{n} \tfrac{\partial^2 \bar{r}}{\partial q_k \partial q_j}\tfrac{\partial \bar{r}}{\partial q_l} dV\right)\dot{q}_k\dot{q}_j\ddot{q}_l + \\
&+ \tfrac{1}{2}\left(\int_V \rho \sum_{k=1}^{n}\sum_{j=1}^{n} \tfrac{\partial \bar{r}}{\partial q_k}\tfrac{\partial \bar{r}}{\partial q_j} dV\right)\ddot{q}_k\ddot{q}_j = E_{ao}(\dot{q}) + E_{a1}(\dot{q},\ddot{q}) + E_{a2}(\ddot{q})
\end{aligned} \tag{4}$$

where:

$$E_{ao}(\dot{q}) = \frac{1}{2}\left(\int_V \rho \sum_{k=1}^{n}\sum_{j=1}^{n}\sum_{l=1}^{n}\sum_{m=1}^{n} \frac{\partial^2 \overline{r}}{\partial q_k \partial q_j}\frac{\partial^2 \overline{r}}{\partial q_l \partial q_m} dV\right)\dot{q}_k\dot{q}_j\dot{q}_l\dot{q}_m \tag{5}$$

contains terms that have no accelerations,

$$E_{a1}(\dot{q},\ddot{q}) = \frac{1}{2}\left(\int_V \rho \sum_{k=1}^{n}\sum_{j=1}^{n}\sum_{l=1}^{n} \frac{\partial^2 \overline{r}}{\partial q_k \partial q_j}\frac{\partial \overline{r}}{\partial q_l} dV\right)\dot{q}_k\dot{q}_j\ddot{q}_l \tag{6}$$

represents the terms in which the accelerations intervene linearly and:

$$E_{a2}(\ddot{q}) = \frac{1}{2}\left(\int_V \rho \sum_{k=1}^{n}\sum_{j=1}^{n} \frac{\partial \overline{r}}{\partial q_k}\frac{\partial \overline{r}}{\partial q_j} dV\right)\ddot{q}_k\ddot{q}_j \tag{7}$$

represents the quadratic terms in accelerations.

GA equations are given by relations:

$$\frac{\partial S}{\partial \ddot{q}_j} = Q_j \quad j = \overline{1,n} \tag{8}$$

where Q_j it represents the generalized force corresponding to the generalized coordinate q_j.

3. Basic Hypothesis

Now consider a finite element modeling a domain of an elastic element from the structure of a multibody system. The aim of the paper is to determine the motion equations of this element. The generalized coordinates will be the displacements of the nodes. The displacements of an arbitrary point of the continuous domain modeled by the finite element shall be expressed using a finite number of nodal coordinates. The finite element shall be related to the mobile reference frame, which participates in the general rigid motion (Figure 1). With $\overline{v}_o(\dot{X}_o, \dot{Y}_o, \dot{Z}_o)$, we shall note the velocity, and with $\overline{a}_o(\ddot{X}_o, \ddot{Y}_o, \ddot{Z}_o)$, the acceleration of the origin of the local reference frame relative to the global reference frame OXYZ. We shall note with $\overline{\omega}(\omega_x, \omega_y, \omega_z)$ the angular velocity and with $\overline{\varepsilon}(\varepsilon_x, \varepsilon_y, \varepsilon_z)$ the angular acceleration. Generally, a multibody system consists of several solids. Every solid is characterized by these two vectors, and the number of pairs of such vectors is equal to the number of distinct bodies. The relations between the components of a vector in the local coordinate system $\{t\}_L$ and the components of the same vector in a global coordinate system $\{t\}_G$ are obtained using a matrix of rotation $[R]$:

$$\{t\}_G = [R]\{t\}_L \tag{9}$$

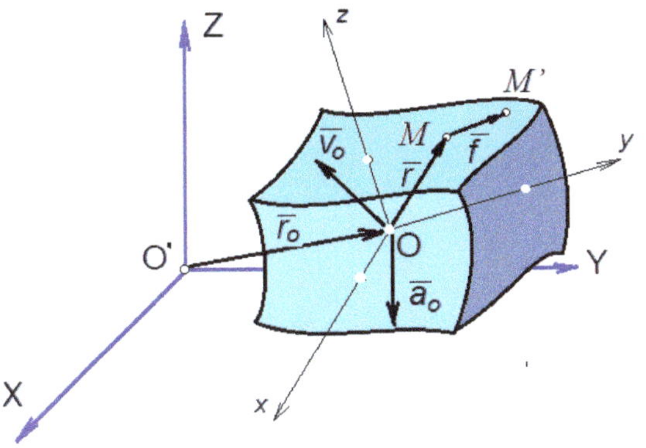

Figure 1. Three-dimensional finite element.

If an arbitrary point M of the finite element has a displacement $\{f\}_L$ changing into M′, we may write:

$$\{r_{M'}\}_G = \{r_O\}_G + [R]\big(\{r\}_L + \{f\}_L\big) \tag{10}$$

where $\{r_{M'}\}_G$ is the position vector of point M' with components expressed in the global coordinate system. The continuous displacement field noted $\{f\}_L$ is approximated, using the finite element method, by relation:

$$\{f\}_L = [N(x, y, z)]\{\delta\}_L \tag{11}$$

where the matrix elements $[N(x, y, z)]$, the shape functions, will depend on the type of the finite element chosen and $\{\delta\}_L$ are the independent nodal coordinates. The velocity of point M′ will be:

$$\{v_{M'}\}_G = \left\{\dot{r}_{M'}\right\}_G = \left\{\dot{r}_O\right\}_G + \left[\dot{R}\right]\{r\}_L + \left[\dot{R}\right][N]\{\delta\}_L + [R][N]\left\{\dot{\delta}\right\}_L \tag{12}$$

and the acceleration will be:

$$\begin{aligned}\{a_{M'}\}_G = \left\{\ddot{r}_O\right\}_G + \left[\ddot{R}\right]\{r\}_L + \left[\ddot{R}\right]\{f\}_L + 2\left[\dot{R}\right]\left\{\dot{f}\right\}_L + [R]\left\{\ddot{f}\right\}_L = \\ \left\{\ddot{r}_O\right\}_G + \left[\ddot{R}\right]\{r\}_L + \left[\ddot{R}\right][N]\{\delta\}_L + 2\left[\dot{R}\right][N]\left\{\dot{\delta}\right\}_L + [R][N]\left\{\ddot{\delta}\right\}_L\end{aligned} \tag{13}$$

These expressions will be used to determine the energy of accelerations. In the future, the index G mark the vector with the components expresses in the global reference system, the index L mark the vector with the components expresses in the local reference frame and the non-indexed vectors are considered to be written in the local coordinate system. The derivatives of the positional matrix $[R]$ occur in the previous relations. These derivatives will define angular velocities and angular accelerations. From the orthogonality conditions written for the positional matrix we have:

$$[R][R]^T = [R]^T[R] = [E] = \begin{bmatrix} 1 & 0 & 0 \\ 0 & 1 & 0 \\ 0 & 0 & 1 \end{bmatrix} \tag{14}$$

where $[E]$ is the unit matrix, the following relation can be obtained:

$$[\omega]_G = \left[\dot{R}\right][R]^T = \begin{bmatrix} 0 & -\omega_{zG} & \omega_{yG} \\ \omega_{zG} & 0 & -\omega_{xG} \\ -\omega_{yG} & \omega_{xG} & 0 \end{bmatrix} \tag{15}$$

which is a skew symmetric matrix angular velocity, with components in global coordinate system, corresponding to the angular velocity:

$$\{\omega\}_G = \begin{Bmatrix} \omega_{xG} \\ \omega_{yG} \\ \omega_{zG} \end{Bmatrix} \tag{16}$$

Obviously, we have, too:

$$[\omega]_L = [R]^T[\dot{R}] = \begin{bmatrix} 0 & -\omega_{zL} & \omega_{yL} \\ \omega_{zL} & 0 & -\omega_{xL} \\ -\omega_{yL} & \omega_{xL} & 0 \end{bmatrix} \text{and } \{\omega\}_L = \begin{Bmatrix} \omega_{xL} \\ \omega_{yL} \\ \omega_{zL} \end{Bmatrix} \tag{17}$$

The angular acceleration operator is defined by:

$$[\varepsilon]_G = [\dot{\omega}]_G = \begin{bmatrix} 0 & -\varepsilon_{zG} & \varepsilon_{yG} \\ \varepsilon_{zG} & 0 & -\varepsilon_{xG} \\ -\varepsilon_{yG} & \varepsilon_{xG} & 0 \end{bmatrix} = \begin{bmatrix} 0 & -\dot{\omega}_{zG} & \dot{\omega}_{yG} \\ \dot{\omega}_{zG} & 0 & -\dot{\omega}_{xG} \\ -\dot{\omega}_{yG} & \dot{\omega}_{xG} & 0 \end{bmatrix} = [\ddot{R}][R]^T + [\dot{R}][\dot{R}]^T \tag{18}$$

and:

$$[\varepsilon]_L = [\dot{\omega}]_L = \begin{bmatrix} 0 & -\varepsilon_{zL} & \varepsilon_{yL} \\ \varepsilon_{zL} & 0 & -\varepsilon_{xL} \\ -\varepsilon_{yL} & \varepsilon_{xL} & 0 \end{bmatrix} = \begin{bmatrix} 0 & -\dot{\omega}_{zL} & \dot{\omega}_{yL} \\ \dot{\omega}_{zL} & 0 & -\dot{\omega}_{xL} \\ -\dot{\omega}_{yL} & \dot{\omega}_{xL} & 0 \end{bmatrix} = [\dot{R}]^T[\dot{R}] + [R]^T[\ddot{R}] \tag{19}$$

It will result after some elementary calculus:

$$[\ddot{R}][R]^T = [\varepsilon]_G - [\dot{R}][\dot{R}]^T = [\varepsilon]_G - [\dot{R}][R]^T\left([\dot{R}][R]^T\right)^T = [\varepsilon]_G + [\omega]_G[\omega]_G \tag{20}$$

whereas:

$$[\omega]_G = -[\omega]_G^T \tag{21}$$

We have, too:

$$[R]^T[\ddot{R}] = [\varepsilon]_L - [\dot{R}]^T[R][R]^T[\dot{R}] = [\varepsilon]_G - \left([R]^T[\dot{R}]\right)^T[R]^T[\dot{R}] = [\varepsilon]_L + [\omega]_L[\omega]_L \tag{22}$$

These relations will be useful in the following considerations.

4. Motion Equations

In the following, the equations of motion for a finite element will be established using the GA formalism. For this, it is necessary to first determine the energy of the accelerations of a finite element. If we use Expression (13) determined previously for the acceleration of a point, the energy of the accelerations for the considered finite element is given by the expression [30]:

$$\begin{aligned}
&E_a = \tfrac{1}{2}\textstyle\int_V \rho a_{M'}^2 dV = \tfrac{1}{2}\int_V \rho\{a_{M'}\}^T\{a_{M'}\}dV = \\
&= \tfrac{1}{2}\textstyle\int_V \rho\Big(\{\ddot{r}_O\}_G^T + \{r\}_L^T[\ddot{R}]^T + \{\delta\}_L^T[N]^T[\ddot{R}]^T + 2\{\dot{\delta}\}_L^T[N]^T[\dot{R}]^T + \{\ddot{\delta}\}_L^T[N]^T\,[R]^T\Big)x \\
&\Big(\{\ddot{r}_O\}_G + [\ddot{R}]\{r\}_L + [\ddot{R}][N]\{\delta\}_L + 2[\dot{R}][N]\{\dot{\delta}\}_L + [R][N]\{\ddot{\delta}\}_L\Big)dV = \\
&\tfrac{1}{2}\textstyle\int_V \rho\Big(\{\ddot{r}_O\}_G^T\{\ddot{r}_O\}_G + 2\{r\}_L^T[\ddot{R}]^T\{\ddot{r}_O\}_G + 2\{\delta\}_L^T[N]^T[\ddot{R}]^T\{\ddot{r}_O\}_G + 4\{\dot{\delta}\}_L^T[N]^T[\dot{R}]^T\{\ddot{r}_O\}_G + 2\{\ddot{\delta}\}_L^T[N]^T\,[R]^T\{\ddot{r}_O\}_G\Big) + \\
&\tfrac{1}{2}\textstyle\int_V \rho\Big(\{r\}_L^T[\ddot{R}]^T[\ddot{R}]\{r\}_L + 2\{\delta\}_L^T[N]^T[\ddot{R}]^T[\ddot{R}]\{r\}_L + 4\{\dot{\delta}\}_L^T[N]^T[\dot{R}]^T[\ddot{R}]\{r\}_L + 2\{\ddot{\delta}\}_L^T[N]^T\,[R]^T[\ddot{R}]\{r\}_L\Big)x \\
&\tfrac{1}{2}\textstyle\int_V \rho\Big(\{\delta\}_L^T[N]^T[\ddot{R}]^T[\ddot{R}][N]\{\delta\}_L + 4\{\dot{\delta}\}_L^T[N]^T[\dot{R}]^T[\ddot{R}][N]\{\delta\}_L + 2\{\ddot{\delta}\}_L^T[N]^T\,[R]^T[\ddot{R}][N]\{\delta\}_L\Big) + \\
&\tfrac{1}{2}\textstyle\int_V \rho\Big(4\{\dot{\delta}\}_L^T[N]^T[\dot{R}]^T[\dot{R}][N]\{\dot{\delta}\}_L + 4\{\ddot{\delta}\}_L^T[N]^T\,[R]^T[\dot{R}][N]\{\dot{\delta}\}_L + \{\ddot{\delta}\}_L^T[N]^T\,[R]^T\,[R][N]\{\ddot{\delta}\}_L\Big)
\end{aligned} \tag{23}$$

The quadratic terms can be easily identified in the accelerations of the independent coordinates $\{\ddot{\delta}\}_L$, noted with E_{a2}:

$$E_{a2} = \frac{1}{2}\int_V \rho\left(\{\ddot{\delta}\}_L^T [N]^T [R]^T [R][N]\{\ddot{\delta}\}_L\right)dV = \frac{1}{2}\{\ddot{\delta}\}_L^T\left(\int_V \rho[N]^T [N]\, dV\right)\{\ddot{\delta}\}_L \tag{24}$$

the terms in which these accelerations appear linear E_{a1}:

$$\begin{gathered} E_{a1} = \int_V \rho\left(\{\ddot{\delta}\}_L^T [N]^T [R]^T\{\ddot{r}_O\}_G + \{\ddot{\delta}\}_L^T [N]^T [R]^T\left[\ddot{R}\right]\{r\}_L + \{\ddot{\delta}\}_L^T [N]^T [R]^T\left[\ddot{R}\right][N]\{\delta\}_L + 2\{\ddot{\delta}\}_L^T [N]^T [R]^T\left[\dot{R}\right][N]\{\dot{\delta}\}_L\right)dV = \\ \{\ddot{\delta}\}_L^T \int_V \rho\left([N]^T \{\ddot{r}_O\}_L + [N]^T ([\varepsilon]_L + [\omega]_L[\omega]_L)\{r\}_L + [N]^T ([\varepsilon]_L + [\omega]_L[\omega]_L)[N]\{\delta\}_L + 2[N]^T [\omega]_L[N]\{\dot{\delta}\}_L\right)dV \end{gathered} \tag{25}$$

and the terms that do not contain at all accelerations and which do not have importance in GA equations, E_{a0}:

$$\begin{gathered} E_{a0} = \frac{1}{2}\int_V \rho\left(\{\ddot{r}_O\}_G^T\{\ddot{r}_O\}_G + 2\{r\}_L^T\left[\ddot{R}\right]^T\{\ddot{r}_O\}_G + 2\{\delta\}_L^T[N]^T\left[\ddot{R}\right]^T\{\ddot{r}_O\}_G + 4\{\dot{\delta}\}_L^T[N]^T\left[\dot{R}\right]^T\{\ddot{r}_O\}_G\right) + \\ \frac{1}{2}\int_V \rho\left(\{r\}_L^T\left[\ddot{R}\right]^T\left[\ddot{R}\right]\{r\}_L + 2\{\delta\}_L^T[N]^T\left[\ddot{R}\right]^T\left[\ddot{R}\right]\{r\}_L + 4\{\dot{\delta}\}_L^T[N]^T\left[\dot{R}\right]^T\left[\ddot{R}\right]\{r\}_L\right)dV + \\ \frac{1}{2}\int_V \rho\left(\{\delta\}_L^T[N]^T\left[\ddot{R}\right]^T\left[\ddot{R}\right][N]\{\delta\}_L + 4\{\delta\}_L^T[N]^T\left[\dot{R}\right]^T\left[\ddot{R}\right][N]\{\delta\}_L + 4\{\dot{\delta}\}_L^T[N]^T\left[\dot{R}\right]^T\left[\dot{R}\right][N]\{\dot{\delta}\}_L\right)dV \end{gathered} \tag{26}$$

The potential energy (internal work) has a classical form:

$$E_p = \frac{1}{2}\int_V \{\sigma\}^T\{\varepsilon\}dV \tag{27}$$

where $\{\sigma\}$ is the stress tensor and $\{\varepsilon\}$ the strains tensor:

$$\{\sigma\} = \begin{Bmatrix} \sigma_{xx} \\ \sigma_{yy} \\ \sigma_{zz} \\ \tau_{xy} \\ \tau_{yz} \\ \tau_{zx} \end{Bmatrix};\ \{\varepsilon\} = \begin{Bmatrix} \varepsilon_{xx} \\ \varepsilon_{yy} \\ \varepsilon_{zz} \\ 2\varepsilon_{xy} \\ 2\varepsilon_{yz} \\ 2\varepsilon_{zx} \end{Bmatrix} \tag{28}$$

Writing the Hooke law as follows:

$$\{\sigma\} = [D]\{\varepsilon\} \tag{29}$$

and of the differential relations between strains and finite deformations:

$$\{\varepsilon\} = [a]\{f\} = [a][N]\{\delta\} \tag{30}$$

where $[a]$ represents the differentiation operator [31] and, using Equations (13), (15) and (16), the internal work becomes [31]:

$$E_p = \frac{1}{2}\int_V \{\delta\}_L^T[N]^T[a]^T[D]^T[a][N]\{\delta\}dV = \frac{1}{2}\{\delta\}_L^T\left(\int_V [N]^T[a]^T[D]^T[a][N]dV\right)\{\delta\}_L \tag{31}$$

where $[k]$ is the classical stiffness matrix:

$$[k] = \int_V [N]^T[a]^T[D]^T[a][N]dV \tag{32}$$

If noted with $\{p\} = \{p(x,y,z)\}$, the vector of body forces per unit volume, then the external work of these is:

$$W = \int_V \{p(x,y,z)\}_L^T \{f\}_L dV = \left(\int_V \{p(x,y,z)\}_L^T [N] dV \right) \{\delta\}_L = \{q^*\}_L^T \{\delta\}_L \tag{33}$$

The concentrated forces $\{q\}_L$ in the nodes of the element will give an external work:

$$W^c = \{q\}_L^T \{\delta\}_L \tag{34}$$

The equations of motion are obtained by applying the GA equations previously presented:

$$\left\{ \frac{\partial E_a}{\partial \ddot{\delta}} \right\}_L - \{Q\}_L = 0 \tag{35}$$

By $\left\{ \frac{\partial E}{\partial \delta_e} \right\}$, we understand:

$$\left\{ \frac{\partial E}{\partial X} \right\} = \left\{ \begin{array}{c} \frac{\partial E}{\partial x_1} \\ \frac{\partial E}{\partial x_2} \\ \vdots \\ \frac{\partial E}{\partial x_n} \end{array} \right\} \text{where : } \{X\} = \left\{ \begin{array}{c} x_1 \\ x_2 \\ \vdots \\ x_n \end{array} \right\} \tag{36}$$

In our case:

$$E_a = E_{a0}(\dot{q}) + E_{a1}(\dot{q}, \ddot{q}) + E_{a2}(\ddot{q}) \tag{37}$$

and:

$$\{Q\}_L = [k]\{\delta\}_L + \{q\}_L + \{q^*\}_L \tag{38}$$

After a series of elementary calculations and rearranging of terms, we get the equations of motion for the finite element considered:

$$\frac{\partial E_{a2}}{\partial \left\{ \ddot{\delta} \right\}_L} = \left(\int_V \rho [N]^T [N] \, dV \right) \left\{ \ddot{\delta} \right\}_L = [m] \left\{ \ddot{\delta} \right\}_L \tag{39}$$

We shall note with $\left[N_{(1)} \right]$, the row i of matrix $[N]$ and with:

$$\begin{gathered} [m] = \int_V \rho [N]^T [N] \, dV = \int_V \rho \left[\begin{array}{ccc} N_{(1)}^T & N_{(2)}^T & N_{(3)}^T \end{array} \right] \left[\begin{array}{c} N_{(1)} \\ N_{(2)} \\ N_{(3)} \end{array} \right] dV = \\ = \int_V \rho \left(N_{(1)}^T N_{(1)} + N_{(2)}^T N_{(2)} + N_{(3)}^T N_{(3)} \right) dV = \\ \int_V \rho N_{(1)}^T N_{(1)} \, dV + \int_V \rho N_{(2)}^T N_{(2)} \, dV + \int_V \rho N_{(3)}^T N_{(3)} \, dV = [m_{11}] + [m_{22}] + [m_{33}] \end{gathered} \tag{40}$$

where:

$$\left[m_{ij} \right] = \int_V \rho N_{(i)}^T N_{(j)} \, dV \tag{41}$$

therefore:

$$[m_{11}] = \int_V \rho N_{(1)}^T N_{(1)} \, dV, \; [m_{22}] = \int_V \rho N_{(2)}^T N_{(2)} \, dV, \; [m_{33}] = \int_V \rho N_{(3)}^T N_{(3)} \, dV \tag{42}$$

$$\begin{array}{c}\frac{\partial E_{a1}}{\partial\{\ddot{\delta}\}_L}=\left(\int_V \rho[N]^T dV\right)\{\ddot{r}_O\}_L+\left(\int_V \rho[N]^T\ ([\varepsilon]_L+[\omega]_L[\omega]_L)\{r\}_L dV\right)+\\ \left(\int_V \rho[N]^T\ ([\varepsilon]_L+[\omega]_L[\omega]_L)[N]dV\right)\{\delta\}_L+2\left(\int_V \rho[N]^T\ [\omega]_L[N]dV\right)\{\dot{\delta}\}_L=\\ \left[m_O^i\right]\{\ddot{r}_O\}_L+\left\{q^i(\omega)\right\}_L+\left\{q^i(\varepsilon)\right\}_L+([k(\omega)]+[k(\varepsilon)])\{\delta\}_L+[c]\{\dot{\delta}\}_L\end{array} \tag{43}$$

where it is denoted:

$$\begin{array}{c}\left[m_O^i\right]=\int_V \rho[N]^T dV;\ \left\{q^i(\varepsilon)\right\}_L=\int_V \rho[N]^T[\varepsilon]_L\{r\}_L dV;\\ \left\{q^i(\omega)\right\}_L=\int_V \rho[N]^T[\omega]_L[\omega]_L\{r\}_L dV; [k(\varepsilon)]=\int_V \rho[N]^T\ [\varepsilon][N]dV;\\ [k(\varepsilon)]=\int_V \rho[N]^T\ [\omega]_L[\omega]_L[N]dV; [c]=\int_V \rho[N]^T\ [\omega]_L[N]dV;\\ \{m_{ix}\}=\int_V \rho\left[N_{(i)}\right]^T xdV\quad;\quad\left\{m_{iy}\right\}=\int_V \rho\left[N_{(i)}\right]^T ydV\quad;\quad\{m_{iz}\}=\int_V \rho\left[N_{(i)}\right]^T zdV\end{array} \tag{44}$$

The term E_{a0} does not contain $\{\ddot{\delta}\}_L$ and, consequently:

$$\frac{\partial E_{a0}}{\partial\{\ddot{\delta}\}_L}=0 \tag{45}$$

Finally, the motion equations can be written in the condensed form:

$$[m]\{\ddot{\delta}\}_L+[c]\{\dot{\delta}\}_L+([k]+[k(\varepsilon)]+[k(\omega)])\{\delta\}_L=\{q\}_L+\{q^*\}_L-\left\{q^i(\varepsilon)\right\}_L-\left\{q^i(\omega)\right\}_L-\left[m_O^i\right]\{\ddot{r}_O\}_L \tag{46}$$

These equations are related to the local system of coordinates. Similar formulas can be obtained, and we consider the global reference frame. The matrix coefficients can be calculated after choosing the shape functions and the independent nodal coordinates for expressing the displacement of a point.

5. Conclusions and Discussions

In the classical Lagrangian formalism, to determine the equations of motion for a three-dimensional finite element, the main step of the calculation is the Lagrangian, based essentially on the kinetic energy of the deformable finite element. The Lagrangian is:

$$L=E_c-E_p+W+W^c \tag{47}$$

where E_p, W, W^c are given, respectively, by the Relations (31), (33) and (34) and the kinetic energy has the expression:

$$\begin{array}{c}E_c=\frac{1}{2}\int_V \rho v_{M'}^2 dV=\frac{1}{2}\int_V \rho\{v_{M'}\}^T\{v_{M'}\}dV=\\ =\frac{1}{2}\int_V \rho\left(\{\dot{r}_O\}_G^T+\{r\}_L^T\left[\dot{R}\right]^T+\{\delta\}_L^T[N]^T\left[\dot{R}\right]^T+\{\dot{\delta}\}_L^T[N]^T[R]^T\right)x\\ \left(\{\dot{r}_O\}_G+\left[\dot{R}\right]\{r\}_L+\left[\dot{R}\right][N]\{\delta\}_L+\ [R][N]\{\dot{\delta}\}_L\right)dV=\\ \frac{1}{2}\int_V \rho\left(\{\dot{r}_O\}_G^T\{\dot{r}_O\}_G+2\{\dot{r}_O\}_G^T\left[\dot{R}\right]\{r\}_L+2\{\dot{r}_O\}_G^T\left[\dot{R}\right][N]\{\delta\}_L+2\{\dot{r}_O\}_G^T\ [R][N]\{\dot{\delta}\}_L\right)d\,V+\\ \frac{1}{2}\int_V \rho\left(\{r\}_L^T\left[\dot{R}\right]^T\left[\dot{R}\right]\{r\}_L+2\{r\}_L^T\left[\dot{R}\right]^T\left[\dot{R}\right][N]\{\delta\}_L+2\{r\}_L^T\left[\dot{R}\right]^T\ [R][N]\{\dot{\delta}\}_L\right)dV+\\ \frac{1}{2}\int_V \rho\left(\{\delta\}_L^T[N]^T\left[\dot{R}\right]^T\left[\dot{R}\right][N]\{\delta\}_L+2\{\delta\}_L^T[N]^T\left[\dot{R}\right]^T\ [R][N]\{\dot{\delta}\}_L+\ \{\dot{\delta}\}_L^T[N]^T[R]^T\ [R][N]\{\dot{\delta}\}_L\right)dV\end{array} \tag{48}$$

Lagrange's equations are:

$$\frac{d}{dt}\left\{\frac{\partial L}{\partial\dot{\delta}_e}\right\}-\left\{\frac{\partial L}{\partial\delta_e}\right\}=0 \tag{49}$$

Leaving aside the other derivatives, which have a low weight in the computing economy, the term most significant in this formulation is the term kinetic energy. In this expression, there are four terms containing the vector $\{\dot{\delta}\}_L$ and four terms containing the vector $\{\delta\}_L$. If we look at the expression of

Lagrange's equations, it follows that we will have four differentiation operations for $\left\{\partial L/\partial\dot{\delta}_e\right\}$, another four for $\{\partial L/\partial\delta_e\}$ and then the derivatives with respect to the time of the terms $\left\{\partial L/\partial\dot{\delta}_e\right\}$.

Using the Gibbs–Appell equations that have the form of Equation (31), we have to differentiate the energy of the accelerations given by the Relations (24)–(26) and (37). The role of kinetic energy is play in these equations by the energy of accelerations $\left\{\ddot{\delta}\right\}_L$. Note that of the three terms of kinetic energy, only two contain the vector of accelerations. The size E_{a2} has a single term that contains this vector and E_{a1} has four terms that contain the vector $\left\{\ddot{\delta}\right\}_L$ and which are differentiated. The total number of differentiation to be made in this case is five.

The conclusion from these two approaches is that the number of operations to be performed when using the Gibbs–Appell equations is lower than when Lagrange's method is applied. The use of this method has the advantage that we are familiar with the calculation of kinetic energy of a solid and to determine the energy of the accelerations it is sufficient to replace the speeds, in the kinetic energy formula, with the accelerations. The use of these equations will result in an economy in the computational effort we make to determine the equations of motion for a single finite element.

An application was made for a rod within a wind pumping mechanism (Figure 2). The mechanism proposed for the study is a mechanism with two degrees of freedom of type "kinematic chain closed by inertia" [32].

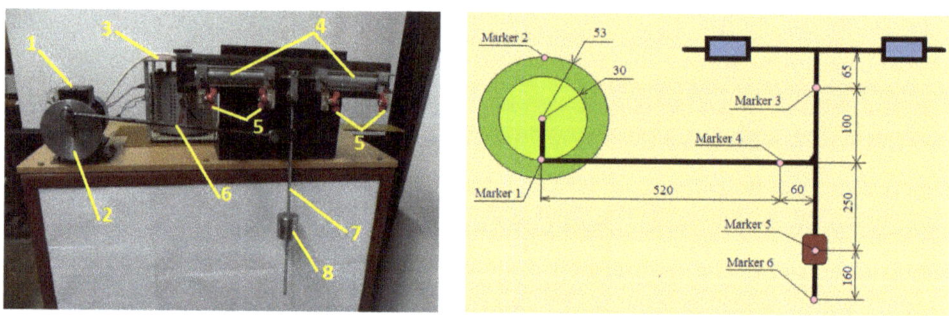

(a) (b)

Figure 2. (**a**) Experimental test stand and (**b**) sketch of the mechanism with the positioning of the markers (dimensions are in mm). 1—engine; 2—rotating disk; 3—invertor; 4—pumps; 5—valves; 6—rod; 7—pendulum rod; 8—weight.

Table 1 presents a comparison between the number of differentiation required when using Lagrange's method and the number of differentiation required if the Gibbs–Appell method is applied. It is noted that the number of these operations is reduced to less than half. For a large number of finite elements used, in the case of complex structures, the reduction can be significant and can reduce the computational effort. We mention that after obtaining the matrix coefficients of the system of differential equations, the next procedures in the two methods become identical, so also the necessary computer times.

Another problem that arises is the accuracy of the results. In the case of finite element analysis of a mechanism that can have large displacements, specific problems arise, very delicate, related to the accuracy of the results obtained and the correctness of the models used. During the present paper, we did not deal with these aspects, especially interesting and exciting. By themselves, these aspects define a field of research. The purpose of the paper was to provide an alternative way of writing the equations of motion, which would have the advantage of a smaller number of operations to perform.

The test mechanism on which the considerations were made, presented in Figure 2, can be considered, at this level as regards the accuracy of the results, accessible through the classical methods. In Figures 3 and 4, we presented the space of the possible positions that can be occupied by the elements of this mechanism, to justify our hypothesis.

Table 1. Comparison between the Lagrange and Gibbs–Appell methods. Number of differentiations.

Number of Finite Elements	Lagrange	Gibbs–Appell
5	288	120
10	528	220
15	768	320
20	1008	420
25	1248	520
30	1488	620
40	1968	820

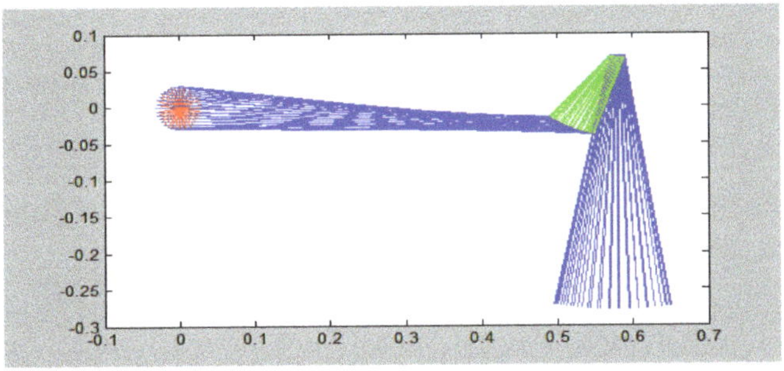

Figure 3. Sketch of the motion of the two degree of freedom system for the angular speed 16 rad/s.

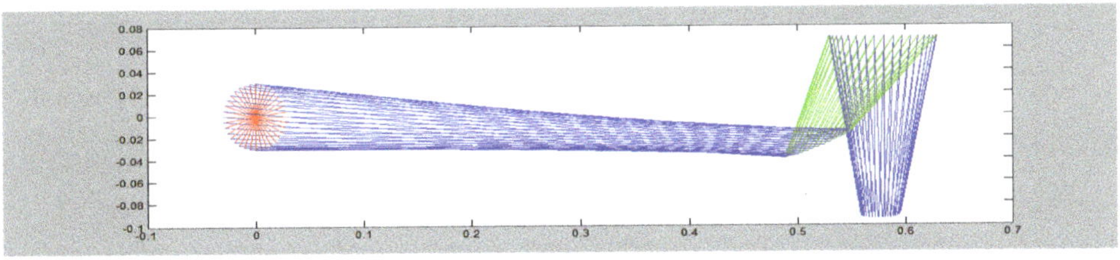

Figure 4. Sketch of the motion of the two degree of freedom system for the angular speed 19 rad/s.

Author Contributions: Conceptualization, S.V. and I.N.; methodology, S.V. and M.L.S.; software, M.M.; validation, S.V., I.N. and M.M.; formal analysis, S.V. and M.L.S.; investigation, M.L.S.; resources, S.V.; writing—original draft preparation, S.V. and I.N.; writing—review and editing, M.L.S.; visualization, M.M.; supervision, I.N. and M.M.; project administration, S.V.; funding acquisition, S.V., I.N., M.M. and M.L.S. All authors have read and agreed to the published version of the manuscript.

Funding: This research received no external funding.

Conflicts of Interest: The authors declare no conflict of interest.

References

1. Erdman, A.G.; Sandor, G.N.; Oakberg, A. A General Method for Kineto-Elastodynamic Analysis and Synthesis of Mechanism. Journal of Engineering for Industry. *ASME Trans.* **1972**, *94*, 1193.
2. Bagci, C. Elastodynamic Response of Mechanical Systems using Matrix Exponential Mode Uncoupling and Incremental Forcing Techniques with Finite Element Method. In Proceedings of the Sixth Word Congress on Theory of Machines and Mechanisms, New Delhi, India, 15–20 December 1983; Volume 472.
3. Bahgat, B.M.; Willmert, K.D. Finite Element Vibrational Analysis of Planar Mechanisms. *Mech. Mach. Theory* **1976**, *11*, 47–71. [CrossRef]

4. Cleghorn, W.L.; Fenton, E.G.; Tabarrok, K.B. Finite Element Analysis of High-Speed Flexible Mechanisms. *Mech. Mach. Theory* **1981**, *16*, 407–424. [CrossRef]
5. Christensen, E.R.; Lee, S.W. Nonlinear finite element modelling of the dynamics of unrestrained flexible structures. *Comput. Struct* **1986**, *23*, 819–829. [CrossRef]
6. Midha, A.; Erdman, A.G.; Frohrib, D.A. Finite element approach to mathematical modeling of high-speed elastic linkages. *Mech. Mach. Theory* **1978**, *13*, 603–618. [CrossRef]
7. Nath, P.K.; Ghosh, A. Steady-state response of mechanism with elastic links by finite element methods. *Mech. Mach. Theory* **1980**, *15*, 199–211. [CrossRef]
8. Vlase, S.; Marin, M.; Öchsner, A.; Scutaru, M.L. Motion equation for a flexible one-dimensional element used in the dynamical analysis of a multibody system. *Contin. Mech. Thermodyn.* **2019**. [CrossRef]
9. Vlase, S. Dynamical Response of a Multibody System with Flexible Elements with a General Three-Dimensional Motion. *Rom. J. Phys.* **2012**, *57*, 676–693.
10. Vlase, S.; Dănăşel, C.; Scutaru, M.L.; Mihălcică, M. Finite Element Analysis of a Two-Dimensional Linear Elastic Systems with a Plane "rigid Motion. *Rom. J. Phys.* **2014**, *59*, 476–487.
11. Deü, J.-F.; Galucio, A.C.; Ohayon, R. Dynamic responses of flexible-link mechanisms with passive/active damping treatment. *Comput. Struct.* **2008**, *86*, 258–265. [CrossRef]
12. Hou, W.; Zhang, X. Dynamic analysis of flexible linkage mechanisms under uniform temperature change. *J. Sound Vib.* **2009**, *319*, 570–592. [CrossRef]
13. Neto, M.A.; Ambrósio, J.A.C.; Leal, R.P. Composite materials in flexible multibody systems. *Comput. Methods Appl. Mech. Eng.* **2006**, *195*, 6860–6873. [CrossRef]
14. Piras, G.; Cleghorn, W.L.; Mills, J.K. Dynamic finite-element analysis of a planar high-speed, high-precision parallel manipulator with flexible links. *Mech. Mach. Theory* **2005**, *40*, 849–862. [CrossRef]
15. Shi, Y.M.; Li, Z.F.; Hua, H.X.; Fu, Z.F.; Liu, T.X. The Modeling and Vibration Control of Beams with Active Constrained Layer Damping. *J. Sound Vib.* **2001**, *245*, 785–800. [CrossRef]
16. Zhang, X.; Erdman, A.G. Dynamic responses of flexible linkage mechanisms with viscoelastic constrained layer damping treatment. *Comput. Struct.* **2001**, *79*, 1265–1274. [CrossRef]
17. Zhang, X.; Lu, J.; Shen, Y. Simultaneous optimal structure and control design of flexible linkage mechanism for noise attenuation. *J. Sound Vib.* **2007**, *299*, 1124–1133.
18. Gibbs, J.W. On the fundamental formulae of dynamics. *Am. J. Math.* **1879**, *2*, 49–64. [CrossRef]
19. Appell, P. Sur une forme générale des equations de la dynamique. *C.R. Acad. Sci. Paris* **1899**, *129*, 459–460.
20. Mirtaheri, S.M.; Zohoor, H. The Explicit Gibbs-Appell Equations of Motion for Rigid-Body Constrained Mechanical System. In *Book Series: RSI International Conference on Robotics and Mechatronics ICRoM*; IEEE: Piscataway, NJ, USA, 2018; pp. 304–309.
21. Mehrjooee, O.; Dehkordi, S.F.; Korayem, M.H. Dynamic modeling and extended bifurcation analysis of flexible-link manipulator. *Mech. Based Des. Struct. Mach.* **2019**, 1–24. [CrossRef]
22. Korayem, M.H.; Dehkordi, S.F. Motion equations of cooperative multi flexible mobile manipulator via recursive Gibbs-Appell formulation. *Appl. Math. Model.* **2019**, *65*, 443–463. [CrossRef]
23. Shafei, A.M.; Shafei, H.R. A systematic method for the hybrid dynamic modeling of open kinematic chains confined in a closed environment. *Multibody Syst. Dyn.* **2017**, *38*, 21–42. [CrossRef]
24. Amini, S.; Dehkordi, S.F.; Fahraji, S.H. Motion equation derivation and tip-over evaluations for K mobile manipulators with the consideration of motors mass by the use of Gibbs-Appell formulation. In Proceedings of the 5th RSI International Conference on Robotics and Mechatronics (IcRoM), Tehran, Iran, 25–27 October 2017.
25. Korayem, M.H.; Dehkordi, S.F. Derivation of dynamic equation of viscoelastic manipulator with revolute-prismatic joint using recursive Gibbs-Appell formulation. *Nonlinear Dyn.* **2017**, *89*, 2041–2064. [CrossRef]
26. Shafei, A.M.; Korayem, M.H. Theoretical and experimental study of dynamic load-carrying capacity for flexible robotic arms in point-to-point motion. *Optim. Control Appl. Methods* **2017**, *38*, 963–972. [CrossRef]
27. Negrean, I.; Kacso, K.; Schonstein, C.; Duca, A. *Mechanics, Theory and Applications*; UTPRESS: Cluj-Napoca, Romania, 2012.
28. Negrean, I.; Crișan, A.-D. Synthesis on the Acceleration Energies in the Advanced Mechanics of the Multibody Systems. *Symmetry* **2019**, *11*, 1077. [CrossRef]
29. Negrean, I.; Crișan, A.-D.; Vlase, S. A New Approach in Analytical Dynamics of Mechanical Systems. *Symmetry* **2020**, *12*, 95. [CrossRef]

30. Vlase, S. Elimination of Lagrangian Multipliers. *Mech. Res. Commun.* **1987**, *14*, 17. [CrossRef]
31. Massonet, C.; Deprez, G.; Maquoi, R.; Muller, R.; Fonder, G. *Calcul des Structures sur Ordinateur*; EYROLLES: Paris, France, 1972.
32. Scutaru, M.L.; Mitrica, B. Dynamical Analysis of the Mechanical System with Two Degrees of Freedom Applied to the Transmission of the Wind Turbine. *Math. Probl. Eng.* **2016**, *2016*, 3821083. [CrossRef]

symmetry

MDPI

Article

The Properties of a Decile-Based Statistic to Measure Symmetry and Asymmetry

Mohammad Reza Mahmoudi [1,2], Roya Nasirzadeh [2], Dumitru Baleanu [3,4] and Kim-Hung Pho [5,*]

1 Institute of Research and Development, Duy Tan University, Da Nang 550000, Vietnam; mohammadrezamahmoudi@duytan.edu.vn or mahmoudi.m.r@fasau.ac.ir
2 Department of Statistics, Faculty of Science, Fasa University, Fasa, Fars 7461686131, Iran; nasirzadeh.roya@fasau.ac.ir
3 Department of Mathematics, Faculty of Art and Sciences, Cankaya University, Ankara 06530, Turkey; dumitru@cankaya.edu.tr
4 Institute of Space Sciences, 077125 Bucharest, Romania
5 Fractional Calculus, Optimization and Algebra Research Group, Faculty of Mathematics and Statistics, Ton Duc Thang University, Ho Chi Minh City 72915, Vietnam
* Correspondence: phokimhung@tdtu.edu.vn

Received: 20 January 2020; Accepted: 4 February 2020; Published: 18 February 2020

Abstract: This paper studies a simple skewness measure to detect symmetry and asymmetry in samples. The statistic can be obviously applied with only three short central tendencies; i.e., the first and ninth deciles, and the median. The strength of the statistic to find symmetry and asymmetry is studied by employing numerous Monte Carlo simulations and is compared with some alternative measures by applying some simulation studies. The results show that the performance of this statistic is generally good in the simulation.

Keywords: symmetry; asymmetry; measure of skewness; decile; Monte Carlo algorithm

1. Introduction

In scientific studies, the researchers can summarize a given dataset using descriptive statistics. The descriptive statistics contain three known tendencies: central tendencies, dispersion tendencies and shape tendencies [1]. The central and dispersion tendencies, such as mean, median, standard deviation and variance deal with the convenience of the dataset [1–5]. The shape tendencies, such as skewness and kurtosis, are related to the distribution of dataset [6–8]. These measures which may be utilized in divergent disciplines consist of the tests of normality and of the lustiness for normal theoretical procedures. Skewness is often utilized to reference to symmetry. Nevertheless, symmetry is not often perspicuously defined, and it is thought that everybody knows it. There are some definitions about symmetry relying on the disciplines that it is utilized in. In literature, any statement related to the symmetry of a structure has to be done with reference to some rules of symmetry—a score, a line or an axis [9]. In the statistical inference, the meaningful score or axis is taken as the center of a distribution. There are several measures employed to quantify the degree of skewness of a distribution. Assume that μ, m, M, σ, μ_3, Q_1 and Q_3, are the mean; median; mode; standard deviation; third centered moment; and the first and the third quartiles, respectively. The statistics introduced for measuring the skewness are Pearson's coefficient of skewness:

$$SK_P = \frac{\mu - M}{\sigma} \tag{1}$$

Pearson's second coefficient of skewness:

$$SK_{P2} = \frac{3(\mu - m)}{\sigma} \tag{2}$$

Yule's coefficient of skewness:

$$SK_{Y} = \frac{(\mu - m)}{\sigma} \tag{3}$$

the standardized third central moment:

$$\gamma_1 = \frac{\mu_3}{\sigma^3} \tag{4}$$

Bowley's coefficient of skewness:

$$SK_B = \frac{Q_3 + Q_1 - 2m}{Q_3 - Q_1} \tag{5}$$

and three Galip's coefficients of skewness:

$$SK_{G1} = \frac{X_{Max} + X_{min} - 2M}{X_{Max} - X_{min}} \tag{6}$$

$$SK_{G2} = \frac{X_{Max} + X_{min} - 2m}{X_{Max} - X_{min}} \tag{7}$$

$$SK_{G3} = \frac{X_{Max} + X_{min} - 2\mu}{X_{Max} - X_{min}} \tag{8}$$

[9–17].

Although there are numerous different measures, and practical elongations of the above coefficients were proposed afterward, the original measures are still employed to this day, especially γ_1(or its variants). It is largely utilized in statistical calculation software.

When we face a dataset containing outliers, we need a measure that can carefully consider these outliers. Therefore, probably, the measures that are based on the extreme values (max and min) such as three Galip's coefficients of skewness; are based on the first and the last quartiles (Q_1 and Q_3) such as Bowley's coefficient of skewness; or are based on the first and the last deciles (D_1 and D_9), should be more effective than other methods. The previous studies indicated that the three Galip's coefficients of skewness had the most power to detect symmetry and asymmetry. But the Bowley's coefficient of skewness acted not so well. There is no deep study about the definition of skewness based on deciles and the comparison between them and other alternatives.

In this work, at first, we consider the definition of skewness based on deciles and then study its asymptotic properties, similar to the approach that was applied in [18–23]. Finally, the power of the considered statistic to detect symmetry and asymmetry is compared with the powers of other measures of skewness.

2. Decile-Based Skewness

Let $X_1, \ldots, X_n$ be a sample from a distribution F on the real line, and we suppose that F is continuous so that all observations are distinct with probability one. We may then arrange the observations in increasing order without ties, $X_{(1)} < \ldots < X_{(n)}$. These variables are called the order statistics, where $X_{(k)}$ is the kth order statistic. For $0 < p < 1$, the pth quantile of F is defined as $x_p = F^{-1}(p)$ and the corresponding sample quantile is defined as $X_{(k)}$ where $k = \lceil np \rceil$, the ceiling of (the smallest integer greater than or equal to np). Let D_1 and D_9 be the first and nine sample deciles (0.1 and 0.9 quantiles), respectively. We consider our statistic for measuring the skewness by

$$SK = \frac{(D_9 - m) - (m - D_1)}{D_9 - D_1} \tag{9}$$

In the following, the asymptotic distribution of the proposed statistic is explored.

Lemma 1. *Let $U_1, \ldots, U_n$ be independent, identically distributed (iid in short) random variables from $U(0,1)$ and $U_{(1)} < \ldots < U_{(n)}$, which are order statistics of $U_1, \ldots, U_n$. If $n \to \infty$, then*

$$\sqrt{n}\begin{pmatrix} U_{\lceil np_1 \rceil} - p_1 \\ U_{\lceil np_2 \rceil} - p_2 \\ U_{\lceil np_3 \rceil} - p_3 \end{pmatrix} \xrightarrow{D} N(\mathbf{0}, \mathbf{\Sigma}) \tag{10}$$

where $0 < p_1 < p_2 < p_3 < 1$, and

$$\mathbf{\Sigma} = \begin{bmatrix} p_1(1-p_1) & p_1(1-p_2) & p_1(1-p_3) \\ p_1(1-p_2) & p_2(1-p_2) & p_2(1-p_3) \\ p_1(1-p_3) & p_2(1-p_3) & p_3(1-p_3) \end{bmatrix} \tag{11}$$

Proof. Assume that $Y_1, Y_2, \ldots$ are iid exponential variables with mean 1 and $S_j = \sum_{i=1}^{j} Y_i$. Additionally, assume that $\sqrt{n}\left(\frac{k_1}{n} - p_1\right) \to 0$, $\sqrt{n}\left(\frac{k_2}{n} - p_2\right) \to 0$ and $\sqrt{n}\left(\frac{k_3}{n} - p_3\right) \to 0$ as k_1, k_2, k_3, and $n \to \infty$. Then by the extension of the results given in [24],

$$\sqrt{n+1}\begin{bmatrix} \frac{1}{n+1}S_{k_1} - p_1 \\ \frac{1}{n+1}\left(S_{k_2} - S_{k_1}\right) - (p_2 - p_1) \\ \frac{1}{n+1}\left(S_{k_3} - S_{k_2}\right) - (p_3 - p_2) \\ \frac{1}{n+1}\left(S_{n+1} - S_{k_3}\right) - (1 - p_3) \end{bmatrix} \xrightarrow{D} N(\mathbf{0}, \mathbf{\Sigma_1}),$$

such that

$$\mathbf{\Sigma}_1 = \begin{bmatrix} p_1 & 0 & 0 & 0 \\ 0 & p_2 - p_1 & 0 & 0 \\ 0 & 0 & p_3 - p_2 & 0 \\ 0 & 0 & 0 & 1 - p_3 \end{bmatrix}$$

Take $g(x_1, x_2, x_3, x_4) = \frac{1}{x_1 + x_2 + x_3 + x_4}[x_1, x_1 + x_2, x_1 + x_2 + x_3]'$; then, by Cramer's theorem [24],

$$\sqrt{n}\begin{pmatrix} \frac{S_{k_1}}{S_{n+1}} - p_1 \\ \frac{S_{k_2}}{S_{n+1}} - p_2 \\ \frac{S_{k_3}}{S_{n+1}} - p_3 \end{pmatrix} \xrightarrow{D} N(\mathbf{0}, \mathbf{\Sigma})$$

Finally, the proof is completed with the reality that the distribution of $\left(\frac{S_{k_1}}{S_{n+1}}, \frac{S_{k_2}}{S_{n+1}}, \frac{S_{k_3}}{S_{n+1}}\right)'$ given S_{n+1} is the same as the distribution of $\left(U_{(k_1)}, U_{(k_2)}, U_{(k_3)}\right)'$. □

Corollary 1. *Let $X_1, \ldots, X_n$ be iid random variables with density and distribution functions f and F, respectively. Additionally, assume that $f(x)$ is continuous and positive in a neighborhood of the quantiles x_{p_1}, x_{p_2} and x_{p_3} with $p_1 < p_2 < p_3$; then,*

$$\sqrt{n}\begin{pmatrix} X_{\lceil np_1 \rceil} - x_{p_1} \\ X_{\lceil np_2 \rceil} - x_{p_2} \\ X_{\lceil np_3 \rceil} - x_{p_3} \end{pmatrix} \xrightarrow{D} N(\mathbf{0}, \mathbf{\Sigma}^*) \tag{12}$$

where

$$\Sigma^* = \begin{bmatrix} \frac{p_1(1-p_1)}{f^2(x_{p_1})} & \frac{p_1(1-p_2)}{f(x_{p_1})f(x_{p_2})} & \frac{p_1(1-p_3)}{f(x_{p_1})f(x_{p_3})} \\ \frac{p_1(1-p_2)}{f(x_{p_1})f(x_{p_2})} & \frac{p_2(1-p_2)}{f^2(x_{p_2})} & \frac{p_2(1-p_3)}{f(x_{p_2})f(x_{p_3})} \\ \frac{p_1(1-p_3)}{f(x_{p_1})f(x_{p_3})} & \frac{p_2(1-p_3)}{f(x_{p_2})f(x_{p_3})} & \frac{p_3(1-p_3)}{f^2(x_{p_3})} \end{bmatrix} \tag{13}$$

Proof. By applying the transformation $g(y_1, y_2, y_3) = \left(F^{-1}(y_1), F^{-1}(y_2), F^{-1}(y_3)\right)'$ to the variables $\left(U_{\lceil np_1 \rceil} - p_1,\ U_{\lceil np_2 \rceil} - p_2,\ U_{\lceil np_3 \rceil} - p_3\right)$ in Lemma 1, the proof will be completed. Be careful that the derivation of g is

$$\dot{g}(y_1, y_2, y_3) = \begin{bmatrix} \frac{1}{f(F^{-1}(y_1))} & 0 & 0 \\ 0 & \frac{1}{f(F^{-1}(y_2))} & 0 \\ 0 & 0 & \frac{1}{f(F^{-1}(y_3))} \end{bmatrix}.$$

□

The asymptotic distribution of SK is provided in the following theorem. This is our major contribution. It is also necessary to infer the skewness of population.

Theorem 1. *Let$X_1, \ldots, X_n$ be iid random variables with density function f. Additionally, assume that $f(x)$ is continuous and positive in a neighborhood of the quantiles $x_{0.1}, x_{0.5}$ and $x_{0.9}$. Then, the asymptotic distribution of the proposed statistic can be illustrated by*

$$T_n = \sqrt{n}\left(SK - \frac{x_{0.9} + x_{0.1} - 2x_{0.5}}{x_{0.9} - x_{0.1}}\right) \xrightarrow{D} N\left(0, \sigma^2\right)$$

where

$$\sigma^2 = \frac{1}{(x_{0.9}-x_{0.1})^4} \left[\frac{0.36(x_{0.9}-x_{0.5})^2}{f^2(x_{0.1})} + \frac{(x_{0.9}-x_{0.1})^2}{f^2(x_{0.5})} + \frac{0.36(x_{0.5}-x_{0.1})^2}{f^2(x_{0.9})} - \frac{0.4(x_{0.9}-x_{0.1})(x_{0.9}-x_{0.5})}{f(x_{0.1})f(x_{0.5})} + \frac{0.08(x_{0.5}-x_{0.1})(x_{0.9}-x_{0.5})}{f(x_{0.1})f(x_{0.9})} - \frac{0.4(x_{0.9}-x_{0.1})(x_{0.5}-x_{0.1})}{f(x_{0.5})f(x_{0.9})} \right] \tag{14}$$

Proof. The proof is simply achieved using Cramer's theorem [24] and taking $g(x_1, x_2, x_3) = \frac{x_1 - 2x_2 + x_3}{x_3 - x_1}$.
□

Corollary 2. *Let$X_1, \ldots, X_n$ be iid random variables from $U(0,1)$; then, the asymptotic distribution of the proposed statistic is given by*

$$\sqrt{n}(SK - 0) \xrightarrow{D} N(0, 1.25) \tag{15}$$

These results can be employed to build an asymptotical confidence interval and to check the hypothesis.

2.1. Asymptotic Confidence Interval

Now, T_n can be utilized as a pivotal quantity to build a confidence interval asymptotic to a population's skewness,

$$\left(SK - \frac{\hat{\sigma}}{\sqrt{n}} Z_{\alpha/2},\ SK + \frac{\hat{\sigma}}{\sqrt{n}} Z_{\alpha/2}\right) \tag{16}$$

where

$$\hat{\sigma}^2 = \frac{1}{(D_9-D_1)^4} \left[\frac{0.36(D_9-m)^2}{f^2(D_1)} + \frac{(D_9-D_1)^2}{f^2(m)} + \frac{0.36(m-D_1)^2}{f^2(D_9)} - \frac{0.4(D_9-D_1)(D_9-m)}{f(D_1)f(m)} + \frac{0.08(m-D_1)(D_9-m)}{f(D_1)f(D_9)} - \frac{0.4(D_9-D_1)(m-D_1)}{f(m)f(D_9)} \right] \quad (17)$$

2.2. Hypothesis Testing

Hypothesis testing related to *skewness* is a crucial issue in practical application. For instance, the assumption $Skewness = 0$ is tantamount to the symmetry. Generally, to test $H_0 : Skewness = \gamma_0$, the test statistic can be

$$T_0 = \sqrt{n}\left(\frac{SK - \gamma_0}{\hat{\sigma}}\right) \quad (18)$$

Similar to the methodology provided in Theorem 1, it can prove that with the null hypothesis, T_0 has, asymptotically, standard normal distribution.

3. Asymptotic Properties of the Proposed Statistic

In this part, many data sets are drawn to analyze the performance of the proposed approach, for distinct symmetric distributions and divergent sample sizes. Firstly, we checked that the given CI and test statistic are truly the asymptotic CI and test statistic. For every parameter, the experiential coverage probability (percentage of runs for which the given CI contains zero (true skewness)) was calculated by relying on 10,000 repetitions using *statistical R 3.6.2 and SPSS 25* software. In addition, for each repetition, the value of the given test statistic is presented and normal Q–Q plots of the given test statistic are provided. The Shapiro-Wilk's normality test is used to confirm the normality of the given test statistic. The experiential coverage probabilities for divergent parameters are illustrated as in Table 1.

Table 1. The experiential coverage probability of the proposed confidence interval.

Distribution	n					
	50	75	100	200	500	1000
Normal (1,5)	0.9732	0.9734	0.9743	0.9744	0.975	0.9761
t(10)	0.9916	0.9928	0.9934	0.9939	0.9942	0.9947
U(0,1)	0.9485	0.949	0.9491	0.9502	0.9521	0.9569

The results show that the experiential coverage probability of proposed approach is more than nominal level (0.95), especially when the sample sizes grow. In the other hand, we can admit the given CI as the asymptotic CI for the skewness of population. Figure 1 and Table 2 show the Q–Q plots for the standard normal distribution and the results of Shapiro-Wilk's normality test in the test statistic, respectively.

Table 2. Shapiro-Wilk's normality test p-value for the given test statistic.

Distribution	n					
	50	75	100	200	500	1000
Normal (1,5)	0.7131	0.7174	0.7899	0.8436	0.9077	0.9213
t(10)	0.433	0.6515	0.781	0.8317	0.9603	0.9945
U(0,1)	0.3144	0.5566	0.6034	0.6219	0.8249	0.9488

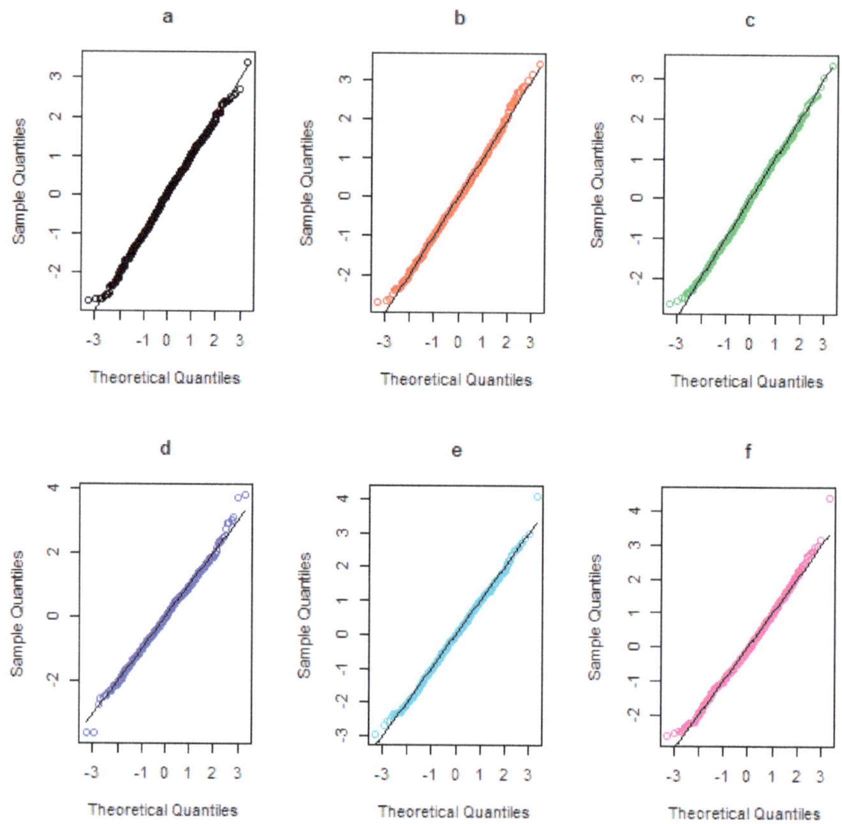

Figure 1. The Q–Q plots versus standard normal distribution. Normal distribution: $n = 50$ (**a**), $n = 1000$ (**b**). *t distribution:* $n = 50$ (**c**), $n = 1000$ (**d**). Uniform distribution: $n = 50$ (**e**), $n = 1000$ (**f**).

It can be then seen that the asymptotic properties are relatively satisfied in all situations (p-value is greater than 5%). Thereafter, it can be seen that our approach is a good choice to build a CI and execute hypothesis testing for the skewness of a population.

4. Comparison with Alternative Measures

To check the performances of the considered statistic, its power to detect asymmetry is compared with the conventional measures of skewness by employing a Monte Carlo simulation. As in Section 3, numerous data sets were drawn to check the performances of the measures, for different asymmetric distributions and different sample sizes using R software. For this purpose, we generated 10,000 samples of size $n = 10,\ 20,\ 50$, from a chi-square distribution with m degrees of freedom, $\left(\chi^2(m)\right)$. We considered three cases: extremely skewed ($m = 1$), moderately skewed ($m = 5$) and slightly skewed ($m = 40$). The powers (at 5% significant level) of different measures to detect asymmetry are summarized in Table 3.

Table 3. The powers of different measures to detect skewness.

Distribution	Measure	n		
		10	**20**	**50**
Extremely Skewed	SK	0.798	0.989	1.000
	γ_1	0.687	0.942	1.000
	SK_{G3}	0.817	0.991	1.000
	SK_{G2}	0.834	0.992	1.000
	SK_{G1}	0.461	0.831	0.997
	SK_P	0.200	0.151	0.130
	SK_Y	0.616	0.869	0.999
	SK_{P2}	0.616	0.869	0.999
	SK_B	0.260	0.403	0.711
Moderately Skewed	SK	0.318	0.597	0.945
	γ_1	0.297	0.530	0.889
	SK_{G3}	0.321	0.564	0.911
	SK_{G2}	0.318	0.623	0.941
	SK_{G1}	0.145	0.397	0.814
	SK_P	0.132	0.108	0.100
	SK_Y	0.207	0.344	0.651
	SK_{P2}	0.207	0.344	0.651
	SK_B	0.123	0.156	0.224
Slightly Skewed	SK	0.144	0.163	0.289
	γ_1	0.129	0.165	0.288
	SK_{G3}	0.135	0.175	0.284
	SK_{G2}	0.143	0.153	0.282
	SK_{G1}	0.116	0.136	0.252
	SK_P	0.106	0.103	0.119
	SK_Y	0.120	0.116	0.180
	SK_{P2}	0.120	0.116	0.180
	SK_B	0.117	0.116	0.135

As preliminary results, based on the maximum power, it can be observed that the performances of SK, γ_1 , SK_{G1}, SK_{G2} and SK_{G3} are approximately similar and are more powerful than other methods for all simulated datasets, and are therefore are very promising. The performances of SK_P, SK_{P2} and SK_Y are approximately similar and have the next best ranks, while SK_B has the worst performance in all situations. In general, the measures that are based on the extreme values (maximum and minimum), such as three Galip's coefficients of skewness, and those based on the first and the last deciles (D_1 and D_9), are more effective than other methods, because of their better performances and easy calculations.

5. Discussion

In this work, at first, we considered the definition of skewness based on deciles, and then studied its asymptotic properties. The results showed that the experiential coverage probability of this measure was more than nominal level (0.95), especially when the sample size was increased. The Q–Q plots versus the standard normal distribution and the results of Shapiro-Wilk's normality test verified the theoretical asymptotic properties. Finally, the power of the considered statistic to detect symmetry and asymmetry was compared with the powers of other measures of skewness. The power study indicated that the performances of decile-based measure and three Galip's coefficients of skewness were approximately similar, and were more powerful than other methods for all simulated datasets, and are therefore are promising for application in practice.

6. Conclusions

We presented a simple measure to find skewness in patterns. The new measure relies on a new definition of skewness that contains many outstanding advantages. The proposed coefficient of skewness could be obviously calculated with only three short statistics; i.e., the first and nine deacons and the median. The strength of the proposed statistic to find symmetry and asymmetry was studied by employing numerous Monte Carlo simulations. The results show that the performance of new statistic is generally very good in the simulation. There are many definitions to describe symmetry and asymmetry. To investigate the skewness in datasets including outliers, we should use the measures that consider the effects of outliers. Therefore, probably, the measures that are based on the extreme values (maximum and minimum), such as three Galip's coefficients of skewness; those based on the first and the last quartiles (Q_1 and Q_3), such as Bowley's coefficient of skewness; and those based on the first and the last deciles (D_1 and D_9), are candidates for application. Other studies showed that Galip's coefficients of skewness are more powerful for detecting symmetry and asymmetry. There is no deep study about the definition of skewness based on deciles and a comparison between them and other alternatives. In this work, at first, we considered the definition of skewness based on deciles and then studied its asymptotic properties. Finally, the power of the considered statistic to detect symmetry and asymmetry was compared with the powers of other measures of skewness. For future works, we suggest readers to use a definition of skewness based on combinations of more deciles, not only the first and the ninth deciles. We think this combination will improve the detection of symmetry and asymmetry.

Author Contributions: Conceptualization, M.R.M., R.N., D.B. and K.-H.P.; data curation, M.R.M.; formal analysis, M.R.M., R.N. and K.-H.P.; investigation, M.R.M., R.N. and D.B.; methodology, M.R.M. and K.-H.P.; project administration, D.B.; supervision, M.R.M.; validation, M.R.M.; visualization, M.R.M.; writing—original draft, M.R.M. and R.N.; writing—review and editing, D.B. and K.-H.P. All authors have read and agreed to the published version of the manuscript.

Funding: This research received no external funding.

Conflicts of Interest: The authors declare no conflict of interest.

References

1. Sprinthall, R.C.; Fisk, S.T. *Basic Statistical Analysis*; Prentice Hall: Englewood Cliffs, NJ, USA, 1990.
2. Manikandan, S. Measures of central tendency: Median and mode. *J. Pharmacol. Pharm.* **2011**, *2*, 214. [CrossRef]
3. Weisberg, H.; Weisberg, H.F. *Central Tendency and Variability*; Sage: Thousand Oaks, CA, USA, 1992.
4. Deshpande, S.; Gogtay, N.J.; Thatte, U.M. Measures of central tendency and dispersion. *J. Assoc. Physicians India* **2016**, *64*, 64–66.
5. Manikandan, S. Measures of dispersion. *J. Pharmacol. Pharm.* **2011**, *2*, 315. [CrossRef]
6. Kim, T.H.; White, H. On more robust estimation of skewness and kurtosis. *Financ. Res. Lett.* **2004**, *1*, 56–73. [CrossRef]
7. Oja, H. On location, scale, skewness and kurtosis of univariate distributions. *Scand J. Stat.* **1981**, *1*, 154–168.
8. Wilkins, J.E. A note on skewness and kurtosis. *Ann. Math Stat.* **1944**, *15*, 333–335. [CrossRef]
9. Murphy, E.A. Skewness and asymmetry of distributions. *Metamedicine* **1982**, *3*, 87–99. [CrossRef]
10. Arnold, B.C.; Groeneveld, R.A. Skewness and kurtosis orderings: An introduction. *Lect. Notes Monogr. Ser.* **1992**, *22*, 17–24.
11. Arnold, B.C.; Groeneveld, R.A. Measuring skewness with respect to the mode. *Am. Stat.* **1995**, *49*, 34–38.
12. Doane, D.P.; Seward, L.E. Measuring Skewness: A Forgotten Statistic? *J. Stat. Educ.* **2011**, *19*, 1–18. [CrossRef]

13. García, V.J.; Martel, M.; Vázquez-Polo, F.J. Complementary information for skewness measures. *Stat. Neerl* **2015**, *69*, 442–459. [CrossRef]
14. Groeneveld, R.A.; Meeden, G. Measuring skewness and kurtosis. *Statistician* **1984**, *33*, 391–399. [CrossRef]
15. Mahmoudi, M.R.; Nasirzadeh, R.; Mohammadi, M. On the Ratio of Two Independent Skewnesses. *Commun. Stat. Theory Methods* **2019**, *48*, 1721–1727. [CrossRef]
16. Tabor, J. Investigating the investigative task: Testing for skewness–An investigation of different test statistics and their power to detect skewness. *J. Stat. Educ.* **2010**, *18*, 1–13. [CrossRef]
17. Tajuddin, I.H. A simple measure of skewness. *Stat. Neerl* **1996**, *50*, 362–366. [CrossRef]
18. Haghbin, H.; Mahmoudi, M.R.; Shishebor, Z. Large Sample Inference on the Ratio of Two Independent. Binomial Proportions. *J. Math. Ext.* **2011**, *5*, 87–95.
19. Mahmoudi, M.R.; Mahmoodi, M. Inferrence on the Ratio of Variances of Two Independent Populations. *J. Math. Ext.* **2014**, *7*, 83–91.
20. Mahmoudi, M.R.; Mahmoodi, M. Inferrence on the Ratio of Correlations of Two Independent Populations. *J. Math. Ext.* **2014**, *7*, 71–82.
21. Mahmouudi, M.R.; Maleki, M.; Pak, A. Testing the Difference between Two Independent Time Series Models. *Iran. J. Sci. Technol. Trans. A Sci.* **2017**, *41*, 665–669. [CrossRef]
22. Mahmoudi, M.R.; Mahmoudi, M.; Nahavandi, E. Testing the Difference between Two Independent Regression Models. *Commun. Stat. Theory Methods* **2016**, *45*, 6284–6289. [CrossRef]
23. Mahmoudi, M.R.; Behboodian, J.; Maleki, M. Large Sample Inference about the Ratio of Means in Two Independent Populations. *J. Stat. Theory Appl.* **2017**, *16*, 366–374. [CrossRef]
24. Ferguson, T.S. *A Course in Large Sample Theory*; Chapman & Hall: London, UK, 1996.

MDPI
St. Alban-Anlage 66
4052 Basel
Switzerland
Tel. +41 61 683 77 34
Fax +41 61 302 89 18
www.mdpi.com

Symmetry Editorial Office
E-mail: symmetry@mdpi.com
www.mdpi.com/journal/symmetry

www.ingramcontent.com/pod-product-compliance
Lightning Source LLC
Chambersburg PA
CBHW061827210926
51850CB00006B/832

* 9 7 8 3 0 3 6 5 5 2 5 8 3 *